The Pennsylvania Academy of Science Publications
Books and Journal

Editor: Shyamal K. Majumdar
Professor of Biology
Lafayette College
Easton, Pennsylvania 18042

1. *Energy, Environment, and the Economy,* 1981. ISBN: 0-9606670-0-8. Editor: Shyamal K. Majumdar.

2. *Pennsylvania Coal: Resources, Technology and Utilization,* 1983. ISBN: 0-9606670-1-6. Editors: Shyamal K. Majumdar and E. Willard Miller.

3. *Hazardous and Toxic Wastes: Technology, Management and Health Effects,* 1984. ISBN: 0-9606670-2-4. Editors: Shyamal K. Majumdar and E. Willard Miller.

4. *Solid and Liquid Wastes: Management, Methods and Socioeconomic Considerations,* 1984. ISBN: 0-9606670-3-2. Editors: Shyamal K. Majumdar and E. Willard Miller.

5. *Management of Radioactive Materials and Wastes: Issues and Progress,* 1985. ISBN: 0-9606670-4-0. Editors: Shyamal K. Majumdar and E. Willard Miller.

6. *Endangered and Threatened Species Programs in Pennsylvania and Other States: Causes, Issues and Management,* 1986. ISBN: 0-9606670-5-9. Editors: Shyamal K. Majumdar, Fred J. Brenner, and Ann F. Rhoads.

7. *Environmental Consequences of Energy Production: Problems and Prospects,* 1987. ISBN:0-9606670-6-7. Editors: Shyamal K. Majumdar, Fred J. Brenner and E. Willard Miller.

8. *Contaminant Problems and Management of Living Chesapeake Bay Resources,* 1987. ISBN: 0-9606670-7-5. Editors: Shyamal K. Majumdar, Lenwood W. Hall, Jr. and Herbert M. Austin.

9. *Ecology and Restoration of The Delaware River Basin,* 1988. ISBN: 0-9606670-8-3. Editors: Shyamal K. Majumdar, E. Willard Miller and Louis E. Sage.

10. *Management of Hazardous Materials and Wastes: Treatment, Minimization and Environmental Impacts,* 1989. ISBN 0-9606670-9-1. Editors: Shyamal K. Majumdar, E. Willard Miller and Robert F. Schmalz.

11. *Wetlands Ecology and Conservation: Emphasis in Pennsylvania,* 1989. ISBN 0-945809-01-8. Editors: Shyamal K. Majumdar, Robert P. Brooks, Fred J. Brenner and Ralph W. Tiner, Jr.

12. *Water Resources in Pennsylvania: Availability, Quality and Management,* 1990. ISBN 0-945809-02-6. Editors: Shyamal K. Majumdar, E. Willard Miller and Richard R. Parizek.

13. *Environmental Radon: Occurrence, Control and Health Hazards*, 1990. ISBN 0-945809-03-4. Editors: Shyamal K. Majumdar, Robert F. Schmalz and E. Willard Miller.

14. *Science Education in the United States: Issues, Crises, and Priorities,* 1991. ISBN: 0945809-04-2. Editors: Shyamal K. Majumdar, Leonard M. Rosenfeld, Peter A. Rubba, E. Willard Miller and Robert F. Schmalz.

SCIENCE EDUCATION IN THE UNITED STATES: ISSUES, CRISES AND PRIORITIES

EDITED BY

S.K. MAJUMDAR, Professor of Biology,
Lafayette College, Easton, PA 18042

L.M. ROSENFELD, Assistant Professor of Physiology,
Jefferson Medical College, Philadelphia, PA 19107

P.A. RUBBA, Associate Professor, Science Education,
The Pennsylvania State University, University Park, PA 16802

E.W. MILLER, Professor and Associate Dean (Emeritus),
The Pennsylvania State University, University Park, PA 16802

R.F. SCHMALZ, Professor of Geology,
The Pennsylvania State University, University Park, PA 16802

Founded on April 18, 1924

**A Publication of
The Pennsylvania Academy of Science**

Library of Congress Cataloging in Publication Data

Bibliography
Index
Majumdar, Shyamal K. 1938-, ed.

Library of Congress Catalog Card No.: 91-60096

ISBN-0-945809-04-2

Printed in the United States of America by

Typehouse of Easton
Phillipsburg, New Jersey 08865

PREFACE

Our age of scientific and technological complexity requires a cadre of scientists, well-trained technicians, and a scientific and technologically literate populace. The educational system of the United States is thus the cornerstone of our future.

Although American science and technology have long been recognized as the premier system in the world, grave concerns are being expressed as to our ability to sustain and improve this system in the future. The major challenge of the day is to provide an adequate supply of well-trained scientific and technological professionals, as well as a scientific and technologically literate citizenry. Unfortunately, there is a growing recognition that in recent years science education in the United States has been deficient in meeting these goals.

While many solutions have been proposed and tried, none has resolved these basic problems of numbers and literacy. The increasing impact of science and technology on all aspects of our lives, as well as population demographics that show women and minority groups that refrain from participation in science will be an expanding portion of the population suggest these problems will intensify. Science education is a vital national issue. A revitalization must occur if our nation is to remain at the forefront of a dynamic scientific and technological world.

The fundamental objective of this volume is not only to describe the state of science education in the United States, but also to evaluate and analyze the present crisis so that solutions can evolve. The papers that comprise its chapters were prepared by a knowledgeable and well-known group of professional scientists and science educators as well as key policy-makers at the local, state and national level. They deal comprehensively with the wide variety of science education issues of the day.

The papers are organized into seven Parts. The first Part contains papers on the present status of science education in the United States. This is followed by Parts 2 and 3 which contain papers devoted to science education K-12 and at the post-secondary level. The role of government, professional societies and industry is considered in Part 4. The papers in Part 5 present a broad view of science education, including science as a liberal art, science education and the electronic media, and perspectives on our science and technology needs. Part 6 includes papers on initiatives such as Project 2061 and NSTA's Scope, Sequence and Coordination Project, as well as non-traditional science education experiences. Critical science education issues involving women, minorities and the economy are discussed in Part 7. The book ends with the editors' reflections on science education in the United States.

The editors hope that this book will be valuable to a wide audience; not only to scientists and educators, but to all individuals who are concerned about the future and science education in the United States. We express our deep appreciation to the contributors for their dedication to the task, the quality of their work and for excellent cooperation as the volume was developed. Gratitude is expressed to Lafayette College, Jefferson Medical College, and The Pennsylvania State University for providing facilities for the editors. Lastly, the editors extend heartfelt thanks to their wives and families for their encouragement and help during the preparation of this book.

Shyamal K. Majumdar
Leonard M. Rosenfeld
Peter A. Rubba
E. Willard Miller
Robert F. Schmalz
March, 1991

ACKNOWLEDGMENTS

The publication of this book was aided by contributions from The Pennsylvania Power and Light Company, Allentown, Pennsylvania; and Consolidation Coal Company, Pittsburgh, Pennsylvania.

Science Education In The United States: Issues, Crises and Priorities

Table of Contents

Part One: State of Science Education

Part Seven: Critical Issues

Science Education in the United States: Issues, Crises and Priorities.

EXECUTIVE OFFICE OF THE PRESIDENT
OFFICE OF SCIENCE AND TECHNOLOGY POLICY
WASHINGTON, D.C. 20506

FOREWORD

D. Allan Bromley
Assistant to President Bush for Science and Technology
Executive Office of the President

During a presentation at a recent meeting of the President's Council of Advisors on Science and Technology—a group of 12 distinguished researchers, academic leaders, and industrialists who advise the President on scientific and technological issues—Michael Boskin, Chairman of the President's Council of Economic Advisors, referred to education as the "ticking time bomb" at the center of the U.S. economy. It is the most serious problem our country faces, he said, because if we do not educate our children adequately, we will find it impossible to compete in the emerging global economy of the 21st century.

This belief is widely shared in the Bush Administration. Since his first days in office, President Bush has emphasized the vital importance of education, saying that "nothing could be more important to current and future generations and to the future of our Nation." In the President's 1991 budget, education was singled out—together with enhancing research and increasing the rate of savings and investment—as a key area in which America must invest today to secure a more prosperous future.

Within this overall emphasis on education, special attention has focused on mathematics and science education because of its role in modern economies. As the economies of the industrialized countries shift toward high-technology manufacturing and services, the jobs available to new entrants into the workforce require higher levels of skill and education than ever before. Projections are that over half of the jobs available by the end of the decade in the United States will require some education beyond high school, compared to 42 percent of currently available jobs. There will be fewer jobs for the poorly educated high school graduate and fewer still for the 25 to 30 percent of students who drop out of high school without graduating.

These economic arguments for improved mathematics and science education are compelling, but I would like to emphasize another rationale for increasing our efforts in this area. It is difficult to make sense of the modern world without knowing something about science and mathematics. For the past several centuries, scientists have been engaged in one of humanity's most profound adventures—increasing our knowledge about ourselves and our world—and in the process they have transformed the way we view ourselves and the world around us. To know nothing about the constituents of the atom, the theory of evolution, or the genesis of the universe is to be intellectually handicapped in modern society.

An appreciation of science and technology is important for another reason. In a democracy such as ours, it is essential that our citizens be able to understand the broad issues that affect them, even though they may lack the background and training to participate in the resolution of those issues. Lacking such understanding, many become alienated from society, and this we simply cannot afford. Literacy—both verbal and numeric—is the foundation on which we necessarily build our future.

REFORM OF PRECOLLEGE EDUCATION

In considering the state of mathematics and science education in the United States, one is immediately struck by a picture of great contrasts. The United States still sets the pace and style for the world in graduate education; in fact, it can be argued that the scientific and engineering education we provide to foreign students who then return home is one of our most valuable exports.

In undergraduate education, although the range of educational quality available is more extreme than elsewhere in the developed world, on average we remain competitive. At the precollege level, however, we have fallen far behind our international competitors.

There remain many excellent public elementary and secondary schools in America, and the system continues to produce many excellent students. It is clear, however, that in overall terms U.S. precollege education is not meeting the needs of society. It is not just science and mathematics that are in trouble. I believe that the problem is a much broader one spanning the entire educational field. Reform of mathematics and science education will not be successful unless the overall reform of precollege education is successful. Just as the federal sector is only one constituency that must work in concert with other groups for mathematics and science reform, the mathematics and science reform effort must be orchestrated with the larger educational reform movement.

Recognizing the crisis in American education, President Bush has made revitalization of the nation's schools a top priority for his administration and has set a number of initiatives in motion. The highlight of these was the Education Summit between the President and the nation's governors held in September 1989, which was only the third time in the history of the country that the President and governors have met to consider a single topic of national importance. Out of this meeting evolved the six national education goals that the President announced in January 1990, three of which directly involve science and mathematics:

Goal 3: By the year 2000, American students will leave grades four, eight and twelve having demonstrated competency in challenging subject matter including English, *mathematics, science,* history, and geography; and every school in America will ensure that all students learn to use their minds well, so they may be prepared for responsible citizenship, further learning, and productive employment in our modern economy. [Emphasis added.]

Goal 4: By the year 2000, U.S. students will be first in the world in science and mathematics achievement.

Goal 5: By the year 2000, every adult American will be literate and will possess the knowledge and skills necessary to compete in a global economy and exercise the rights and responsibilities of citizenship.

Achieving these ambitious goals—along with the other goals of ensuring that students start school ready to learn, raising the high school graduation rate to 90 percent, and ridding every school in America of drugs and violence—will require substantial changes in our educational system, in our attitudes toward teaching and learning, and in our culture. We must return to objective standards of performance and learning. We must reinvolve parents in the education of their children and maintain discipline so that students can learn. In this volume you will read much about ways to achieve these objectives, such as the restructuring of schools , the professionalization of teaching, curriculum development, and enhanced accountability.

Perhaps most important, however, we must ensure that all sectors of society are involved and working together to achieve educational excellence. Our teachers, students, industry, academia, state and local governments, and the federal government—Congress and the Executive Branch—all have important and varied roles to play.

The problems we face in mathematics and science education are complex but not intractable. We must build on the successes that have already been achieved around the country in establishing magnet schools for science and mathematics, promoting the hiring of teachers who are trained in science and technology, and rewarding excellent teachers and students. These changes will occur largely at the state and local level, but the federal government can catalyze changes by removing obstacles to innovation and by identifying and working to disseminate successful approaches.

The federal government can also catalyze a form of education that is often overlooked in discussions of mathematics and science education—continuing education. The effective half-life of a formal scientific or technical education is shortening rapidly and is now in many cases less than 5 years. Thus, formal education should, above all else, train students how to learn and how to continue learning by drawing on all of the resources of today's information revolution. Schooling must instill in people the desire to view education as a vital and continuing part of their life—not something that is endured and is over and done with.

Finally, the federal government has a unique responsibility at the opposite end of the educational process. The federal government has accepted the responsibility of working with the states and localities to ensure that when children become old enough to start school, they are both physically and mentally prepared for the experience. Such is far from the case at present—particularly in urban areas. As a result, President Bush has proposed a 36 percent increase in funding for Head Start to $1.9 billion, which would enable the program to serve up to 70 percent of poor four-year-olds. Other programs for preschoolers would also be augmented.

THE FEDERAL ROLE

A large number of federal agencies are involved in the effort to revitalize precollege education in the United States. In mathematics and science education, the National Science Foundation has launched a Statewide Systemic Initiative that will encourage and support proposals seeking broad-based, fundamental changes at the state and local level. The President has proposed significant funding increases for the Department of Education's Dwight D. Eisenhower

Mathematics and Science Education Program, which provides grants to states, school districts, and colleges to increase the quality of mathematics and science teaching to improve access of all students to quality instruction. The Department of Energy is mobilizing its national laboratories and other research facilities to bring cutting-edge science to America's teachers and students. NASA's "Science and Technology Literacy for the 21st Century" plan sets a new course for its educational activities extending throughout the 1990s. And the National Institutes of Health is increasing its training and research grants in the life sciences. For the five agencies mentioned above, the President's budget for fiscal year 1991 proposes over $1 billion for programs whose primary purpose is improvement in mathematics and science education—an increase of 26 percent over the previous fiscal year.

The effects of these programs will be limited, however, unless they are closely coordinated across agency lines and unless they work in concert with reforms in the states, local school districts, and schools, colleges, and universities. Central to this coordination will be the new Committee on Education and Human Resources, chaired by Secretary of Energy James Watkins, within the Federal Coordinating Council for Science, Engineering, and Technology. The charge to the committee is to promote more efficient use of the expertise that exists in federal agencies, avoid needless duplication of effort, identify areas of new program opportunities, and make more efficient use of limited federal resources in efforts related to science, mathematics, and engineering education across the Administration. Senior agency administrators from sixteen federal departments and agencies are involved in this effort.

Complementing the advice the President receives from within government will be the advice of the private sector. The President's Council of Advisors on Science and Technology has responded to the President's request that mathematics and science education be a major issue that PCAST addresses. PCAST is currently creating a panel, chaired by Peter Likins, the President of Lehigh University, to study and report on mathematics and science education.

There are also more broadly focused efforts under way within the White House to reform precollege education. The Assistant to the President for Economic and Domestic Policy, Roger Porter, provides coordination for a sustained, high-level focus on Administration efforts specifically related to the national education goals and other commitments made at the Education Summit. The Domestic Policy Council has established a Working Group on Education chaired by Secretary of Education Lauro Cavazos. The President has set up an Education Policy Advisory Committee of representatives from labor, business, the media, and all levels of education to recommend ways to mobilize national efforts to realize the goals. And, most recently, the President and the nation's governors have created a National Education Goals Panel to determine the indicators used to measure and report on progress toward achieving the national education goals.

It is too early to issue a report card on the new institutional entities that have been established to enhance the federal role in precollege education. Nor will it be easy to turn around the quality of education in a decentralized system of 15,000 school districts in 50 states, over 40 million students, and 2.8 million teachers and administrators. But the nation cannot address its long-term economic and social needs without responding to this serious national problem. As President Bush has often said when referring to education, "It's time to take action."

The Pennsylvania Academy of Science is to be commended for "taking action" via it's initiative in assembling such a diverse group of authors: educators, scientists, economists, corporate and foundation leaders as well as representatives of local (school boards), state (governors), and federal governments (Congress as well as The Executive Branch). Out of such cooperative efforts, we can look for success.

D. Allan Bromley
Assistant to the President for Science
and Technology
Executive Office of the President

INTRODUCTION

Robert P. Casey
Governor
Commonwealth of Pennsylvania

COMMONWEALTH OF PENNSYLVANIA
OFFICE OF THE GOVERNOR
HARRISBURG

As Governor, I congratulate the Pennsylvania Academy of Science for sponsoring the publication of *Science Education in the United States: Issues, Crises and Priorities.* Since 1924, the Academy has not only served the citizens of Pennsylvania, but the nation and the world, in examining critical scientific issues in our society.

The timing of this publication could not be more appropriate. Less than one year ago, the nation's Governors and the President adopted six national education goals that aim by the year 2000 to reestablish the United States as the world leader in student achievement. One of those goals calls for American students to lead the world in math and science expertise.

That is no small order. This goal is the most ambitious among all six because the level of math and science education of our nation, as a whole, lags behind that of most industrialized nations. This stands in sharp contrast to America's unique and unmatched collection of academic and research institutions.

But it is a goal we must resolve to meet. Not simply because the ability of our nation and our children to be competitive internationally depends upon it. But also because ultimately the quality of American life—economic, social and political—cannot be sustained without well-educated, literate citizens who have the ability to succeed in a complex society that is being constantly transformed by science and technology.

We simply cannot accept or afford to leave science to the select few who are able naturally to appreciate its breadth and master its challenges. But that is what is occurring all too often in our nation today. The imperative before us, then, is to make science education an integral part of every American student's academic career at all educational levels.

There is no reason we cannot fulfill this task. Because science pervades every discipline and virtually every aspect of our lives. Science has unquestionably shaped history; and our political and social history has in turn shaped science. The language of science and technology constantly finds its way into the language and commerce of every day life. Science, no less than any discipline and possibly more so, is one critical way we define ourselves and understand our world, from the composition of life's smallest elements and the mysteries of distant stars, to the ethical dimensions of genetic engineering and life-sustaining medical technology.

It is telling that Benjamin Franklin, a predecessor of mine as Pennsylvania's chief executive and a founder of our nation, was not only a statesman and a philosopher, but also a man of science.

"The good education of youth," Franklin wrote, "has been esteemed by wise men in all ages as the surest foundation of the happiness both of private families and of Commonwealths."

Franklin's words ring as true today as they did two hundred and fifty years ago. Our economic vitality, the quality of our democratic society depend centrally on the education we provide to our young people; on our ability to examine our past and peer into our future.

And science is nothing if not a way to examine our origins and peer into our future. Perhaps now, more than at any other time in our history, a strong foundation in science is a prerequisite for finding the public and private "happiness" of which Franklin spoke.

Over the past four years, we have laid a foundation of education reform in Pennsylvania that positions our state to take advantage of the new national focus on improving American education and, in particular, science achievement. We have increased spending on basic education by nearly $1 billion; established programs that have strengthened the teaching profession; and provided financial rewards to schools that show significant academic improvement.

In so doing, the Commonwealth has also sought to emphasize the importance of science education. In addition to a college loan forgiveness program for all new teachers that choose to teach in economically hard-pressed rural and urban schools, the Commonwealth provides college loan forgiveness to attract more math and science teachers. The Governor's School for the Sciences, located at Carnegie Mellon University, provides advanced work in the sciences for highly talented high school students each summer.

The Commonwealth also supports programs that encourage partnerships between higher education institutions and local school districts designed to improve the quality of science instruction. The Commonwealth Elementary Science Teaching Alliance, administered through the Franklin Institute in Philadelphia, provides science curricula and other assistance to more than fifty school districts. And the Pennsylvania Science Teacher Education Program targets $1 million each year to assist elementary and secondary science teachers in updating their skills. These represent just some of the initiatives state government has undertaken in recent years.

But clearly there is much more that we must do if we are to improve the level of science achievement among our students. We need new efforts to integrate math and science education into the basic education curriculum of our public schools—efforts which must involve teachers themselves in the process of redefining the way they teach. Teacher education programs must give new emphasis to ensuring that all teachers have a strong math and science foundation and to attracting more teachers to the fields of math and science. Women and minorities represent the fastest growing segment of our workforce. Therefore, we must take dramatic steps to improve their math and science education levels and encourage them to pursue studies in math, science and engineering.

This is not a job that state government or the federal government can accomplish alone. Meeting the goals we have set for ourselves and our children will require an unprecedented partnership among government, educators, business and parents. And it will require the long-term commitment of political, human and financial resources necessary to get the job done.

This publication is part of a ten-year effort by the Pennsylvania Academy of Science to focus public attention on key scientific issues facing our state, region and nation. It presents in one volume some of the best thinking of the key figures who shape science education in America, from the nation's top academics and educators to policymakers at the state and federal levels.

By taking stock of the current situation and identifying the challenges we face as a nation in science education, this anthology can help us chart the future of science education that will prepare our children for the 21st century.

Robert P. Casey
Governor
Commonwealth of Pennsylvania
December, 1990

DEPARTMENT OF ENERGY
WASHINGTON, DC 20585

MESSAGE

James D. Watkins
Admiral, U.S. Navy (Retired)
Secretary of Energy

When President Bush convened the Education Summit in September, 1989, he and the Governors focussed the attention of America on the importance of education. The National Education Goals developed as a result of that meeting include as perhaps the most ambitious goal of all: "By the year 2000, U.S. students will be first in the world in science and mathematics achievement."

Such an achievement will require a great commitment from the American people—commitment to initiating innovation and reform in math and science education, and commitment to sustaining the multitude of reform efforts now underway in the States and the private sector. This achievement will also require continued thoughtful effort by individuals like those who produced this volume—top public officials and policy makers, educators, corporate executives and scientists—coupled with a commitment to action.

The entire nation has awakened to a new sense of urgency in education and the Federal government, under the direction of the President, has been challenged to assume a new leadership role. As one result of the Education Summit, Dr. Allan Bromley, the President's Science and Technology Advisor, established the Committee on Education and Human Resources, which I chair, under the Federal Coordinating Council on Science, Engineering, and Technology (FCCSET). Within the next month, the Committee will have completed its first tasks: an inventory of the Federal Government's science education activities and the preparation of the first coordinated interagency budget submission for math and science education. While the Federal Government contributes only 6% of the total funds spent on education in this country, the $1.48 billion (FY 1990) that it spends specifically on math and science education (kindergarten through graduate school) can be leveraged with State and private sector funds to achieve at least three of the National Education Goals by the year 2000.

One area in which the Federal government, with its hundreds of world-class research facilities, is uniquely positioned to make a contribution is teacher preparation and enhancement. The Department of Energy, for example, spent $3.2 million on precollege teacher preparation and enhancement programs in FY 1990. This money is leveraged, however, by the more than $30 billion invested in facilities, equipment and technically trained personnel within our nation-wide laboratory system. Similar facilities and personnel exist within NASA, the Departments of Agriculture, Health and Human Services, Interior, Defense and many others.

Federal involvement here addresses one fundamental flaw in the American educational system—the disconnect that exists between cutting-edge science and what is learned in the classroom. Almost nowhere in 20th century America are students given a glimpse of 21st century science, or the technical innovation that will be a part of their everyday lives and workplaces. Why? Because the teachers who must convey this vision of discovery and accomplishment to our children are often poorly prepared by their own educational experiences to do the job.

To solve this problem, it is vital that we begin to treat math and science teachers as respected members of America's scientific community, and prepare them just as diligently. The Federal Government, in partnership with the States and the private sector, can and must reach out to the nation's teachers, make them our number one priority, and bring them together with science in every setting—

in federal laboratories, in field research programs, and in the private sector workplace—so that they can return this knowledge to the classroom, the very essence of technology transfer.

Our FCCSET Committee on Education and Human Resources has proposed exactly that. We have established the reform of mathematics and science teaching as our first priority, and will look for innovative ways to reach every teacher in America by the middle of this decade, in partnership programs with business, State and local governments, and community-based organizations.

One of the next tasks I am hopeful that the Committee will undertake is the compilation of a directory of all federal facilities, programs and materials available on a regional basis for use by teachers and students in mathematics and science. I would like to put this directory in the hands of every teacher, corporate executive and State education policy maker in the nation. Our goal is to make these individuals aware of the federal resources available to them, so that as State and regional reform efforts are constructed, these resources can be included in the mix.

As Secretary of Energy, I can also see the importance of math and science education from an angle even more immediate than the often-cited long-term issues of international competitiveness. The Department of Energy employs over 70,000 scientists, engineers and technicians, making it one of the single largest employers of scientists and engineers in the world today. And there are shortages today of professionals in the fields necessary to carry out the Department's mission. In important areas like environmental restoration or waste management—certain to be one of the fastest growing employment areas in the next decade—we have jobs that go begging.

This problem of manpower shortages, both current and projected, becomes even more acute when one considers that the bulk of those entering the workforce in the near future will come largely from groups now grossly underrepresented in technical fields, including women and minorities.

We need to send a message to America's youth that you don't need to have a Ph.D. to be successful in science and technology—we need technicians, and we need them badly. But those technicians must be able to read well, compute accurately, and have a mastery of their technical area. We need to tell our youth that we in the Federal Government and the private sector will help them along the way, and that there will be jobs waiting for them at the end of their education, if they will only stick with us.

The obvious solution to the problem of having an inadequate pool of qualified professionals is to increase the number of students entering math and science fields, i.e., to keep the "science pipeline" filled. Most children are lost to math and science by the time they reach seventh grade, many of them even before this. To keep students in the pipeline we must keep them excited about science, through hands-on laboratory research activities. Students learn more and find greater satisfaction when they discover for themselves. Hands-on science will turn students on where rote memorization from out-of-date textbooks will surely turn them off.

The Department of Energy is greatly expanding its precollege programs, again with a special emphasis on teachers, on hands-on research experiences, and on attracting underrepresented groups into science and technology. Just ten days after the Charlottesville Education Summit, I co-chaired the Math/Science Education Action Conference in Berkeley, California with Nobel laureate Glenn T. Seaborg. The purpose of the conference was to develop a model showing how a federal agency could utilize its resources in innovative ways, working with States and the private sector, to help reach the National Education Goals.

Our conference report was issued in May 1990, along with the announcement of several new regional initiatives. These initiatives, such as the Academy for Math and Science Teachers in Chicago, The Bay Area Science and Technology Collaboration in Oakland, the "adoption" of the Native American schools in the New Mexico and Arizona, and similar initiatives in rural Tennessee, the Yakima Valley in Washington State, Idaho, and elsewhere, all represent collaborations between one or more National Laboratories, educators, corporations, and in some cases, States. We are opening our doors, sending our scientists out into the community, and bringing students and teachers in, to see first hand the science of tomorrow.

One of our primary goals, as we attempt to reach the 15,000 math and science teachers in Chicago, the 2,700 math and science teachers in Oakland, and so many others through our new initiatives, is to raise the basic level of scientific literacy in this country. A successful democracy requires an informed citizenry. At an ever increasing rate, decisions on public policy matters demand a technical or scientific base of knowledge. But, according to one study, approximately 94% of this country's population is scientifically illiterate. More than half of all Americans do not know that the Earth moves around the sun and that this process takes one year. This does not allow us to look with confidence to our future in a world concerned with how to protect the environment, with what sort of research to perform and fund, and with how to interpret the results of those studies, much less headlines and news stories on scientific issues.

Another aspect of math and science education about which there is very little discussion is one that I find personally important: one of the most pleasant, rewarding experiences in life is the understanding of the basic processes in nature. There are not many better feelings than discovering the solution to a complex problem or understanding for the first time the forces driving a system. All scientists and engineers will recognize the feeling. Very young children also understand it, as they are "natural scientists," exploring and testing their world.

I think of this as the "aesthetic value" of math and science education. It kept me and hundreds of like-minded colleagues in the laboratories and library stacks late at night, always searching for more. Math and science can provide deep and satisfying insights into life, but too often in modern society, these insights are never experienced.

"Science" and "knowledge" are synonymous in many languages of the world. Here in the United States, however, we seem to have connected "science" with other concepts—"difficulty," "dryness," "elitism," or even "irrelevance to everyday life." Basic misconceptions about the scientific process abound.

To quote from our Berkeley conference report, "Our mission is to ensure that the history books of the 21st century will not tell of a once great nation that declined and fell because it lost its passion for science." Ours is an age of scientific and technological advancement unlike any other in history. Will the promise of this new renaissance be fulfilled? Will an informed citizenry, having mastered science and technology, find in the 21st century freedom from disease, hunger and hardship now felt by millions in our world? The answers lie with our teachers, our students, and with the investment in human capital we either are or are not willing to make at this critical point in time.

America must continue to lead this technological expansion, this new renaissance, but not by leaving behind the great proportion of its population. There is room for everyone, and a mission for all. The great message of science—of discovery waiting to happen, and accomplishment open to all—is what we must find new ways to convey to our school children, and represents the fundamental passion behind the education revolution of the 1990's. Those who cannot see this may have missed the true spirit of the age in which they live.

James D. Watkins
Secretary of Energy
Department of Energy
Washington, D.C. 20585

OFFICE OF THE PRESIDENT
LAFAYETTE COLLEGE
EASTON, PA 18042

REMARKS

Robert I. Rotberg
President
Lafayette College

Scientists are educated in liberal arts colleges before they go on to obtain advanced and specialized degrees. During the course of this century, undergraduate liberal arts colleges have given first degrees to more MDs and PhDs than universities, on a per capita basis. Although colleges like Lafayette, Swarthmore, Oberlin, Carleton, Williams, Amherst, and Bowdoin can claim only a handful of Nobel laureates, they have nurtured and trained a legion of professionals in the health and biomedical sciences, in the mathematical, physical, and earth sciences, and in chemistry. They are doing so today, under increasingly difficult circumstances.

At a time when the doing of science increasingly demands larger and more expensive technology and bigger and more costly laboratory teams, undergraduate colleges struggle to afford what it takes to continue to be excellent. It is in the undergraduate classrooms and laboratories that the Nobel winners of the next several decades will be stimulated to apprentice themselves to science; equipping those laboratories and providing high-quality faculty with research experience for the classrooms is becoming harder and much more expensive.

The equation is evident: without the maintenance of scientific excellence in the undergraduate colleges of the United States, big science will lack recruits and thus will suffer. Without big science we compete nationally in the global marketplace of ideas and commerce less well, and our American quality of life inevitably faces decline. Thus it behooves all of us to find ways to nurture undergraduate science and the education of undergraduate scientists.

Simply asking the undergraduate colleges to teach harder will not do. Those who teach must have opportunities to do their own research, even on the periphery of big science, and must have the facilities and the support to involve undergraduates in ongoing research. Hands on collaboration has been demonstrated to be a crucial ingredient in the making of young scientific leaders.

As undergraduate colleges we may also have a critical role to play in the broader scientific and technical education of all of our young people. In this decade and the next we have a responsibility to prepare them for a world increasingly influenced by science and technology. No one could have anticipated the number of major scientific advances in the second half of this century, especially in the past decade. Progress is never linear, but the pace of scientific change is accelerating ever faster, and the liberal arts to which we have long introduced our national leaders have now become infused with science. Indeed, training our young people to be able to appreciate and do at least some science, and focus their minds cognitively, may be as important today as introducing them to our cultural heritage has always been.

The undergraduate college will continue to play a critical part in the education of America's scientists and engineers. It is the place in which our best and brightest minds are directed toward lifetime pursuits. It will thus be essential to strengthen our national support for science in undergraduate colleges, and to appreciate—as a matter of national policy—that what happens in the up-to-date college laboratory contributes directly to the development of big science in the larger research universities, and in industry. Big science can hardly do without undergraduate science.

Robert I. Rotberg
President
Lafayette College
Easton, PA 18042

mathematics and science education, although the other four also have at least indirect implications for math and science achievement.

Goal 3:

By the year 2000, American students will leave grades four, eight, and twelve having demonstrated competency in challenging subject matter including English, mathematics, science, history, and geography; and every school in America will ensure that all students learn to use their minds well, so they may be prepared for responsible citizenship, further learning, and productive employment in our modern economy.

Goal 4:

By the year 2000, U.S. students will be first in the world in science and mathematics achievement.

It is ambitious, perhaps rash, to say that we can make these gains by the year 2000. But it is essential that we do so, because the economic competitiveness of this Nation requires that our workforce has strong grounding in mathematics and science.

No longer can we believe that our industries will remain viable without workers that are well prepared to address scientific and technological issues. There is perhaps no issue more difficult and yet more crucial than to vastly increase the population of capable, motivated, and successful science and mathematics learners.

The problems we face revolutionizing science and mathematics education are well known. The following are only a few of the well-documented deficiencies.

Over the past few years, we have all read reports about the poor performance of American students in mathematics and science. When compared with students in twelve other countries, our high school students finished 9th in physics, 11th in chemistry, and last in biology[1] In a recent study of five countries and four Canadian provinces, U.S. 13-year-olds came in last in mathematics and near last in science.[2]

Nor are our students learning enough in absolute terms. Only about half of our high school juniors can do junior high math—tasks such as calculating the area of a rectangle or estimating 87 percent of 10. Just one in four Black and Hispanic 17-year-olds performs at the junior high level in mathematics.[3]

In science, fewer than half of our 17-year-olds seem to know enough "to perform . . . jobs that require technical skills or to benefit substantially from specialized on-the-job training," according to the National Assessment of Educational Progress. It tells that fewer than half appear to be adequately prepared in science "for informed participation in the nation's civic affairs." Only about half of the majority of White juniors can evaluate experiments and interpret texts and graphs. Fewer than 15 percent of Black and Hispanic juniors can do these things, and only one third of females.[4]

These problems are not new, as it has been shown in previous international studies that U.S. students, as a whole, perform at a mediocre level. But there was the consolation that our best students—the future mathematicians, scientists, and engineers—were as good as those in any other country. But this just isn't true anymore. In mathematics, our top 13 percent generally fell into the bottom 25% in comparison to other countries.[5] We can no longer assume that the science, mathematics, and engineering professions will continue to have the new talent they require.

Another concern about the "pipeline" of scientific talent is the lack of enrollments in advanced math and science classes. If students do not take chemistry, physics, advanced algebra, and trigonometry, they will simply not be able to pursue scientific and technological careers. While enrollments are beginning to increase, less than 50 percent of our 1987 high school graduates took chemistry of any type; only about 20 percent took physics.[6] Given the number of students we are preparing at the secondary school level, it is not surprising that nearly half of the doctorates in some fields of engineering are awarded to foreign students.[7] What makes this situation more critical is that with the worldwide demand for scientists and engineers, it will be increasingly difficult to retain these doctorates in the United States.

Of particular concern is to encourage minorities and women to enter science and mathematics fields. In the past, majority males have been the predominant group in these occupations; only about 25% of the doctorates awarded have been awarded to females. Our changing demographics clearly argue that unless we successfully reach out to other groups, current shortages will only become worse. Expanding the participation of minorities and women is not just the right thing to do. It is the only choice.

One of the factors that led to these problems is the nature of instruction our students are receiving. Hands-on learning has been espoused by science and mathematics education groups for years, and research suggests that using objects and "manipulatives" is an effective way to teach mathematics and science concepts. Yet student experiments and hands-on assignments have actually *declined* in recent years.[8] A third of our seventh graders say they never get to do experiments.[4]

Another factor is the quality and quantity of teachers available in science and mathematics. Many elementary teachers say they are uncomfortable teaching science and mathematics. While in college, many of the elementary teachers in our schools today completed only one or two courses—if any—in these two disciplines; further, almost 50% have had no additional college coursework in the past ten years.[9] Can you imagine trying to inspire students in a subject that is unfamiliar to you? If children do not enjoy science in the early grades (some say by the third grade), they are less likely to be willing to invest the hard work and discipline that success in later science study demands.

There are problems in teaching at the secondary level as well. During the 1980s at least two-thirds of the states were unable to hire enough mathematics and science teachers. Sixty to 70 percent of junior high and high school principals say they have difficulty hiring teachers for physics, chemistry, and computer science classes.[8] Nor is there relief in sight. During the next five years, we will need to develop as many new mathematics and science teachers as the number we currently have in our schools.[10]

Changing these conditions is the responsibility of each of us. Here are just a few of those involved and a partial list of the tasks required.

- STUDENTS must understand essential scientific and mathematical knowledge and skills.
- TEACHERS of science and mathematics must possess a thorough understanding of their subjects, even at the elementary grades.
- PARENTS must expect their children to take demanding science and math courses and encourage them to do their best.
- SCHOOLS must introduce mathematical concepts and scientific principles early and actively in a student's education experience.
- SCHOOL DISTRICTS must establish high expectations for student achievement and ensure that assessments of student learning reflect the goals of learning for understanding and thinking.
- POSTSECONDARY INSTITUTIONS must expand their efforts to recruit, retain, and graduate more students—especially women and minorities—from mathematics, science, and engineering programs and from mathematics and science teaching programs.
- STATES must review, revise, and coordinate the essential elements of mathematics and science learning: curriculum, assessment, teacher certification standards, and criteria for the selection of teaching materials and equipment.
- FEDERAL AGENCIES must ensure that programs are well-coordinated and consistent with national learning goals in science and mathematics.
- BUSINESS and INDUSTRY must make clear statements of their expectations of education, at all levels, and assist educational institutions through partnerships, public advocacy, and corporate leadership practices.
- THE PUBLIC must demand improved science and mathematics programs and provide the financial and vocal support for them.

ENCOURAGING SIGNS OF CHANGE

As awareness increases of the magnitude of the mathematics and science education deficiencies, calls for change have been made and actions has been taken. Certainly, the establishment of national education goals in these areas signals an awareness of the problem and a willingness to pursue answers.

Another important source of leadership in this move to rethink and reform mathematics and science education has been the professional associations and science societies. The willingness of the science, mathematics, and education communities to come together to address a common problem has been outstanding. Some of the more striking of these efforts are briefly noted.

- The National Council of Teachers of Mathematics (NCTM) released their *Curriculum and Evaluation Standards for School Mathematics.*[11] These *Standards* reflect a consensus of the mathematics and mathematics education community of how the mathematics curriculum must be altered. Emphasis is placed on active mathematics learning, based on students' natural curiosity. NCTM will shortly release a companion *Standards* volume on teaching mathematics.

- The Mathematical Sciences Education Board (MSEB) of the National Academy of Sciences has launched a national effort to reform mathematics. Through reports, conferences, working groups, and other mechanisms, MSEB plans to implement new approaches to mathematics learning, such as the NCTM *Standards*. The MSEB has also put into place an impressive program and issued a report on the future of mathematics, *Everybody Counts!*[12]

- The American Association for the Advancement of Science (AAAS) has initiated Project 2061. Names for the year that Halley's Comet will next be in the vicinity of Earth, it aims to determine the long-term science and mathematics learning needs of our students. The project has identified a small number of common themes, such as constancy and scale, on which the future curriculum would be based. Their project is described in *Science for All Americans.*[13]

- The National Science Teachers Association (NSTA) has proposed a major curricular reform of science.[14] Rather than the sequence of biology, chemistry, and physics through the high school years, NSTA has proposed that all sciences be taught at each grade level in an integrated manner. Pilot implementations of this plan are now under way in several states, including California and Texas.

Along with these, and the numerous reports that have been issued, there have been other important activities that should be noted.

- A number of states have upgraded their curriculum requirements in mathematics and science to guarantee that students have been exposed to more content in these areas.

- Assessment plays the important roles of monitoring progress and increasing visibility of educational goals. The National Assessment of Educational Progress (NAEP) has begun a state-by-state assessment of mathematics achievement. The first of these assessements, at the eighth grade, is currently being conducted. In many instances, States have also increased the assessment of student achievement in math and science.

- Several states have adopted alternative certification of teachers. With shortages of qualified mathematics and science teachers, it is essential that we take advantage of the pool of well-educated mathematicians, scientists, and engineers who might be interested in teaching. New Jersey, the first State to adopt alternative certification, reports that a number of mathematics and science teachers have been identified using this approach, and that the percentage of minority applicants has been increased.[15]

- Funds for science and engineering education at the National Science Foundation have increased from under $100 million five years ago to the current request for $251 million.
- The U.S. Congress and most state legislatures are debating how best to devise solutions for the mathematics and science crisis. Many bills have been submitted reflecting proposed solutions. While no consensus has yet been reached, the message is clear that the problem has been recognized and can no longer be ignored.

A U.S. DEPARTMENT OF EDUCATION RESPONSE TO THE NEEDS OF SCIENCE AND MATHEMATICS EDUCATION

Before discussing the Department of Education's role in mathematics and science education, it is important to recognize that the Department is only a small part of the support for mathematics and science education, providing just 6 percent of total education funding. The Department cannot be the solution, nor can the rest of the Federal government, but we can contribute to solutions. The following are ways that the Department can contribute.

First, the Department must maximize the impact of the resources already directed toward science and mathematics education. There is almost $300 million in the current budget request for the Department that can be directly associated with mathematics and science education programs. However, this is a small portion of what actually is directed toward science and mathematics in the budget. For example, the Department will have an investment of more than $1 billion in mathematics in our Chapter 1 Program of the Elementary and Secondary Education Act. The program provides grants to local education agencies to support educational programs for the disadvantaged. Approximately 45 percent of the students served by these programs receive mathematics instruction. A similar analysis can be done with many of the large, formula grant programs that the Department operates. Thus, these funds are substantial and can make a difference.

The initial step is to *inventory* the mathematics and science activities currently being supported. This is not an easy task, because the Department is not structured by content areas, as are most secondary schools and postsecondary institutions. Rather it is organized according to administrative structures, such as elementary and secondary education, or specialized areas, such as vocational education. Also, many programs, and projects within them, include mathematics and science activities in combination with other activities. What proportion should be considered math and science is often arbitrary.

The Department has organized a Task Force on Science and Mathematics Education and charged it with undertaking this inventory of activities. The inventory has been completed, and it is now possible to begin an analysis of each of the programs to determine if the money is being used effectively. For example, many programs support the development of usable models and materials that might well be made available to other projects. Identification of these types of exemplary products and practices is one way to maximize investment. Another way is to promote coordination across programs and educational units. *Second,* the Department will identify the needs of special groups and content areas, in particular determining how mathematics and science issues impinge on these needs. Some examples of these activities are:

- *Hispanic Task Force.* This task force was formed to advise the Working Group on Education of the Domestic Policy Council, chaired by the Secretary of Education. Its purpose is to assess participation of Hispanics in Federal education programs, identify barriers that may limit participation in education programs, and suggest goals and strategies for the education of Hispanics. A broad range of topics will be explored that have relevance to science and mathematics education, including early school success programs, adequately trained bilingual teachers in all content areas, and prevention of dropouts.
- *Metric Education Task Force.* The Metric Education Task Force is exploring ways to promote metric education in schools at all levels and to ensure compliance of the Education Department with Federal laws. The coming changes in the European economic system will bring with it requirements for conversion to the metric system. U.S. schools must begin to prepare for new expectations regarding the teaching of metrics.
- *Indian Nations at Risk Task Force.* This task force is charged with reviewing data on the quality of learning in the Nation's tribal schools; comparing curricula, standards, and expectations regarding the education of Indian and non-Indian students; studying program quality; and reviewing major changes that have occurred in the education of Indians. Included under these topics will be discussions of how to make mathematics and science relevant to Indian culture and experience so that Indian students can be appropriately educated.

Third, the Department will explore other ways to address urgent mathematics and science education concerns. Some changes have already been proposed and/or adopted; others will need further investigation and analysis. As these opportunities for change are considered, the following concerns must be evaluated:

1. Science and mathematics must be taught to *all* students. While it is crucial for us to encourage students to become scientists and engineers, we must not forget that science and mathematics literacy are essential for everyone.

2. We must be able to leverage other resources with ours. What we invest must multiply itself many times if we are to make a difference. An obvious example is in teacher development, where what is learned can be multiplied in classroom after classroom.
3. We must emphasize those areas where we play a unique role and have a proven track record of success. Clearly research, dissemination, and collection of statistical information are areas where we have proven successful. Another area is our increasing collaboration with schools, school systems, states, and others.

With these criteria in mind, some of the activities the Department is undertaking or considering follow:

For Teacher Professional Development

- The Administration has proposed to emphasize the preparation and continued professional development of elementary and middle school teachers in order to increase their level of knowledge and skills in mathematics and science. This will be done by substantially expanding the Dwight D. Eisenhower Science Education Program by $94 million, a 69 percent increase.
- Another possible option is to support local school districts in establishing professional development schools for teachers that provide education for a cadre of well-prepared science and mathematics teachers, including those who enter the profession through alternative certification. Special emphasis would be placed on developing teachers from minorities, women, or disabled.

For Research and Assessment

- The Department will sponsor national research centers to focus on improved teaching and learning in science and mathematics. These centers will also address appropriate uses of advanced information technologies as tools in the acquisition and application of mathematical and scientific knowledge and skills.
- Expansion of the National Assessment of Educational Progress to accommodate state-by-state achievement comparisons has previously been discussed. Consideration should also be given to increasing support for the NAEP to permit testing with measures that are appropriate for *authentic* assessment of mathematics and science achievement. This will require the use of measures that stress analytical thinking, problem solving, and active, hands-on performance.

For School Improvement

- The Department will assess the feasibility of providing a network of technical assistance with special emphasis on mathematics and science instruction. This assistance would be delivered through the Department's existing research/practice collaboratives, regional laboratories, ERIC Clearinghouses, and other Department-supported institutions.

- The Department will ask every program to identify ways in which science and mathematics activities could be encouraged and used to reinforce state and local systemic reform initiatives.

These are only some of the actions being considered by the Department. Nothing will be certain until all deliberations have been completed and bugetary contraints weighed. The examples are given to indicate that the Department is currently evaluating its options and intends to make a strong commitment to science and mathematics education.

Fourth, the Department will collaborate with other government agencies that are committed to improving mathematics and science education. Perhaps the most significant step has been the formation of an Education and Human Resources Committee under the Federal Coordinating Council for Science, Engineering, and Technology (FCCSET). The President's Science Advisor, Dr. D. Allan Bromley, has appointed Secretary of Energy James Watkins as chair, with the Under Secretary of Education and the Assistant Director for Education and Human Resources of the National Science Foundation as vice chairs. This committee will encourage and coordinate programs and policies related to science, mathematics, engineering, and technological education, training, and human resource development.

In addition to the FCCSET Committee, the Secretary of Education and the Director of the National Science Foundation (NSF) have established formal mechanisms for the coordination of mathematics and science education programs between the two agencies. Collaboration of the two agencies under this mechanism has already reached beyond coordination to the development of cooperative initiatives and relationships. Areas in which agreements in principle to collaborate (subject to receipt of positive peer reviews) include:

Project 2061 Phase II curriculum development projects.
- Joint funding of a mathematics assessment conference by the National Academy of Science's Mathematical Science Education Board.
- Joint funding of educational television programs.
- Joint funding of National R&D Centers for mathematics and science education.

Other areas of collaboration with NSF and with other Federal agencies are being pursued. Each agency brings complementary strengths and expertise that make the joint activity much stronger than individual initiatives could be.

CONCLUDING STATEMENT

Much remains to be done to make the national educational goals a reality by the year 2000. There will be interesting implications for our institutions when these goals are achieved. For example, what will this mean for our colleges and universities when they are faced with an incoming student body that is far better prepared in the areas of science and mathematics? Here are a few of the problems that will then have to be faced.

- Institutions would be able to discontinue the large number or remedial courses now being offered. In their place, more demanding science and mathematics courses would need to be offered that would require different staffing patterns.
- The pool of minority and women students ready to enroll in mathematics and science courses would be much larger. Overall enrollments would markedly increase, forcing a reassessment of facilities and equipment needs.
- Likewise, the pool of minority students from which to draw prospective teachers would also increase. Internship placements, collaborations with schools, and ongoing professional development opportunities for teachers would all have to be expanded.

Businesses and industries would also have to make the following adjustments to absorb a better prepared workforce:

- Supervisory staff would need to be reeducated to guide and assist these new workers. In addition, the skill levels of older staff would need to be improved in order to keep up. In some instances, a new pattern of management and staff expectations would need to be developed.
- Equipment might need to be upgraded to take advantage of the capabilities of these new employees.

For now, this is a mental exercise. But it could be reality soon. These possibilities could present daunting challenges for our institutions, but they certainly would be more interesting and rewarding than the problems we now face. Let us hope that these will be the challenges we will face when we enter the next century, and let us work together to guarantee this future.

We wish to thank Senta Raizen, Director, Center for Improving Science Education, and Kirk Winters, Special Assistant in the Office of Educational Research and Improvement, U.S. Department of Education, for kindly agreeing to review and comment on this chapter.

REFERENCES

1. International Association for the Evaluation of Educational Achievement. 1988. *Science Acheivement in 17 Countries: A Preliminary Report.* Pergamon Press, Oxford, England.

2. Lapoint, A.E., Mead, N.A., Phillips, G.W. 1989. *A World of Differences: An International Assessment of Mathematics and Science*. Educational Testing Service, Princeton, N.J.
3. National Assessment of Educational Progress. 1988. *The Mathematics Report Card: Are We Measuring Up?* Report No. 17-M-01. Educational Testing Services, Princeton, N.J.
4. National Assessment of Educational Programs. 1988. *The Science Report Card: Elements of Risk and Recovery*. Report No. 17-S-01. Educational Testing Service, Princeton, N.J.
5. McKnight, C.C., et al. 1987. *The Underachieving Curriculum: Assessing U.S. School Mathematics from an International Perspective*. Stipes Publishing, Champaign, IL.
6. Westat, Inc. 1988. *Tabulation for the Nation At Risk Update Study as Part of the 1987 High School Transcript Study*. U.S. Department of Education, National Center for Education Statistics, Washington, D.C.
7. National Science Board. 1989. *Science and Engineering Indicators — 1989*. NSB 89-1. U.S. Government Printing Office, Washington, D.C.
8. Weiss, I.R. 1987. *Report of the 1985–86 National Survey of Science and Mathematics Education*. Research Triangle Institute, Research Triangle Park, N.C.
9. Weiss, I.R. 1988. "Course Background Preparation of Science Teachers in the U.S.: Some Policy Implications." In A.B. Champagne (ed.), *Science Teaching: Making the System Work*. American Association for the Advancement of Science, Washington, D.C.
10. Shavelson, R., McDonnell, L., and Oakes, J. 1989. *Indicators for Monitoring Mathematics and Science Education: A Sourcebook*. Rand Corporation, Santa Monica, CA.
11. National Council of Teachers of Mathematics. 1989. *Curriculum and Evaluation Standards for School Mathematics*. Reston, VA.
12. National Research Council. 1989. *Everybody Counts: A Report to the Nation on the Future of Mathematics Education*. National Academy Press, Washington, D.C.
13. American Association for the Advancement of Science. 1989. *Science for All Americans*. Washington, D.C.
14. National Science Teachers Association. 1989. *Essential Changes in Secondary School Science, Scope, Sequence, and Coordination*. Washington, D.C.
15. New Jersey State Department of Education. 1989. *Provisional Teachers Program: Fifth Year Report*. Trenton, N.J.

Science Education in the United States: Issues, Crises and Priorities. Edited by S. K. Majumdar, L. M. Rosenfeld, P. A. Rubba, E. W. Miller and R. F. Schmalz. ©1991, The Pennsylvania Academy of Science.

Chapter Two

THE EDUCATION SUMMIT, NATIONAL GOALS, AND ACCOUNTABILITY: PRESIDENT BUSH AND THE GOVERNORS ANNOUNCE A NEW ERA

JOHN ASHCROFT

Governor of Missouri
Executive Office
State of Missouri
P.O. Box 720
Jefferson City, MO 65102

John Ashcroft
Governor of Missouri

"We did not meet for yet another conference, more speeches and white papers. No. We forged a National compact in education reform." (President Bush reflecting on the Education Summit in remarks at the National PTA Legislative Conference, March 5, 1990).

INTRODUCTION

Inviting the nation's governors to a "Summit" on education last fall, President Bush declared his intentions for "more than dialogue—a new sense of direction." "Developing a national strategy that includes new directions for education is pivotal to the Nation's well-being," he observed, stating his belief that without "tradition-shattering" strategies we would fail to fix the serious problems in our education system.

This paper describes that meeting, its outcomes, and its promise. It elaborates on the most promising result of the Summit—the agreement to establish national goals and to be held publicly accountable for their accomplishment. Concluding sections encourage educators and policy-makers to draw on evidence about what works in education and to support experimentation and assessment of multiple, diverse reform efforts.

THE NATIONAL EDUCATION SUMMIT, CHARLOTTESVILLE, VIRGINIA, SEPTEMBER 27–28, 1989

Two Days of Discussion and Debate: The governors and President Bush agreed to casual, candid discussions. For two days, with only minimal staff and no outside speakers, they deliberated:

- *Teaching, Revitalizing a Profession* (chaired by Governors Branstad of Iowa and Gardner of Washington);
- *The Learning Environment* (chaired by Governors Hunt of Alabama and Mabus of Mississippi);
- *Governance: Who is in Charge?* (chaired by Governors Carruthers of New Mexico and Clinton of Arkansas);
- *Choice and Restructuring* (chaired by Governors McKernan of Maine and Sinner of North Dakota);
- *A Competitive Workforce and Education* (chaired by Governors Campbell of South Carolina and Schaefer of Maryland); and
- *Postsecondary Education: Strengthening Access and Excellence* (chaired by Governors Ashcroft of Missouri and Andrus of Idaho).

They captured the sense of the agreements hammered out during the summit in a *Joint Statement*, which included the following:

"The President and the nation's Governors agree that a better educated citizenry is the key to the continued growth and prosperity of the United States. Education has historically been, and should remain, a state responsibility and a local function, which works best when there is also strong parental involvement in the schools. And, as a Nation we must have an educated workforce, second to none, in order to succeed in an increasingly competitive world economy."

To develop this "better educated citizenry," the President and Governors agreed to

1. establish a process for setting national education goals;
2. seek greater flexibility and enhanced accountability in the use of Federal resources to meet the goals . . . ;

3. undertake a major state-by-state effort to restructure the education system; and

4. report annually on progress in achieving the goals.

The agreement elaborates on each of these promises. It notes, for example, the need to build a "broad-based consensus" around the goals. It recognizes the importance of including diverse groups inside and outside education in defining them. *It makes clear the intention that they be ambitious targets* regarding

- children's readiness to start school;
- students' performance on international achievement tests, especially in math and science;
- reduction of the drop-out rate and improvement of academic performance especially among at-risk students;
- functional literacy of adult Americans;
- the level of training necessary to guarantee a competitive workforce;
- the supply of qualified teachers and modern technology; and
- the establishment of safe, disciplined, and drug-free schools.

In this pledge to improve the effectiveness of "the federal/state partnership," the Joint Statement urges *examination and removal of counter-productive rules and regulations.* It argues for "swapping red tape for results" and acknowledges that the Federal financial role in education—although still important—is limited, having two functions: to promote educational equity and to provide research and development for programs that work and for good information on the performance of students, schools, and states.

This commitment to a state-by-state restructuring of education pledges systems of accountability focusing on *results* rather than on compliance with rules and regulations. *It advocates decentralizing authority so educators are empowered to address goals and be held accountable for accomplishing them.* It calls for rigorous instructional programs to ensure that every child can acquire the knowledge and skills required in an economy in which our citizens must be able to think for a living. It pledges to develop first-rate teachers and professional environments for them that reward success and sanction failure. It encourages active, sustained involvement from parents and the business community.

In addition, *governors promise as elected chief executives to be held accountable for progress in meeting the national goals.* They note their intention to establish clear performance measures once the goals are defined and then to issue annual report cards on the progress of students, schools, states, and the federal government.

The Education Summit proved demanding but stimulating. Each participant worked conscientiously to forge an agreement that would be both meaningful and enforceable. The Joint Statement concludes,

"Over the last few days we have humbly walked in the footsteps of Thomas Jefferson. We have started down a promising path. We have entered into a compact—a Jeffersonian compact—to enlighten our children and the children of generations to come. The time for rhetoric is past; the time for performance is now."

The Merits of This Plan: The three elements of the agreement—national performance goals, freedom to devise diverse strategies to address them, and accountability for results—hold considerable promise for improving the Nation's education system.

It emphasizes results, not processes, as a way to foster creativity and commitment. This approach is the only way to stimulate America's considerable talents and preserve the rich tradition of diversity that has served the nation so well historically.

In fact, the United States holds a distinct advantage compared to its international competitors. It is free of the mandates of a bureacratic, central education ministry with requirements for uniformity and standardization. It is free to be innovative and resourceful—to try to solve long-standing problems in new and diverse ways. The Summit heralded and promoted the continuance of America's traditional strength of diversity and flexibility.

This strategy matches that of the scientific method, itself. It emphasizes multiple hypotheses, the testing of them, and the implementation and further testing of those that work.

America's honor roll of great scientists demonstrates the advantages of multiple, diverse talents building on different backgrounds and education. Think of the contributions of Einstein, Salk, Fermi, Oppenheimer, Franck, and Curie, for example[1,2] Indeed, during the first fifty years of the compilation of the "American Men of Science," one-sixth of the nearly 2600 starred in the listing (for special contributions to their fields) were foreign-born. In addition, many were foreign-educated.[3] In 1988, roughly half of the recipients of U.S. patents were of foreign origin.[4]

Had the President and governors chosen, rather, to mandate a reform *process* instead of targets, they would've risked neutralizing the great surges of intellect seen across the Nation. Mandating a uniform process to solve education problems is a certain way to flatten, homogenize, and squelch creativity. Instead the Summit's participants agreed to identify objectives, assess results in reaching them, and stand accountable for those results. Speaking specifically of science education, Myron Atkin, former dean of Stanford's school of education, endorses this notion: "The challenge in improving [it] is to avoid the extremes of top-down curriculum pronouncements, on the one hand, and unmonitored grassroots and opportunistic strategies on the other."[5]

Assessing Progress and Reporting: In their 1990 summer meeting, the Nation's governors established the "Education Goals Panel" to be comprised of five governors, four members of the Executive Branch, and four Congressional leaders. The panel will determine measures for each goal, define the baselines against which to evaluate progress, plan the format for the annual report, and describe the federal government's actions to implement their part of the agreements reached at the Summit.

THE NATIONAL GOAL IN MATH AND SCIENCE

Five months after the Summit, the governors and White House published the National Education Goals, sticking to their original promise to be ambitious and declaring, "These goals are about excellence."[6] The National Goals read, as follows:

Goal 1: Readiness. By the year 2000, all children in America will start school ready to learn.

Goal 2: School Completion. By the year 2000, the high school graduation rate will increase to at least 90 percent.

Goal 3: Student Achievement and Citizenship. By the year 2000, American students will leave grades four, eight, and twelve having demonstrated competency over challenging subject matter, including English, Mathematics, Science, History, and Geography, and every school in America will ensure that all students learn to use their minds well, so they may be prepared for responsible citizenship, further learning, and productive employment in our modern economy.

Goal 4: Mathematics and Science. By the year 2000, U.S. students will be first in the world in mathematics and science achievement.

Goal 5: Adult Literacy and Lifelong Learning. By the year 2000, every adult American will be literate and will possess the knowledge and skills necessary to compete in a global economy and execise the rights and responsibilities of citizenship.

Goal 6: Safe, Disciplined, and Drug-free Schools. By the year 2000, every school in America will be free of drugs and violence and will offer a disciplined environment conducive to learning.

Upon their publication, one education writer noted, "The goals [the governors] set were ambitious, and political veterans could not recall any previous occasion on which the White House and the National Governors' Association had agreed on a set of goals for anything, much less an enterprise that is generally considered to be a local affair."[7]

What does the National Goal in math and science require, and where does the U.S. stand now with respect to it? Three subsidiary objectives were specified:

1. Math and science education will be strengthened throughout the system, especially in the early grades.
2. The number of teachers with a substantive background in mathematics and science will increase by 50 percent.
3. The number of U.S. undergraduates and graduate students, especially women and minorities, who complete degrees in mathematics, science, and engineering will increase significantly

The National Goal is that U.S. students be *first internationally* by the year 2000. In contrast, current achievement rankings are low: in 1988, 13-year-olds in the United States scored last in math and near the lowest in science among twelve countries and Canadian provinces; in 1986 even those U.S. students who had taken high school biology, chemistry, or physics courses scored last in those subjects among fourteen industrialized countries.[8]

One objective requires schools to *strengthen courses,* especially at the early grades. Currently they offer elementary students (K-3) roughly twenty minutes a day in science and math; about twenty-seven minutes per day to those in grades 4-6. And contrary to research findings about effective science and math instruction, having students read a textbook continues to be the predominant classroom approach.[9]

The second objective is to *increase the numbers of science and math teachers with substantive backgrounds* by fifty percent. How many does that mean? Current estimates of the teaching force in the fields vary, but rough estimates show approximately 230,000 teachers employed nationwide in math in grades 5-12 and 130,000 in science.[10,11,12] A 1986 survey found that roughly one-quarter of them have taken the college courses recommended by their professional organizations as the appropriate "substantive background."[13]

The third objective calls for increases in *degrees awarded—especially to women and minorities—in science and engineering.* Their numbers have increased recently. Data from the National Science Foundation show women receiving just over 144,000 of the roughly 400,000 undergraduate and graduate degrees in these fields in 1985[14]; the U.S. Department of Education estimates that roughly 31,000 undergraduate and graduate degrees in life science, computer systems, and engineering were awarded that year to minority students.[15]

What Accounts for These Troubling Findings? After examining the current state of affairs in science and math education, several prominent groups offer insights into the causes of the difficulties. They note that too often our schools take curious, fascinated young students and, rather than nurturing this natural curiosity, "extinguish it with catalogs of dreary facts and terms." Indeed, by the third grade, half of all students don't want to take any more science[16], and their feelings about math are similar: in 1981, 9-year-olds on the National Assessment of Educational Progress rated math as the best liked of five academic subjects; 13-year-olds rated it as the second-best liked; 17-year-olds rated it as the least liked.[17]

Even those students who do take advanced math and science courses score poorly on international tests. International comparisons show that few U.S. students take the higher level courses (6%, 1%, and 1%, respectively, in biology, chemistry, and physics compared to 12%, 16%, and 11% in those fields in Japan, for example), and those who do too often fail to master the more challenging aspects of the disciplines. This low "yield"—the low proportion of young people taking these courses and the little they learn in the them—sends a clear signal of the need to improve the way science and math are taught![8]

In addition, further analysis of the test scores suggests where among our students the improvements should focus. While top U.S. math students score as well or better than those of other countries (in algebra, for example, our top score exceeds everyone else's), U.S. students in the middle two quartiles score lower than the same groups from other countries![9]

Similar patterns emerge in science. Although analyses similar to the ones in math are not widely available, the problems of middle and lower students are evident in tabulations of the proportion of American students scoring at each of five proficiency levels. Almost 96% of U.S. students "know everyday science facts" (the lowest proficiency level); 78% demonstrate the next higher level of "understanding and applying simple science principles"; then fewer than half—42%—reach the next level, "using science procedures and analyzing data." (Twelve percent and one percent, respectively, accomplish the top levels, "understanding and applying intermediate knowledge and principles" and "integrating information and experimental evidence.") In contrast, most—8 of 11—other countries have the majority of their students demonstrating at least the third level of proficiency; that is, they can use scientific procedures and analyze data.[20]

The most promising reforms, then, for achieving the national goal by the year 2000 will pay considerable attention to improving how the middle and lower U.S. students perform. As noted by one analyst, "The curriculum offered to our highest . . . classes is not a core problem when seen cross-culturally, but the curriculum offered to our lower . . . classes is."[21]

ATTAINING THE GOAL

Solid Information to Build On: Several researchers and commissions have assessed education in mathematics and science. Although their investigations proceeded independently, their findings are strikingly similar. Table 1 displays the characteristics of effective programs in both disciplines—clear, well-defined curriculum; hands-on educational experiences; attention to needs of students at all ability levels; inquiry-based instruction; and a focus on the core of important material.

As compelling as these findings are, however, to reap the full benefits of science and math reforms they must be placed in schools that are effective. Research documents that good schools:
- benefit from strong instructional leadership by their principals,[37,38,39]
- operate with enough autonomy to encourage those principals to manage the site effectively,[40]
- maintain a climate conducive to learning, free of discipline and vandalism problems,[41,42,43,44]
- emphasize basic academics skills on a schoolwide basis,[45,46]

- hold high expectations for all teachers, students, and parents,[47,48,49]
- use a system for monitoring and assessing pupil performance tied to instructional objectives,[50]
- recognize academic successes publicly and schoolwide,[51]
- maximize learning time,[52] and
- encourage collegial efforts among teachers for planning, advising each other on strategies, and offering remediation where necessary.[53]

Illustrative Efforts Underway: In addition to the solid base of information available about effective practices in science, math, and good schools, education policy can be informed by an impressive array of reform efforts currently underway. A few illustrations are included below, and more comprehensive information is available from professional groups overseeing various initiatives.

TABLE 1

What Research Tells Us About Effective Science and Math Programs

In science, effective programs—	In math, effective programs—
Address a curriculum that is not necessarily age-specific, but at least is sequential and well-defined[22]	Follow well-defined courses with a clear syllabus[23]
Stress hands-on, experiential learning[24]	Use varied ways of instruction with younger grades, especially, manipulating physical objects[25,26]
Attend to learning needs of all children, all grades, all performance levels[27]	Advocate common math experiences for all students, not minimal math for the majority, advanced math for few[28]
Dismiss the "layer cake" of high school courses in biology, chemistry, and physics, addressing, instead, the connections among them, the "scientific habits of mind"[29]	
Start with students' questions about phenomena; engage students in active use of hypotheses[30]	Encourage student-centered practice, not teachers as authorities transmitting knowledge[31]
Focus on a core of ideas and skills with the greatest scientific merit, not comprehensive smatterings of facts—depth not breadth[32]	Develop "broad-based math power" in students instead of merely inculcating routine skills[33]
	Rely less on paper and pencil tasks, more on calculators and computers[34]
Gear content to real life[35]	Emphasize topics relevant to students' present and future needs, not just what they have to know for later courses[36]

As noted earlier, science and math educators are encouraged to offer courses to younger children, allow them "hand-on" experiences, and build on their natural curiosity. The Monsanto Fund, headquartered in St. Louis, has funded efforts at Maryville College to design and aid the implementation of "Academies of Developmental Learning" as early childhood centers in three St. Louis schools. The academies will offer experiential, hands-on learning experiences focusing on "logical-mathematical and physical (science) knowledge" and will be designed to have the students' parents serve as team-members with the instructional staff.

The involvement of parents is encouraged, also, in an elementary school in suburban Indianapolis through its "PACTS—Parents and Children for Terrific Science" program. In a lab at the nearby high school, thirty teams of students and parents do experiments one evening each week under the supervision of teachers. The school reports improved performance and attitudes on the part of all participants.

Research tells us also to approach both science and math as integrated sets of principles, not conglomerations of facts and figures. Three schools in Houston, 100 in California, and sites in Philadelphia, San Francisco, San Diego, two counties in Georgia, and McFarland, Wisconsin are implementing a new curriculum developed by the American Association for the Advancement of Science that emphasizes principles across disciplines and "scientific habits of mind."

University staff are aiding the reforms as well. A University of Iowa team helps districts revamp their curricula so that students in grades 4-9 come to class with questions, design ways to discover the answers, and proceed with their investigations. Similarly, a physicist at Caltech, arguing that the best science program is one "with no textbook at all," helps schools in Pasadena develop "kits" or choose commercial ones to address the objectives of their science curriculums.

These collaborative efforts between schools and university faculty help teachers continue learning, too, and often rekindle their enjoyment of their disciplines. University City, Missouri has a formal program for summer institutes aimed at improving teacher morale and practice in science. Focusing on teachers in the lower elementary grades, "SEER—Science Education for Equity Reform" offers all teachers a formal course in physics under the direction of the American Institute of Physics.

In their efforts to reform science and math education to be more inquiry-based, policy-makers and educators can learn from efforts underway nationwide to transform all schooling in this manner. The Coalition of Essential Schools, headed by Dr. Ted Sizer at Brown, forms the core of sites nationwide that are redesigning schools. They emphasize a central intellectual purpose, define simple goals so each student masters a limited number of essential skills, and ensure that those goals are expected of all students. In addition, they personalize teaching and learning as much as possible, building on the "student-as-worker." Before students are graduated, they must show evidence that they have mastered the essential areas. Numerous science and math curricular reforms build on this approach.[54,55]

Addressing needs in science and math of learners at all achievement levels will require considerable realignment of attention. While residential academies for top students show great promise in several states,[56] similar opportunities must be offered to middle and lower students. Consider, for example, the potential for learning if juniors or seniors in our urban high schools were offered the opportunity for a year-long residential experience at some of our underutilized state colleges and university campuses. Great benefits would result from raising their aspirations as well as from improving their study habits and academic performance.

Similarly, note the advantages to offering teachers summer experiences in business and industry related to their fields or in residential university studies. McDonnell Douglas Corporation in St. Louis hires science, math, and computer science teachers for summer employment after which they redesign curriculum and make professional presentations about the new instructional areas. While such programs exist in many states, the need for vast upgrades in teacher knowledge, skills, and often morale, calls for more to be done.

DISCUSSION AND CONCLUSIONS

Americans have much to celebrate in their education system in general, and in science and math instruction, in particular. Bright, talented students and teachers who are conscientious and productive are numerous; and parents, businesses, and government leaders offer tremendous support for education.

The need for extensive improvements is also evident, however, to attract more students to science and math and to improve what they learn in those areas.

The deliberations and the agreements forged among the nation's governors and the President have laid a solid groundwork for these improvements. As President Bush remarked later about the Education Summit,

"We've moved in concert to bring a sense of direction to education reform. We've held the first-ever summit with the nation's governors. And we've set ambitious goals for our students, our schools, and ourselves—rallying points for the progress we all know is greatly needed now." (on the occasion of presenting the National Teacher of the Year Award, April 4, 1990.)

The sense of direction has its foundation in the national goals the governors pledge to address together. The promise for attaining them lies in diverse initiatives around the nation using the best minds to solve problems. And annually the governors and White House will assess the progress and report those findings to all Americans.

At one time, offering inadequate education was a moral travesty as our schools failed to help all citizens fulfill their potential. Now the problem is broader: not only do poor schools fail to fulfill the trust citizens place in them, but they also threaten our nation's economic survival as our businesses are handicapped with underprepared employees. Knowledge is available about how to proceed; policy-makers and educators must muster the will to do so.

REFERENCES

1. The Editors of Fortune Magazine. 1961. *Great American Scientists: America's Rise to the Forefront of World Science.* Prentice-Hall, Englewood Cliffs, New Jersey.

2. Manning, Kenneth R. 1983. *Black Apollo of Science: The Life of Ernest Everett Just.* Oxford University Press, New York.

3. Visher, Stephen Sargent. 1947. *Scientists Starred in American Men of Science.* John Hopkins Press, Baltimore, MD.

4. National Science Board. 1989. *Science & Engineering Indicators.* U.S. Government Printing Office, Washington, D.C.

5. Atkin, J. Myron. 1990. On 'Alliances' and Science Education. *Education Week.* April 11.

6. National Governors' Association. 1990. *National Education Goals.* Washington, D.C., February 25.

7. Fiske, Edward B. 1990. Lessons. *New York Times,* February 28.

8. National Science Board. *op. cit.*

9. *Ibid.*

10. U.S. Department of Education. 1989. *The Condition of Education: Postsecondary Education.* Government Printing Office, Washington, D.C.

11. Weiss, I.R. 1987. *Report of the 1985–86 National Survey of Science and Mathematics Education.* Research Triangle Institute: Research Triangle Park, N.C.

12. National Science Teachers Association. 1990.Registry of Science Teachers in the United States. Washington, D.C.

13. Weiss, I.R. 1988. *Course Background Preparation of Science and Mathematics Teachers in the United States.* Special Report Prepared for the National Science Foundation, Washington, D.C.

14. National Science Board. *op. cit.*

15. U.S. Department of Education. *op. cit.*

16. _____ . 1990. Not Just for Nerds. *Newsweek.* April 9: 52–64.

17. International Association for the Evalation of Education Achievement. 1989. *The Underachieving Curriculum: Assessing U.S. School Mathematics from an International Perspective.* Stipes Publishing Company, Cham-

paign, Illinois.

18. *Ibid.*
19. Westbury, Ian. 1990. Personal Correspondence. School of Education, University of Illinois, Urbana-Champaign, May 3.
20. National Science Board. *op. cit.*
21. Westbury, Ian. 1989. The Problems of Comparing Curriculums across Educational Systems. In Alan C. Purves (ed.). *International comparisons and Educational Reform.* Association for Supervision and Curriculum Development, Alexandria, Virginia.
22. _____ . *Newsweek. op. cit.*
23. International Association for the Evaluation of Education Achievement. *op. cit.*
24. _____ . *Newsweek. op. cit.*
25. International Association for the Evaluation of Education Achievement. *op. cit.*
26. U.S. Department of Education. 1986. *What Works: Research about Teaching and Learning.* Government Printing Office, Washington, D.C.
27. National Council on Science and Technology Education. 1990. *Project 2061: Science for All Americans.* American Association for the Advancement of Science, Washington, D.C.
28. Mathematical Sciences Education Board. 1990. *Reshaping School Mathematics: A Philosophy and Framework for Curriculum.* National Academy Press, Washington, D.C.
29. National Council on Science and Technology Education. *op. cit.*
30. *Ibid.*
31. Mathematical Sciences Education Board. *op. cit.*
32. National Council on Science and Technology Education. *op. cit.*
33. Mathematical Sciences Education Board. *op. cit.*
34. *Ibid.*
35. _____ . *Newsweek. op. cit.*
36. Mathematical Sciences Education Board. *op. cit.*
37. Cohen, Michael. 1981. Effective Schools: What the Research Says. *Today's Education,* April-May: 46–50.
38. _____ . 1984. What Makes Great Schools Great. *U.S. News & World Report,* August 27.
39. Purkey, Stewart C. and Smith, Marshal S. 1982. *Ends Not Means: The Policy Implications of Effective Schools Research.* Wisconsin Center for Education Research, School of Education, University of Wisconsin, Madison, Wisconsin.
40. *Ibid.*
41. Cohen. *op. cit.*
42. Sewall, Gilbert T. 1984. Great Expectations, Successful Schools. *Education Week.* February 29.

doctorates awarded in 1988! There is also considerable concern with the under-representation of minorities (particularly Blacks, Hispanics and Native Americans) in all the sciences, and of women in the physical sciences and engineering. This presents not only a serious equity issue, but also a formidable problem in meeting anticipated needs for skilled personnel in science- and technology-related fields, given the profound demographic shifts anticipated in the population.

Student Performance in Science

The most recent evidence concerning the science competence of U.S. students compared to students elsewhere comes from two sources: the second international study of science achievement conducted by the International Association for the Evaluation of Educational Achievement (IEA) during 1983 to 1986, and an international assessment of science and mathematics achievement conducted by Educational Testing Service (ETS) in 1988. Although these tests are largely in short-answer format and do not probe many important goals of science education, the results are nevertheless dismaying. Some of the most startling findings from these two large-scale studies are:

From the IEA study:[2]

- At age 10 (grades 4/5), U.S. students rank eighth among 15 countries, with Japan, Korea, Finland, and Sweden ranking the highest.
- At age 14 (grades 8/9), U.S. students are among the lowest scoring, being third from the bottom among 17 countries together with Thailand, and only Hong Kong and the Philippines scoring lower. The highest scoring countries were Hungary and Japan. The bottom 25 percent of U.S. children did particularly badly, with the lowest scoring doing no better than chance.
- Even for the relatively small (and presumably science-interested and science-able) 12th grade population taking chemistry and physics, U.S. students managed to rank only eleventh and ninth, respectively, out of 13 countries. In biology, they did worse, ranking last.

From the ETS study:[3]

- More than 70 percent of the 13-year-olds in British Columbia and Korea can use scientific procedures and analyze scientific data, while only about 40 percent can do so in the U.S.
- U.S. 13-year-olds fall well below the mean performance achieved by students in twelve countries and Canadian provinces, with only Ireland, French Ontario, and French New Brunswick joining the U.S. in this low ranking.

There is hardly more reassuring news from the most recent national assessment of science achievement conducted by the National Assessment of Educational Progress (NAEP) in 1986.[4] The report summarizing the findings notes that test scores of 13- and 17-year-olds were still below those achieved in 1969, the year of the first science assessment, whereas 9-year-olds had, after sharing in the general test score decline of the 1970s, just about returned to their 1969 levels. The report concludes (p.6): "More than half the nation's 17-year-olds appear to be inadequately prepared . . . to perform competently jobs that require technical skills . . . Only 7 percent . . . have the prerequisite knowledge and skills thought to be needed to perform well in college-level science courses . . . the probability that many more students will embark on future careers in science is very low."

With respect to the achievement of Hispanic and Black students in science, the NAEP findings are also discouraging: the average proficiency of 13- and 17-year-olds from these population groups remains some four years or more behind their White peers; and only about 15 percent of Black and Hispanic 17-year-olds demonstrated the ability to analyze scientific procedures and data, compared to nearly half of the White students. There also continues to be a gender gap in achievement in the physical sciences, which increases as students move through school, becoming very large by eleventh grade, with girls lagging considerably behind boys in physics.

The Underrepresentation of Women and Minorities in Scientific Fields.

These differences in achievement do not bode well for increasing the participation of women and minorities in science and technology. Of all factors, prior achievement is most clearly related to future prospects, including placement in academic tracks, enrollment in science and mathematics courses, and increase in competence. Thus, even though the achievement gap between Blacks and Whites has been narrowing somewhat in the last two decades and women have increased their participation in science-related occupations, science education will have to proceed quite differently if the current underrepresentation of minorities and women is to be remedied.

What is the extent of this underrepresentation, and why is it of concern? In 1986, only 13 percent of employed scientists, mathematicians, and engineers were women, and Hispanics and Blacks each constituted about 2 percent of the scientific workforce, even though these three groups made up 49 percent, 7 percent, and 3 percent, respectively, of all professional workers.[5] Statistics on current enrollment in graduate programs in the sciences, which can be taken as an indicator of the number of future science professionals, give little indication of change in the near future. Although women have increased in the proportion of advanced science degrees earned, a preponderance of these degrees is in psychology and the social sciences—a tendency that also holds for Blacks and Hispanics.

For several decades, this underrepresentation in the prestigious and well-paying science/technology fields has been of concern because it indicates continuing problems with lack of equal opportunity. Today, however, the concern has broadened into a more comprehensive manpower issue, namely whether the country will have a workforce adequate to the challenges of the 21st century. There are two reasons why the manpower issue has taken on increasing significance: the perceived need for a more highly skilled workforce, and the demographic shifts in the population.

Increased Skill Requirements. The notion of an adequately educated person, that is, one who can competently address personal, family, civic, and work responsibility in modern society, is embodied in the concept of "literacy." Originally, the term referred to the ability to read and write; presently, the notion of numerical literacy was added. Historically, the definition of what constitutes a literate person has encompassed more and more education and an escalating standard of performance. Today, requirements are still increasing, triggered in part by job demands imposed by a changing U.S. economy. While there is some debate about the mix of skills that will be needed in the next several decades, there is little question that computers and associated information technologies have brought about new opportunities and demands.[6] The changes in both the manufacturing and service industries due to computerization and the creation of a global marketplace are making demands for workers who can be reflective and flexible. Many are likely to have responsibilities that will require the kind of higher order thinking described by Resnick:[7] capability for non-algorithmic solving of complex problems; knowing that optimal solutions cannot always be specified in advance; ability to take several perspectives in order to develop better approaches; awareness that multiple solutions often are possible; ability to appraise the costs and benefits of alternatives and make reasoned judgments; willingness to obtain additional information when necessary to reduce uncertainty; and seeking to impose order and meaning on ill-defined problems and seemingly random information. The kind of non-routine problem solving encountered with increasing frequency in the world of work (and in one's personal and civic life as well) requires effortful performance; it also requires a great deal more than traditional literacy skills. Yet, as the testing data indicate, the schools continue to fail to provide many youngsters from minority groups or from backgrounds of poverty with even minimum preparation let alone the kind of science and mathematics education that would enable them to obtain and hold rewarding jobs in an increasingly technology-driven economy.

Demographic Shifts. A recent report by the Task Force on Women, Minorities, and the Handicapped in Science and Technology[8] estimates that a shortage of 560,000 scientists and engineers may develop by the year 2010 if present trends population is seeing a significant increase in the cultural diversity of the youth population, so that many more school systems will join the large urban systems in having "minority majorities" as well as increasingly heterogeneous minority populations.[9] Thus, by the year 2000, 85 percent of the new entrants into the workforce will be women or minorities. Putting it crassly, the country will no longer be able to rely on White males to fill its needs in science- and technology-related occupations. For example, to meet the needs for individuals with doctoral degrees in these fields, postsecondary institutions will have to produce three times as many White women with relevant doctorates, 12 times as many Hispanics, and 20 times as many Blacks.[10].

ISSUES IN SCIENCE EDUCATION

Science education appears to be particularly ineffective for women, minorities and poor children, but in fact is not doing well by anyone. Students' attitudes as they progress through school tell the story. Many students seem to find science difficult, boring, and irrelevant.[11] The National Assessment of Educational Progress (NAEP)[12] found that, in third grade, about 70 percent of the students thought what they were learning in science was useful in every-day life and that they would use science in the future, but by age thirteen, only about half of the students thought so, and little more than a third thought that science would be important in their life work. On the whole, most students appear to be unenthusiastic about their science learning, and their attitudes seem to decline as they progress through school.

Elementary School

The problems start in the early grades. At a time when children are full of curiosity and eager to investigate the world around them—with all the inclinations of a scientist, as Carl Sagan has noted—they meet teachers who, for the most part, are insecure in science and schools that devote little time to it.

Curriculum and Instruction. Despite the exhortations in the national reports on educational reform that have been issued over the last decade (see, for example, *A Nation at Risk*[13] and *Educating Americans for the 21st Century*[14]), science has not become a "basic." Even in California, which has made concerted reform efforts in all the major subject areas, the implementation of science education reform in elementary school has lagged far behind mathematics and writing reforms.[15] Less than 20 minutes a day, on average, is spent on science in grades K-3; a little more than half an hour a day is spent in upper elementary school.[16] Moreover, science is not taught as a way of knowing, a way of responding to interesting questions about the natural world and human beings. Of the time actually devoted to science, lecture and discussion dominate three-fourths of the time in K-3 and almost wholly (nearly 90 percent) in upper elementary school; textbooks are the focus for instruction. Even though teachers and principals report that they believe that "hands-on" science activities promote effective science learning, few teachers actually find the time or opportunity to provide activities-based instruction to their students.[17,18] Most science textbook series and instructional practices reflect the belief that science learning consists of memorizing information dispensed by the teacher or through the textbook, with the students taking a largely passive role.[19]

Assessment. Current assessment practices aggravate the problem. As science education reform proceeds, more and more states are mandating science assessments in 4th (sometimes in 3rd or 5th) grade.[20] In some sense, this is a hopeful sign, since it is a tired but nevertheless true maxim that what gets tested gets taught. Hence, the increase in testing can be seen as an indication that science is assuming a more important role in the curriculum. The problem is the nature of the testing, which generally consists of multiple-choice questions designed, at best, to probe for recall of factual information and some limited deductive reasoning skills.[21] Because they are machine scorable and—if well designed and administered—give unequivocal numerical results, such tests are efficient for assessments involving large numbers of students, as in district-wide, state-wide, national or international assessments. Obviously, knowledge of factual information is important; even more important is the organization of information and the ability to apply it—matters less easily probed by multiple-choice tests. Moreover, such tests cannot probe at all the laboratory skills that students need to acquire: reading a real thermometer (not a temperature scale reproduced on paper), connecting a wire to a terminal, focusing a microscope, observing and recording the behavior of an animal, or measuring the amount of rainfall and its acidity. Even more important are the science-thinking skills involved in when and how these procedures are appropriately employed to answer a question that has arisen in the course of an investigation. More broadly, by the end of the elementary years, children should have acquired some facility in generating a hypothesis regarding some observed phenomena, designing an experiment that is a valid test of that hypothesis, and collecting, analyzing, and presenting data. If their science instruction involved genuine investigations, they should also have become more skilled at the type of problem-solving described by Resnick. Unfortunately, most of these science and generic thinking skills cannot be assessed with much validity through multiple-choice or other short-answer paper-and-pencil tests of the kind used in large assessments.

It is disappointing, therefore, that teachers tend to imitate the practices of large-scale assessments, even though they have other assessment strategies open to them. Tests and quizzes given by many teachers to assign grades to individual students focus largely on recall of vocabulary words and isolated facts, restatement of concepts out of context, and rote solving of the repetitive problem sets found at the end of units in the textbooks;[22] the possibility that there might be more than one "right" answer hardly arises. These, then, are the messages sent to the students about what really matters in science, since what is tested is, quite understandably, taken to represent what the teacher considers important to learn. No wonder, most children turn away from science as soon as they can, unless they are fortunate in their schooling circumstances and teachers, or special factors in their home or community environment keep them engaged with the subject.

Middle School

Conditions hardly improve at the middle level of schooling. Recent reports have recognized the great potential for cognitive development and consequently the educational needs of young adolescents.[23,24,25] Although the reforms suggested generally have not addressed the issue of how to teach particular subject matter, they have stressed the importance of school programs that encourage exploration and personal development through student-centered instruction, flexible scheduling and variable group sizes, interdisciplinary curricula and team teaching, and evaluation that involves students in self-appraisal and reflectiveness about their own work.

Curriculum and Instruction. The recommendations by science educators generally follow these prescriptions.[26,27] They emphasize learning science in settings that are interdisciplinary and related to society and daily living, using inquiry to identify and solve problems, and learning about current problems that illustrate the interdependence of science and technology. Hands-on science investigations are to be stressed as means for engaging studens as well as developing their understanding of science as a way of knowing; students are to be encouraged to discuss and explain their work and apply their newly gained knowledge. In this instructional paradigm, the role of the teacher changes from being an expert who delivers the codified science knowledge to be learned to being a model who enters into the scientific inquiry and discussion and coaches students to do so as well.

The reality that most students aged 10-14 experience in the science classroom is far different. Few specific programs that incorporate this style of instruction have been developed, fewer still that have successfully investigated how the acquisition of competence in critical thinking and complex problem solving can be facilitated by formal school experiences in science.

By contrast, the organizational arrangements, science curricula, science training of teachers, and working conditions that obtain in most schools all promote the kind of traditional instruction and learning based on memorization that turns so many students away from science. For example, 80 percent of the teachers assigned to 7th-9th graders are members of a subject-area department and teach that subject (say, earth sciences or biology) to intact classes—no interdisciplinary or team teaching there. The curriculum focuses almost exclusively on academic preparation and ignores most of the proclaimed developmental goals of middle-school. Teachers lack adequate preparation in science; more than a quarter are teaching out of field,[28] and even those qualified in science rarely have had training in working with young adolescents. In addition, the physical environments in which the teachers of young adolescents work tend to be more difficult than in elementary or senior high schools: buildings that discourage integrated, cross-disciplinary programs, inadequate facilities and equipment, and lack of supplies and materials suitable for individual or small-group instruction.[29]

Assessment. Raizen et al.[30] note that "nowhere [in middle-school science] are the gaps larger and the obstacles greater than in assessment. Current practices of assessing science learning with 'objective,' paper-and-pencil instruments focused on the mastery of basic science facts, using individual assessments exclusively, placing students in individual competition for grades, and measuring the quality of programs by aggregating students' test scores stand in stark contrast to assessment strategies that serve the type of science instruction we and others envision for young adolescents." The problems discussed at the elementary level with the sort of testing that characterizes most classroom and large-scale asessments today become even more troubling at the middle level. Early adolescence should be a time for exploration and experimentation, a time when science as a mode of disciplined inquiry should become particularly meaningful, yet current tests send the message that science is a static body of facts, principles, and procedures to be mastered and recalled on demand. As young adolescents grow in their capacity for analysis and abstract thought, such messages about the nature of science as embodied in tests become increasingly important in affecting students' attitudes.

A second problem is that most tests are designed and used to rank-order students against each other, with the inevitable result that there are a few winners and many that are perceived to be mediocre or losers. Their rankings become obvious to students as they move through the middle grades. As a result, most students will become persuaded that science is too difficult for them. They will opt to be tracked out of taking any science courses beyond those required for high school graduation and thereby limit their choices regarding postsecondary education or future careers. Although there is no clear evidence on when the decision process starts that leads to the very small retention rate of students in science, testing in the middle grades seemingly designed to convince students that only a very few are smart enough to pursue science likely is a critical factor. If assessment could be changed so as to demonstrate to students that they are making good progress in their development of science knowledge, investigative and laboratory skills, and science thinking and reasoning skills, quite apart from how they rank compared to other students, it might encourage rather than discourage their further engagement with the subject in high school and beyond.

High School

Interest in any science continues to wane as students move through high school. For example, of four million students who were high school sophomores in 1977, some 730,000 expressed an interest in science and engineering; by their senior year, this number had dropped to 590,000. As they became college freshman, only 340,000 said they intended to major in relevant fields, and 206,000 actually obtained baccalaureate degrees in these fields. About 9,700 or *0.24* percent of the 1977 high school sophomores are expected to obtain PhDs in science and engineering fields by 1992, twelve years after high school graduation.

There is some possibility that this bleak picture may be changing for the better. The education reform reports of the early 1980s all called for a more rigorous high school curriculum. In response, most states mandated increases in high school graduation requirements, particularly in mathematics and science. As a result, enrollments in science courses rose in all fields between 1982 and 1987: the proportion of students who had taken a full year of biology by high school graduation rose from 75 percent to 90 percent, from 31 percent to 45 percent in chemistry, and from 14 percent to 20 percent in physics.[31] Unfortunately, there is some evidence that the increases in science course enrollments occasioned by increased high-school graduation requirements consist at least in part of basic, general, or remedial courses,[32] designed to allow average or below-average students to fulfill their science requirements. In fact, interest in majoring in the more quantitative sciences (mathematics and statistics, physical sciences, engineering, and computer science) has been declining over the last several years, at least among college-bound seniors taking the Scholastic Aptitude Test (SAT): their intent to major in these sciences peaked in 1983 at 20 percent and had declines to 13 percent by 1988.[33,34]

Curriculum and Instruction. Obviously, student learning and interest is not merely a matter of more time spent on a subject. The quality of the curriculum and the effectiveness of instruction are critical factors. No matter what their intent, high school curricula in science in the main appear to be constructed on a "trickle-down" theory: Subject-matter experts decide what general knowledge is needed to provide an adequate base for their discipline, and this is then codified to produce high school (or introductory college) textbooks. Since scientific knowledge keeps growing at an exponential rate, textbooks get heavier and heavier, with more topics included but treated in less depth in every revised edition. Texts for the less able students water down the treatment of topics even more. Reviewing the results, the American Association for the Advancement of Science[35] has called the present curricula in science "overstuffed and undernourished." As in the earlier grades, the emphasis is on learning answers out of textbooks and lecture notes rather than exploring questions through doing scientific investigations. Applications and issues are discussed after basic science information and principles have been presented, rather than allowing questions and problems requiring science knowledge to arise from important issues and applications. Even worse, information too often is memorized as unconnected facts taken out of context, with little discernible reason apparent to the student for having to know them other than the teacher's statement that they will be needed later. And laboratory exercises, when they are part of instruction at all, are largely procedural, resembling the following of recipes in a cookbook, to ensure that the expected results are obtained.

However, to teach science differently, so that questions arise out of interesting and important problems, and to lead students of varying capability in *bona fide* investigations requires considerable science knowledge and understanding on part of the teacher. Yet, in 1986, 72 percent of high school principals reported difficulty in hiring qualified physics teachers, and 63 percent reported difficulty in hiring qualified chemistry teachers, although only 38 percent reported such difficulties with respect to biology teachers.[36] The problems were particularly acute in rural schools, where half the principals reported difficulties in biology, as contrasted to only 13 percent of suburban principals who reported difficulties in this field. Even though having teachers certified in their field does not necessarily guarantee effective science instruction, it is a necessary condition that does not obtain in many schools today.

Assessment. Little needs to be added here to what has already been said about the content validity of science tests. The problem is threefold: First, the proxies—generally multiple-choice or other short-answer items—used to assess students' science knowledge and their ability to do science are of questionable worth, that is, the relationship of scores on tests composed of such items to the abilities exhibited by individuals who are good in science is tenuous, at best. Second, insofar as tests are taken to be important indicators of what students are to learn, the current tests tend to distort the curriculum. The reason is as follows: If tests have high stakes, i.e., important consequences for students, teachers, or schools, it is only to be expected that teachers will teach to the test to ensure their students' success, and students will study what they will get graded on. Therefore, when tests focus as narrowly as most of them currently do on memorization of a multitude of science facts and principles and their rote application, the effect inevitably is to skew curriculum and instruction toward broad but superficial coverage of subject matter without "wasting time" on investigations that might have unpredictable results. And third, both the format and the content of most of the current tests misrepresent the nature of science and in this way affect students' attitudes toward either further study of or engagement with science, let alone considering science- and technology-related careers.

THE BASIS FOR REFORM

Over thirty years ago, this country embarked on a major effort to reform science education in response to the technological feat performed by the USSR in launching Sputnik, the first earth satellite. The National Science Foundation (NSF) took the lead in this effort, funding curriculum improvement projects, teacher training, special opportunities for high school students, and a variety of ancillary programs aimed at precollege science education. The NSF-sponsored programs involved collaborations among research scientists and college and high school teachers; the traditional educational administrative structures were largely bypassed, at least initially. The high school curricula emerging from the first set of reform projects were intended to provide more up-to-date science for students then studying science; they were rigorous and accompanied by carefully designed laboratory experiences intended to give students an appreciation of science as a way of knowing. Summer and year-long institutes for teachers focused on upgrading their science knowledge; as the reform effort proceeded, some of the institutes were built around the knowledge necessary for teaching the improved curricula and new laboratory activities. At the height of the reform movement, a majority of high school students studying biology and chemistry were exposed to the new materials; and student achievement was relatively high compared to test performance in the 1970s and 1980s. Within a few years, it was realized that reform, if it was to be successful, had to start before the high school level. Consequently, curricula intended for all students, not just the science-able, were developed for the middle and elementary grades. Training of teachers proved to be a more difficult problem, given the large number of elementary school teachers, most of whom had minimal preparation in science. Nevertheless, the reforms at this level left behind a legacy of outstanding curriculum materials and science activities still in use and being built upon today.

Other priorities overtook education in the middle 1960s, most particularly efforts to have schools serve various hitherto segregated and neglected populations more adequately. At the federal level, major new programs were launched to provide compensatory education for children falling behind in the basics (generally reading and arithmetic), for children with limited proficiency in English, for students with disabling conditions, and to help schools with desegregation decrees mandated by the courts. Also, concerns arose with respect to the inequities involved in financing schools largely through local taxes, given the unequal tax base among poor and wealthy districts within the same state. The resulting actions to bring about some equalization of school financing led to a major realignment, with states now supplying, on average, half the funds for public schools (though this varies from 7 percent in New Hampshire to 69 percent in California and 90 percent in Hawaii, the only state to have a state-wide school system).[37]

Then and Now

With the country declared once more to be in a state of crisis in science education, 25 years after the last reform effort, one might be justified in asking what expectations for success there are for the current round of reforms. Indeed, in some ways, the goal is much more ambitious, striving not only to enlarge the scientific and engineering manpower pool but also to ensure that all high school graduates are sufficiently scientifically literate to cope with their work, civic, and family responsibilities. There are, however, a number of factors present that lend strength to the current reform efforts:

Improved Understanding of Science Learning. Research by teams of scholars from the science disciplines traditionally taught in school (physics, chemistry, biology, the earth sciences) and from cognitive science has demonstrated that individuals, including even quite young children, build concepts in their heads about the phenomena they observe, for example, about light, electricity and simple circuits, heat and temperature, force and motion, the nature of matter, and the earth in space.[38] Their experiences engender strongly held beliefs, which often conflict with the sometimes counterintuitive canonical explanations of science. Formal science teaching many times does not dislodge these beliefs but simply creates parallel systems used by students for school problems, while they refer to their own concepts (sometimes called "naive theories") outside of school.

Changing these concepts is a long-term process which needs to take account of the learner's ideas. Driver et al.[39] suggest that effective science teaching should (a) give students opportunities for making their own ideas explicit, (b) introduce discrepant events accompanied by questioning and discussion with the teacher and among themselves in small groups, (c) help students generate a range of explanations of events which they should be encouraged to evaluate for themselves, and (d) provide practice to students in using their newly formed ideas in a range of situations to establish their generalizability and power.

The Potential of Computer and Communication Technology. Although very much underutilized at present, the recent advances made in computer and associated technologies could greatly expand the range of science instruction found in the average classroom. Raizen[40] points out the following educational improvements these new technologies make possible: tailoring instruction to the individual needs and learning styles of the student, creating new learning environments by simulating or modeling situations impossible to bring into the classroom, representing knowledge in several different ways, providing calculating and graphing power, making large data bases accessible, creating learner networks apart from the physical location of the learner, and motivating learning through providing an open and non-judgmental learning environment under the control of the learner. Unfortunately, the obstacles to the realization of this potential are many: schools locked into inappropriate hardware, lack of suitable software except for experimental programs, failure to establish an effectively functioning educational market in this area, and reluctance on part of many educators and teachers fully to integrate computers and associated technologies into instruction. These obstacles have yet to be addressed in a concerted manner.

Lessons from Previous Efforts. A critical lesson from the reform efforts of the 1960s is the need for a systemic approach. All parts of the educational process must be addressed—goals and objectives, curriculum, instruction, assessment, teacher preparation and staff development, school organization and the teaching context, leadership and policy formulation. By the same token, people at every level of the system must be involved—policymakers, administrators in state and local educational agencies, school board members, principals, parents, and representatives of business and higher education, as well as teachers and scientists.

Particularly noteworthy are two characteristics of the current reform effort: the increased importance of the state role and greater involvement on part of the business and higher education communities. Most reform initiatives of the 1980s have been initiated and carried out at the state rather than the federal level. These reforms have encompassed, first, the already noted increases in academic standards for students and raises in teachers' salaries accompanied by revamped accreditation standards, and, more recently, the development of educational goals and curriculum frameworks, together with standards and accountability mechanisms designed to encourage schools to meet the goals. With respect to the involvement of business and higher education (including museums and research institutions), Atkin and Atkin[41] note that there are now some 500 sizable partnerships between schools and businesses or institutions of higher education and the like active in improving mathematics and science education at the precollege level. Though not all of these have focused goals, and some of the coalitions may not be stable over the long run, they are a demonstration of the concern as well as the willingness to get involved in the schools, and in science education specifically, on part of the broader community.

Staying Power. Perhaps in part because of the much greater extent of state and community involvement, this round of educational reform appears to be maintaining momentum well beyond what might have been predicted. Increasingly, people have become restive with the shortcomings of their schools and are willing to invest in them, as the increases in teacher salaries—a nation-wide phenomenon—demonstrate. No doubt, the continuing attention to education is fueled by the wide dissemination of the data on international comparisons of student achievement cited above, which raise concerns about this country's ability to maintain its competitive status as a world economic and political power. In fact, with the recent proclamations by the President and the National Governors' Association of national goals in education, including several in science, the nearly ten-year old reform effort seems to be gaining momentum.

Recommendations for Reform

The directions that improvements in science education need to take are fairly evident. Building on the lessons learned, the improvements need to be planned and carried out with attention to all components of science education and all parts of the educational system; they need to attend to current knowlege about the nature of science learning and performance; they need to create coherent approaches suitable for a variety of students as these move through school; they would do well to consider the feasibility of making computers and communication technology an integral part of the instructional environment; and they need to concentrate on improving science education for the populations currently turning away from science.

Curriculum and Instruction. As in the 1960s, much attention is being given to the curriculum. NSF is supporting the development of improved science curricula at the elementary, middle school, and high school levels; California, followed by several other states, has developed a detailed curriculum framework for science;[42] the American Association for the Advancement of Science (AAAS) has made recommendations about the science knowledge all graduating high school students ought to have;[43] the National Center for Improving Science Education has recommended science concepts and instructional models appropriate for elementary and middle schools;[44,45] the National Science Teachers Association (NSTA) has recommended a scope and sequence for science courses to coordinate the high school curriculum.[46] Consensus is building on the goals of science education, the major conceptual themes to be conveyed by the curriculum, the emphasis on depth rather than breadth in the treatment of these themes, and the need for coherence and appropriate transitions between levels of schooling. There also is agreement on the nature of effective instruction: an integrated approach to science teaching in the lower grades; balancing traditional methods with new approaches including a mixture of instructional groupings, group reporting and debates in the classroom, and the use of computers for simulations and as a laboratory and computational tool; active student engagement through problem-solving, laboratory investigations, field studies, and long-term projects; and flexibility in scheduling and school organization to foster these instructional strategies. The challenge is to:

- *Develop model science programs that embody current recommendations on curriculum and instruction, with particular attention to the uses of the computer.*
- *Elaborate these models in some detail to make them useful to schools and teachers interested in improving their science progams.*

Even though there are several promising curriculum efforts under way, some of them related to the AAAS and NSTA initiatives, they are too few in number and, for the most part, still in the early stages of development. Schools need effective program components now, together with training for teachers and administrators that will help them in formulating goals appropriate to their setting and in selecting programs consonant with their goals.

Assessment. Even well formulated goals and programs matched to the goals will be undermined by the kind of testing that currently characterizes both large-scale, externally mandated assessments and assessments conducted by teachers for their own purposes. Raizen et al.[47] have argued that assessment exercises should closely resemble instructional activities and differ from them only in their purpose. Teachers need to understand the validity of assessment modes other than traditional tests and quizzes: evaluating students' portfolios of work, including laboratory notes; assessing records of student performance in class and in the laboratory; judging presentations of individual or group work made to outside audiences. But in order to make such alternative assessment strategies viable, they must be systematized, and public standards for rendering judgments on the basis of performance must be developed and understood.

Unfortunately, much less has been invested in reforming assessment than in reforming curriculum. Although testing is a billion-dollar industry, almost all the money goes into purchasing, administering, and scoring tests, very little into developing better tests or more valid assessment strategies. For example, the National Assessment of Educational Progress (NAEP) successfully piloted the use of some hands-on exercises for assessment,[48] but these exercises have not been incorporated into NAEP science assessments. A series of laboratory tasks has been made part of the 4th-grade science assessment in the state of New York; this subtest has served to demonstrate that few of the students have been exposed to hands-on science. Connecticut, together with several other states, is designing science investigations and accompanying scoring protocols to assess the science knowledge and performance of twelfth grade students. California is revising its science tests to make them more consonant with the state curriculum framework. But these are scattered efforts that leave testing in most states, districts, and classrooms unaffected. Coherent reform of assessment needs to proceed along the following lines:

- *Develop more complex short-answer and essay questions to probe students' understanding of concepts and science reasoning;*
- *Design performance tasks suitable for assessing students' laboratory and science investigation skills; and*
- *Exploit the computer's potential for presenting challenging science tasks and interpreting complex performance.*

Once appropriate modes have been developed, they must be introduced into assessments at all levels of the system. This will entail educating teachers as well as principals and administrators on the use and interpretation of a wide range of assessment strategies so as to reduce reliance on overly narrow tests. In addition, reporting procedures must be developed that will be as persuasive to policymakers, parents, and the public in general as the currently reported numerical test scores.

Teacher Education and Staff Development. Teacher development must be viewed as an ongoing process, with preservice and inservice preparation building on each other and integrated with internships and practice teaching. Formal educational experiences must be augmented by opportunities for teachers to work together as they learn and implement new practices.[49] School districts and institutions of higher education need to plan jointly on providing prospective and practicing teachers with the science and pedagogic knowledge they will need for effective instruction at a given level. Above all, teachers need to be taught as they will be expected to teach, through active experiences that include science investigations as well as opportunities to reflect on alternative instructional strategies. Teacher education programs—preservice or inservice—that exhibit these features are rare and will be difficult to implement, even if created, given the incentive systems that operate within institutions responsible for teacher education and staff development. Therefore, there is great need to:

- *Develop teacher education programs that reflect the nature of science, incorporating science knowledge and understanding with truly investigative laboratory and field experiences;*
- *Integrate education in science with education about the learning of science and with teaching practice; and*
- *Develop strategies for implementing reforms in teacher education programs that will take account of and successfully deal with the values and incentives currently governing academic and teacher training departments in institutions of higher education.*

A recent report by the AAAS[50] points out that natural science faculties will have to deal with the difficulties of engendering a multidisciplinary understanding of science, which is especially important for elementary and middle school teachers. They also will have to change their teaching style from lecturing to large groups of students to the mentor/apprenticeship style of working together that characterizes graduate study. And, they will have to work in close cooperation with educators to ensure the integration of the science knowledge and the pedagogical knowledge essential for teachers to acquire.

Teaching Context and School Organization. However, even the best trained teachers having available excellent programs will be impeded in their efforts if they are not provided with a supportive teaching environment. Teachers must be provided with the resources, including time, materials, and facilities needed to provide good science programs. For example, at the elementary level, a system for providing and replenishing hands-on materials is critical. Psychological support on the part of principals, curriculum specialists, and assessment directors also is necessary to encourage teachers in their new roles as individuals who not only do science together with their students but also approach their own teaching in a reflective manner. To make this possible, school organization, structure, and scheduling must be flexible, and teachers must be given both authority and responsibility for their instruction. Developing such an environment entails:

- *Coherent state and district policies on curriculum, assessment, and staff development consistent with the enunciated goals in science education;*
- *Provision of resources, organizational support, and time for professional activities (planning, discussion with peers, attending meetings of scientific and educational societies, etc.); and*
- *An ongoing program of staff development integrated with the school system's goals and programs in science education.*

Increasing the Participation in Science of Underrepresented Groups. The reader may note that no recommendations have been made regarding special initiatives to increase the participation of minorities and females in science. The reason is that there is no persuasive research evidence indicating differences among groups in the way cognition develops or individuals from these groups learn science.[51] Effective science programs are effective for everyone, as was demonstrated over 25 years ago with the then new elementary science programs. However, given that resources for improving science education will be limited, the following recomendation is perhaps the most critical of all:

- *The highest priority for implementing improvements in science education should be given to schools that serve high proportions of minority and poor students to encourage their increased participation in science.*

This is not to say that special programs to encourage greater involvement with science on the part of currently underrepresented groups should not be initiated and supported. For too many students already in school, the fundamental reforms that need to take place in science education will come too late, and the country can hardly afford to write off the current generation of students while mapping out the future. While current interventions to increase participation have not been evaluated systematically,[52] they do indicate approaches that appear to be successful and need to be implemented more broadly.

The reform agenda laid out in this paper is ambitious. It cannot be accomplished through "quick fixes" of the kind mandated in the early 1980s, like increases in high school graduation or college entrance requirements. It will take the cooperation of institutions and individuals having their own separate and at times conflicting interests. And it will take a commitment of resources matched to the complex and difficult development and implementation issues that must be successfully resolved. But there is reason for optimism: Concern of the public and private sectors is high, and the knowledge and prototypes needed for effective development exist. As with all innovation and reform, wide-spread adoption and implementation will represent the greatest challenge, but if the lessons of the 1960s are understood and applied, the country should see a considerably improved science education system serving the next generation of students.

REFERENCES

1. National Science Foundation/Science Resources Studies. 1988. *NSF Survey of Earned Doctorates*. Author, Washington, DC.
2. International Association for the Evaluation of Educational Achievement (IEA). 1988. *Science Achievement in Seventeen Countries; a Preliminary Report*. Pergamon Press, New York, NY.
3. Lapointe, A.E., N.A. Meade and G.W. Phillips. 1989. *A World of Differences*. Educational Testing Service (ETS), Princeton, NJ.
4. Mullis, I.V., and L.B. Jenkins. 1988. *The Science Report Card*. Educational Testing Service (ETS), Princeton, NJ.
5. National Science Board. 1989. *Science and Engineering Indicators — 1989*. NSB 89-1. U.S. Government Printing Office, Washington, DC.
6. Murnane, R.J. 1988. Education and the productivity of the work force: Looking ahead, pp. 215-243. In E. Litan, R.Z. Lawrence, and C.J. Schultze (Ed.) *American Living Standards*. The Brookings Institution, Washington, DC.
7. Resnick, L.B. 1987. *Education and Learning to Think*. National Academy Press, Washington, DC.
8. The Task Force on Women, Minorities, and the Handicapped in Science and Technology. 1988. *Changing America: The New Face of Science and Engineering*. Author, Washington, DC.
9. H.L. Hodgkinson (Ed.) 1985. *All One System: Demographics of Education, Kindergarten through Graduate School*. The Institute for Educational Leadership, Inc., Washington, DC.
10. The Task Force on Women, Minorities, and the Handicapped in Science and Technology, *op. cit.*
11. Goodlad, J.I. 1984. *A Place Called School*. McGraw-Hill Book Co., New York, NY.
12. Mullis and Jenkins, *op. cit.*
13. The National Commission on Excellence in Education. 1983. *A Nation at Risk: the Imperative for Educational Reform*. U.S. Department of Education, Washington, DC.
14. The National Science Board Commission on Precollege Education in Mathematics, Science and Technology. 1983. *Educating Americans for the 21st Century*. National Science Foundation, Washington, DC.
15. Marsh, D.D. and A.R. Odden. 1990. State-Initiated Curriculum Reform in Elementary School Mathematics and Science Programs. A paper presented at the annual meeting of the American Educational Research Association in Boston. The paper will appear as a chapter in A. Odden (Ed.) *Policy Implementation* to be published by SUNY press.
16. Weiss, I.S. 1987. *Report of the 1985-86 National Survey of Science and Mathematics Education*. National Science Foundation. SPE-8317070. U.S.

Government Printing Office, Washington, DC.

17. Weiss, I.S. 1978. *Report of the 1977 National Survey of Science, Mathematics and Social Studies Education.* The National Science Foundation. Supt. of Doc. No. 083-000-00364-0. U.S. Government Printing Office, Washington, DC.
18. Mullis and Jenkins, *op. cit.*
19. Novak, J. 1988. Learning Science and the Science of Learning. *Studies in Science.* (1988): 77-101.
20. Blank, R. and P. Espenshade. 1988. *State Education Indicators on Science and Mathematics.* Council of Chief State School Offices, Washington, DC.
21. Raizen, S.A. and J.B. Baron, A.B. Champagne, E. Haertel, I.V. Mullis, J. Oakes. 1989. *Assessment in Elementary School Science Education.* The National Center for Improving Science Education. The NETWORK, Inc., Andover, MA.
22. Dorr-Bremme, D.W. and J.L. Herman. 1986. *Assessing Student Achievement: A Profile of Classroom Practices.* Center for the Study of Evaluation, University of California, Los Angeles, CA.
23. California State Department of Education. 1987. *Caught in the Middle.* Author, Sacramento, CA.
24. Carnegie Council on Adolescent Development. 1989. *Turning Points: Preparing American Youth for the 21st Century.* Author, Washington, DC.
25. National Middle School Association. 1982. *This We Believe.* Author, Columbus, OH.
26. Yager, Robert E. 1988. Criteria for Identifying and/or Developing Exemplary Science Programs in Middle Schools. *Science in the Middle School* XI(1): 9-14.
27. Rakow, S.J. 1988. Images of Science Through the Eyes of Middle School Students. *Science in the Middle School* XI(1): 1-8.
28. Weiss, 1987, *op. cit.*
29. Weiss, 1987, *op. cit.*
30. Raizen, S.A. and J.B. Baron, A.B. Champagne, E. Haertel, I.V.S. Mullis, J. Oakes. 1990. *Assessment in Science Education: The Middle Grades.* The National Center for Improving Science Education. The NETWORK, Inc., Andover, MA.
31. Westat, Inc. 1988. *Tabulations for the Nation at Risk Update Study as Part of the 1987 High School Transcript Study.* National Center for Education Statistics, U.S. Department of Education, Washington, DC.
32. Educational Testing Service (ETS). 1989. *ETS Policy Notes* 1, no. 3 (June). Author, Princeton, NJ.
33. Grandy, J. 1989. *Trends in SAT Scores and Other Characteristics of Examinees Planning to Major in Mathematics, Science, or Engineering.* Research Report No. RR-89-24. Educational Testing Service (ETS), Princeton, NJ.

34. Clune, W.H. 1988. *The Implementation and Effects of High School Gradua-
    tion Requirements: First Steps Toward Curriculum Reform.* Wisconsin
    Center for Education Research, University of Wisconsin-Madison,
    Madison, WI.
35. American Association for the Advancement of Science (AAAS). 1989.
    Science for All Americans. A Project 2061 Report on Literacy Goals in
    Science, Mathematics, and Technology. Author, Washington, DC.
36. Weiss, 1987. *op. cit.*
37. National Center for Education Statistics. 1988. *Digest of Education
    Statistics 1988.* Office of Educational Research and Improvement (OERI),
    U.S. Department of Education, Washington, DC.
38. Driver, R. and E. Guesne, A. Tiberghien (Eds.). 1985. *Children's Ideas in
    Science.* Open University Press, Philadelphia, PA.
39. *Ibid,* p. 200.
40. Raizen, S.A. 1988. *Improving Educational Productivity Through Improv-
    ing the Science Curriculum.* A Center for Policy Research in Education
    (CPRE) Report. Rutgers, The State University of New Jersey, New
    Brunswick, NJ.
41. Atkin, J.M. and A. Atkin. 1989. *Improving Science Education Through
    Local Alliances.* A Report to the Carnegie Corporation of New York. Net-
    work Publications, a Division of ETR Associates, Santa Cruz, CA.
42. California State Department of Education. 1989. *Science Framework for
    California Public Schools.* Pre-publication draft. Author, Sacramento, CA.
43. American Association for the Advancement of Science. *op. cit.*
44. Bybee, R.W. and C.E. Buchwald, L.S. Crissman, D. Heil, P.J. Kuerbis, C.
    Matsumoto, J.D. McInerney. 1989. *Science and Technology Education for
    the Elementary Years: Frameworks for Curriculum and Instruction.* The
    NETWORK, Inc., Andover, MA.
45. Bybee, R.W. and C.E. Buchwald, S. Crissman, D.R. Heil, P.J. Kuerbis, C.
    Matsumoto, J.D. McInerney. 1990. *Science and Technology Education for
    the Middle Years: Frameworks for Curriculum and Instruction.* The
    National Center for Improving Science Education, Andover, MA.
46. Aldridge, B.G. 1989. *Essential Changes in Secondary School Science: Scope,
    Sequence and Coordination.* National Science Teachers Association,
    Washington, DC.
47. Raizen, S.A. et al., 1989, 1990, *op. cit.*
48. National Assessment of Educational Progress (NAEP). 1987. *Learning by
    Doing.* Report No.: 17-HOS-80. Educational Testing Service (ETS),
    Princeton, NJ.
49. Loucks-Horsley, S. and C.K. Harding, M.A. Arbuckle, L.B. Murray, C.
    Dubea, M.K. Williams. 1987. *Continuing to Learn: A Guidebook for
    Teacher Development.* The Regional Laboratory for Educational Improve-
    ment of the Northeast and Islands, Andover, MA.

50. American Association for the Advancement of Science (AAAS). 1990. *The Liberal Art of Science: Agenda for Action.* The Report of the Project on Liberal Education and the Sciences. Author. Washington, D.C.
51. Oakes, Jeannie. 1990. *Lost Talent: The Underparticipation of Women, Minorities, and Disabled Persons in Science.* The RAND Corporation, Santa Monica, CA.
52. Ibid.

Science Education in the United States: Issues, Crises and Priorities. Edited by S. K. Majumdar, L. M. Rosenfeld, P. A. Rubba, E. W. Miller and R. F. Schmalz. ©1991, The Pennsylvania Academy of Science.

Chapter Four

SCIENCE ACHIEVEMENT IN THE UNITED STATES AND SIXTEEN OTHER COUNTRIES

RODNEY L. DORAN[1] and WILLARD J. JACOBSON[2]

[1]Professor of Science Education
Department of Learning and Instruction
University at Buffalo
Buffalo, NY 14260
and
[2]Professor Emeritus of National Sciences
Teachers College
Columbia University
New York, NY 10027

INTRODUCTION

During the past several years an unprecedented number of reports on American education have been prepared by researchers, professional associations, government funded task forces and commissions. Some of these reports cite their own and/or other sources of data. Much of the data relate to interest, enrollment, and achievement in mathematics and science.

From the titles and headlines created from the recent reports most people admit there is room for improvement in the mathematics and science achievement of school age students in the United States. Phrases such as "a rising tide of mediocrity", "the under achieving curriculum", and "are we measuring up" are indicators of concern obtained from longitudinal studies and international comparisons.

This chapter will summarize some results from the Second IEA Science Study (SISS) and discuss several implications raised by the study. This and many other cross national studies have been coordinated by the International Association for the Evaluation of Educational Achievement (IEA). IEA is a consortium of research institutes and researchers from many countries.

The SISS, which began in 1981, was built on an earlier study in which data was collected in 1970. This earlier study is now referred to as the First IEA Science Study (FISS). The major international report of the FISS was entitled *Science Education in Nineteen Countries: An Empirical Study* by Comber and Keeves.[1] For more information about the procedures used in the SISS, please consult article by Malcolm Rosier, who was the International Coordinator of the study.[2] A preliminary report of the SISS was published in 1988 with results from 17 countries who had submitted their data to the IEA.[3] Most of the international data cited in this chapter is from the preliminary report. An indepth report of the U.S. study has been prepared by Jacobson and Doran.[4]

The findings from the SISS tests when administered to U.S. samples will be presented in several sections. The first three sections will describe the results from the testing of grade 5, grade 9 and advanced science students. Then a section with the results from the process testing and results when analyzed by gender will be presented. In each section, some comparisons with the international context will be made and changes over time and between population will be cited where available.

The data was collected during seven weeks, from late March to early May of 1986. The cooperation of the schools and teachers with our request for testing time was admirable. The following table includes the numbers of schools and students tested at each population level.

TABLE 1

Number of Schools and Students Tested by Population

| | Grade 5 | | Grade 9 | | Advanced Science | | |
	Set A	Set B	Set A	Set B	Bio	Chem	Phy
Schools	123	123	119	119	43	40	34
Students	1298	1286	1136	1112	630	540	474

SCIENCE ACHIEVEMENT IN THE FIFTH GRADE

In the United States, over 2,500 fifth graders from 123 schools formed the SISS sample. The mean age of these US students in the spring of 1986 was 11.3 years. When the mean achievement scores of the countries testing at grade 5 were ranked, the U.S. mean score (55%) was eighth of fifteen countries. These scores were based on the 24 item international core test administered in all countries. As one can see from the following table, the countries with the highest mean scores were Japan, Korea and Finland. Note that the mean scores of many countries were quite similar. The mean scores of four countries (Canada [Eng], Italy, Australia and Norway) are within 2% of the U.S. mean scores.

Twenty-one of the items in the grade five core test were also used in the 1970 study. This set of "bridge" items provide a valuable time comparison for U.S. elementary science education. The 1986 mean on these bridge items of 56.3% was virtually the same as in 1970, 55.9%. When these items were categorized into life science and physical science sub-tests, comparisons of performance between 1986 and 1970 were virtually the same (less than 1% difference). When these same items were classified into process and non-process sub-tests, a slight difference was detected. The 1986 mean score on the process items (N = 11) in 1986 was 1.8% higher than in 1970, while the mean on the 1986 non-process items (N = 10) was 1.2% lower than in 1970. "Process" items required students to perform a skill such as classify objects, state hypotheses, control variables or design experiments.

In the U.S. the item on which the students had the highest percentage correct (79%) asked students to identify the shadow of a tree at noon. While this concept is part of most school instructional programs, this observation can be made in many everyday life experiences. By contrast, the most difficult item (26% correct) for U.S. fifth graders was one which required the reading of a table of temperatures and inferring the time a cool wind begins to blow.

In summary, the U.S. fifth graders were near the middle of the 15 countries who tested at this grade level. There were small gains from 1970 on items classified as "process" items. They may have been the results of the emphasis on science processes in the elementary science programs developed in the 1960s and 1970s.

SCIENCE ACHIEVEMENT OF THE NINTH GRADE

Ninth graders (or 14 years-olds) in 16 countries were tested with an achieve-

TABLE 2

Grade Five Science Achievement in 15 Countries

Japan	64.2
Korea	64.2
Finland	63.8
Sweden	61.3
Hungary	60.0
Canada (Eng)	57.1
Italy	55.8
U.S.	55.0
Australia	53.8
Norway	52.9
Poland	49.6
England	48.8
Singapore	46.7
Hong Kong	46.7
Philippines	39.6

ment test of 30 items. In the U.S. over 2,200 students in 119 randomly selected schools responded to the SISS tests. In the U.S. and the other countries testing at grade 9, almost all of the eligible students are still enrolled in full-time schooling (except Thailand, 60%). The mean age of the U.S. ninth graders tested in the spring of 1986 was 15.4 years.

When compared with the other 15 countries, the U.S. mean score (55%) was near the bottom of the ranking, with five other countries; England, Italy, Thailand, Singapore and Hong Kong. The highest country mean score was from Hungary (72.3%) with the ninth graders from Japan and the Netherlands also scoring high.

The U.S. and many countries try to provide equal opportunities for all to achieve in science. When this is happening, most of the variance of test scores should be between students in schools and there should be little variance between schools. In the U.S. the difference in science achievement between schools was 29% of the total between student variance. By contrast in Norway, Japan and Finland, the between school variance was less than 5% of the total between student variance. On the other hand, in Singapore, 56% of the total variance was due to the school variable. The between school variance in 13 countries was less than that in the United States. All ninth grade students in U.S. schools do not have equal opportunity to learn science.

Nineteen items of the 1986 test had been selected from the 1970 test. These "bridge" items provide a good measure of U.S. science achievement across time. The 1986 mean score of ninth grade bridge items of 49.2% was lower than the 1970 mean on these items of 53.8%. Each of these samples was composed of over 2,000 students from randomly selected schools. When these "bridge" items

TABLE 3

Grade Nine Science Achievement in 16 Countries

Hungary	72.3
Japan	67.3
Netherlands	66.0
Canada (Eng)	62.0
Finland	61.7
Sweden	61.3
Poland	60.3
Korea	60.3
Norway	59.7
Australia	59.3
England	55.7
Italy	55.7
Thailand	55.0
Singapore	55.0
U.S.	55.0
Hong Kong	54.7

were categorized into life science/physical science and process/non-process groups, the 1970-1986 declines were approximately the same for these content and process sub-tests.

With the U.S. ninth grade sample, the item on which they had the highest percentage correct was one in which they had to locate a number in a table and then identify the thermometer scale which indicated that temperature. This skill was one that these students had learned in school or through everyday experience and used often as 88% of the sampled students chose the correct answer.

The item on which U.S. ninth grade students had the greatest difficulty (20% correct) was one in which they had to determine the tension in a string between two hanging weights. While the item looks like a simple addition of the weights below that point on the string, this concept is usually taught in U.S. physics courses. This item was difficult in other countries as well. Only in Hungary and Japan did more than 50% of ninth graders choose the correct response.

The middle/junior high school years are the time when students should develop science concepts and skills for use in life and in future science-related study and work. The poor performance of U.S. ninth graders builds a very weak foundation for high school science courses and personal applications.

ACHIEVEMENT IN ADVANCED SCIENCE COURSES

The third population assessed in the SISS was composed of those students in the last years of secondary school who are still studying science. In the U.S. we chose to define our sample as all students enrolled in a second year/advanced course in biology, chemistry or physics after having successfully completed a full, first year course in that subject. The large majority of these students were 12th graders, but a few 11th graders in the U.S. sample were "accelerated" students and scored higher than their 12th grade cohorts. The presence of a few 10th grade students in the biology testing did influence the low U.S. biology score. Many of these U.S. "science specialist" students were in courses labelled Advanced Placement, Second Year or College Credit. Indeed, these students are the "cream of the crop" in the U.S. science programs. U.S. students were tested *only* in the content area which they were currently studying. The testing was conducted near the end of the school year.

TABLE 4

Enrollment in Advanced Science Courses by Grade Level

Course	Grade Level		
	10	11	12
Advanced Biology	20%	26%	54%
Advanced Chemistry	0%	33%	67%
Advanced Physics	0%	10%	90%

The U.S. advanced science sample was composed of three sub-samples in biology, chemistry and physics. The following table includes the percentages of students in each sub-sample and the breakdown by grade level.

The performance of U.S. advanced science students in each specialty area was poor. In biology, the U.S. mean score (37.9%) was more than 5% lower than the next country Italy, ranking 13th of 13 countries. An even more sobering comparison is the percentage of schools that score below the lowest performing school in the highest ranking country. In biology, the highest ranking country was Singapore. The mean of the lowest scoring school in Singapore was 57%. In the U.S. 98% of the schools had a lower mean score than the lowest scoring school in Singapore.

In chemistry, the U.S. mean score (37.7%) was ranked 11th of the 13 countries testing. The U.S. chemistry score was less than half of the highest ranking country, Hong Kong (77%). The U.S. physics mean score (45.5%) was 10th of the 14 countries testing.

There was a great variation in the proportion of the 18 year-old cohort completing secondary school in the participating countries. It ranges from 90% in the U.S. to 17% in Singapore. However, the percentages of U.S. high school graduates who completed a second year course in biology, chemistry or physics are 16%, 4% and 1% respectively. These highly selected U.S. students did not perform well when compared with the students tested in the other countries.

From an analysis of the number and the nature of the science and mathematics courses completed by these U.S. advanced science students, a number of interpretations are possible. For the advanced biology students, it is clear that being in an Advanced Placement class or having had at least four years of science or math was associated with higher scores. The following table contains the data from these sub-groups. It appears that the amount and kind of courses may be linked with achievement. It also was true that there was considerable difference between schools. In the U.S. 40% of the total variance of biology scores was associated with the between schools effect. Student scores seem to be associated with the schools which the students attended.

TABLE 5

Scores of Sub-Groups of Advanced Biology Students

	N	Mean Score in Percent
All Advanced Biology Students	674	39.4
Advanced Placement Biology	102	49.6
Non-advanced Placement Biology	572	37.6
Students with at Least Four Years of Science	308	44.4
Students with Less Than Four Years of Science	365	35.2
Students with at Least Four Years of Math	226	45.0
Students with Less Than Four Years of Math	419	36.0

In chemistry, those students enrolled in Advanced Placement classes received higher scores than those not in AP classes. The number of years of science and math classes was not a major predictor of chemistry scores. As with biology, a great deal of the variance of chemistry achievement (48%) was associated with the between schools effect. In countries such as Norway and Finland, the proportion of between-school variance was only 10%.

TABLE 6

Scores of Some Sub-Groups of Advanced Chemistry Students Tested

	N	Mean Score in Percent
All Advanced Chemistry Students	564	34.9
Advanced Placement Chemistry	117	46.9
NonAdvanced Placement Chemistry	445	31.7

In physics (as in biology and chemistry) those students enrolled in an Advanced Placement class had a higher mean score than those students not in AP classes. There also was a difference based on those who had at least 4 years of science and those who had not that number of science credits. On the U.S. physics scores, 38% of the variance was attributed to the between schools effects. The U.S. was joined by Hungary, Poland and Japan with a high proportion of between school variance. In these countries, a students' chance to achieve in physics depended to a large extent on the school attended.

TABLE 7

Scores of Sub-Groups of Advanced Physics Students

	N	Mean Score in Percent
All Advanced Physics Students	500	45.5
Advanced Placement Students	280	49.0
NonAdvanced Placement Students	220	41.3
Students with at Least Four Years of Science	381	47.4
Students with Less than Four Years of Science	119	40.2

SCIENCE PROCESS LABORATORY SKILLS

The emphasis on inquiry skills, experimentation, and laboratory activities in science programs has been high in the U.S. for several decades. Curriculum development projects and, subsequently, commercial materials have incorporated these goals into instructional products that are available for school use.

Researchers have cited the impact of these programs on gains in student cognition, attitudes, science skills, and creative outcomes.

In the First IEA Science Study students in England and Japan responded to an optional laboratory practical test (in Grade Nine). Comber and Keeves concluded that these laboratory practical tasks were "measuring some attributes quite distinct from those measured by the written examination, and that these attributes are only probed to a limited extent by the pencil and paper type of "practical items." (1, p. 104).

As part of the Second IEA Science Study (SISS), six countries administered science laboratory skills tests to students in Grades Five and Nine. These countries were Hungary, Israel, Japan, Korea, Singapore, and the U.S.

The skills tests were developed by the researchers in the six participating countries. Students were asked to manipulate equipment and materials, observe, reason, record data, and interpret results. These specific skills were classified into three general categories: Investigating, Performing, and Reasoning. "Performing" included observing, measuring, manipulating; "Investigating" included the planning and design of experiments; and "Reasoning" included interpreting data, formulating generalizations, building and revising models. The tests were based on science content and equipment believed to be common to the respective grade levels, with the biology, chemistry, and physics areas being represented.

The following lists are descriptions of the tasks in each of the process tests used in Grades Five and Nine.

Process (Lab) Tests Used in Grade Five, Second IEA Science—US

A1. Describe and explain color change of bromthymol blue solution after blowing through a straw. (Chemistry)
A2. Cite at least three similarities and differences of two plastic animal specimens. (Biology)
A3. Determine if four objects are electrical conductors by testing in a battery-bulb circuit. (Physics)
B1. Predict and measure the temperature of the mixture of equal amounts of hot and cold water. (Physics)
B2. Observe and explain the dissolving of coffee crystals in water. (Chemistry)
B3. Determine which seeds contain oil by rubbing them on paper. (Biology)

Process (Lab) Tests Used in Grade Nine, Second IEA Science Study—US

A1. By testing with battery-bulb apparatus, determine the circuit within a "black box." (Physics)

A2. Using phenolphthalein and litmus paper, prepare and execute a plan to identify three solutions as to being acid, base, or neutral. (Chemistry)

A3. Using iodine solution, prepare and execute a plan to determine the starch content of three unknown solutions. (Biology)

B1. Using a spring scale and graduated cylinder, determine the density of a metal sinker. (Physics)

B2. Explain movement rates and separation of water soluble dots in paper chromatography activity. (Biology)

B3. Using a sugar test tape and iodine solution, identify three unknown solutions as to presence of starch and/or sugar. (Chemistry)

There were few differences between the U.S. boys and girls on the science process skills tests. This is in sharp contrast to the gender differences on the paper and pencil achievement tests in science. The results on the skills tests by gender are given in Table 8. On both subtests (A and B), each given to half of the Grade Five students tested, the mean scores for boys and girls were not statistically different. At the Grade Nine level, the mean score on Subtest A for boys was 3% higher than for girls, but the mean scores on Subtest B were not significantly different.

Performance (by boys and girls) on some of the individual tasks did vary. Generally, boys did better on tasks with a physical science content and girls did better on tasks for the life sciences. For instance, the items with the greatest differences favoring the boys were two items involving electrical circuits (one at each grade level) and one item requiring the measurement of mass and volume and the calculation of density. Girls achieved higher than boys on an item testing seeds for oil content and on an item involving paper chromatography.

TABLE 8

*Laboratory Process Test for Girls and Boys
(Percent Scoring Correctly)**

| | Percent Scoring Correctly | | | |
| | Grade 5 | | Grade 9 | |
Item	Girls	Boys	Girls	Boys
Subtest A				
A1	50.0	52.2	70.7	76.9
A2	64.7	61.3	46.0	47.0
A3	62.1	67.0	44.0	46.3
Mean (A)	60.1	61.1	53.1	56.1
Subtest B				
B1	44.5	47.2	38.2	44.6
B2	44.4	45.2	71.8	67.5
B3	60.4	52.7	45.3	44.0
Mean (B)	49.8	48.5	49.0	49.8

*Weighted Scores, since each item has a different point value.

GENDER AND SCIENCE ACHIEVEMENT

In a number of studies, including the FISS, boys generally had higher achievement scores in science than girls. Since science achievement was studied in two grades and in three advanced science subjects in the Second International Science Study in the United States, there was an excellent opportunity to study the comparative science achievement of boys and girls in schools in the U.S. and around the world. Table 9 shows the percent female enrollment and the achievement scores of boys and girls in the U.S. at Grade 5, Grade 9, second year biology, second year chemistry, and second year physics. In the United States boys had somewhat higher scores in all grades and in all subjects. Differences between males and females in science achievement range from 5.2% in Grade 5 to 7.4% in second year physics.

The differences in science achievement between males and females were reported in a number of countries as part of this study. At Grade five, the male-female achievement difference (in percent) ranged from 0.8% in the Philippines to 7.9% in Norway. The U.S. difference of 5.2% was relatively high with only Finland, Korea and Norway having higher gender-related differences in achievement.

In Grade Nine the differences between the scores of boys and girls tended to be somewhat greater than in Grade Five. The smallest difference was in Hungary when the mean percent difference between scores of boys and girls was 2.3%, while the greatest difference was in the Netherlands with a difference of 7.7%. The United States ranked 13th of 17 countries with a mean percent difference between the scores of boys and girls of 6.2%. Only Singapore, Norway, Korea, and the Netherlands had larger differences between the scores of boys and girls.

Biology is the only subject in which girls did better than boys in science achievement in some countries. In Sweden, Hong Kong, and Australia, the girls had higher science achievement scores than boys. In the other 10 countries the boys had higher biology science achievement scores than the girls. The greatest difference in biology achievement between boys and girls was in Singapore, where

TABLE 9

Enrollment and Achievement by Gender and Grade Level or Course

Grade/ Course	Percent Enrollment	Perfect Scoring Correctly		
		Females	Males	Difference
Five	51.0%	54.6	59.8	5.2%
Nine	48.1%	56.4	62.5	6.2%
Biology 2	58.5%	40.2	45.4	5.2%
Chemistry 2	43.6%	35.7	42.8	7.1%
Physics 2	21.4%	40.2	47.6	7.4%

there was a 7% difference between boys' and girls' science achievement. U.S. students had a difference of 5.2% between the boys' and the girls' biology achievement scores, with only Italy, Norway and Singapore having larger differences.

In advanced chemistry the differences between the boys science achievement score and the girls' science achievement score ranged from 0.7% in Sweden to 12.5% in Hungary. Of the 14 countries that tested in chemistry, the U.S. ranked tenth in the differences between boys' and girls' chemistry achievement with a 7.1% difference.

Boys scored higher than girls in physics in all countries. The differences range from 0.6% in England to 8.9% in Poland. Of the 14 countries that tested in physics, U.S. students ranked ninth with a difference in scores between boys and girls of 7.4%.

It was expected that there would be smaller gender-related differences in science achievement in the biological life sciences than in the physical sciences. The items in the achievement tests for Grades Five and Nine were classified as to whether they were physical science or life science. The difference between boys and girls in science achievement was greater on the physical science items than on the biological or life science items. In the secondary school, boys had higher science achievement than girls in both content areas. However, the difference between scores of boys and girls in biology was smaller than in chemistry or physics.

It has been suggested that female students would do better if they had access to or were taught by female science teachers. In Grade Five, 69% of the teachers were female. The gender of the teacher seemed to have no significant influence upon student achievement in Grade Five. In Grade Nine, 32% of the teachers were female, but the male teachers seemed to have a more positive impact on student achievement than female teachers. Interestingly, in biology, female teachers had a slightly greater influence on the science achievement of both boys and girls than did male teachers. At the advanced level in chemistry, students of male teachers had higher achievement. Unfortunately, there were too few female teachers in advanced physics[3] to provide sufficient data for valid comparisons.

DISCUSSION

These findings of low international rankings and declines (or no change) in achievement since 1970 have been of great concern to American educators and policy makers. If these were the only results with such conclusions, one could ignore the findings as isolated or criticize the study on various grounds. However, these results have been substantially supported by the 1985-86 science survey by the National Assessment of Educational Progress [NAEP][5] and an International Assessment of Educational Program conducted by NAEP in 1986.[6] No single reason can be found that explains these disturbing results, however, a number of possible reasons have been identified.

1. *Selectivity of students tested.* This factor is potentially very important, because not all countries are committed to keeping as large a proportion of young people enrolled in school for so many years. Through the grade 9 level, however, all countries tested still enrolled 95-99% of the age cohort in full-time schooling (except for Thailand—60% enrolled). The enrollment pattern changes drastically after grade 9 (age 14) in many countries.

A measure of "selectivity" might be the percentage of the age cohort (grade 12 or 17 years-olds) in each country enrolling in the advanced science courses. In the U.S., 1.5% of high school graduates have completed second year courses in physics and 3.5% have completed a similar experience in chemistry. These figures are lower (therefore more selective) than most of the countries tested. In the case of biology, the statistics obtained from several U.S. sources vary considerably (from 8% to 16%). These percentages are much higher than chemistry and physics, but still less than many countries.

From these data, the U.S. low positions cannot simply be explained away as a function of the degree of selection of students in the various countries.

2. *The U.S. "Layer Cake" Curriculum.* In the U.S., students normally study one science course per year. The pattern usually includes biology in 10th grade, chemistry in 11th grade and physics in 12th grade. About 80% of U.S. high school graduates have completed one year of biology, about 35% one year of chemistry and only about 15% complete one year of physics. Students who pursue the advanced (second year) science courses, either beginning the sequence early ("accelerate") or "double-up"—taking two science courses during some year(s).

By contrast, students in most other countries study two or three science courses each year for the several years of their high school program. By studying several courses at the same time, more integration and connection of key concepts with the several disciplines is possible. It is hypothesized that such integration would enhance comprehension of science concepts.

3. *Intensity of Study (Time on Task) of Science.* This idea is related to the above discussion and may be the critical factor, rather than the layer cake curriculum. Data to answer this question are not available now, but must be included in future surveys. Questionnaires would need to provide easy ways for students to indicate how many hours per day they spend studying science and mathematics (in and out of school). The idea of "intensity" of study may be of interest as a national parameter as well as for individual students.

An important aspect of time on task is the amount of time spent on homework. In the U.S. students "self report" that they spend much less time on homework than watching television. This is especially true at grade 5 and 9, however, even the advanced science students spend more time watching TV than doing homework.

There is extensive research support for the claim that "time on task" correlates positively with achievement. However, just increasing the "quantity" of homework is not a viable solution. Students need to experience homework stressing higher level skills and utilizing a variety of examples and methods. In other words, just three more of the same worksheets will not help—actually it may hurt!

4. *Specialized vs. General Focus of education.* In the U.S. all high school students (even those "majoring" in science) must study a full range of academic subjects (math, english, social studies, languages) as well as electives, such as health, physical education, career skills, and fine arts. The result of this very general focus to education is that a relatively small proportion of student's academic life can be spent on science. In some other countries, students "majoring" in the sciences would enroll in 2 or 3 science courses and perhaps also 2 math courses each year—comprising half of their academic load.

One must realize that the U.S. "general education" high school program is clearly linked with our democratic ideals and philosophy, attempting to provide a broad education to all citizens. Therefore, most secondary schools are comprehensive, academic schools. The U.S. system will allow late educational decisions and changes in career options more easily than narrower programs of study. However, if we naively thought that our general system of secondary education was preparing specialists at the same level of achievement as countries with their very focussed programs, we now know that is not the case.

When we examined the performance of sub-groups of U.S. advanced science students, we noted some logical trends. A sub-group of special significance was the set of students who were enrolled in Advanced Placement courses. For each science discipline (biology, chemistry, physics) the mean of the AP group was significantly higher than of the group of students not enrolled in an AP course. The "AP effect" could be from the selection of students for the course, the exam, the motivation for possibly receiving college credit, the quality of the instructors assigned to the course, or the carefully structured curriculum. Similarly, sub-groups with "more science" and with "more math" performed better than those with less science and with less math.

The last element of the "general" U.S. science programs has to do with the resultant breadth of U.S. science curriculum. There is a pressure to "cover everything" in a first year, high school introductory course. This pressure creates textbooks that are virtual encyclopedias and cause very superficial treatment of all topics. Teaching a smaller set of concepts in greater depth ("less is more") may produce better results, in achievement and attitudes toward science.

5. *Impact of Testing.* Many people have argued that testing is having a negative impact on education in general, and science and math education, in particular. Appropriate assessment can provide important feedback for students, teachers, parents, administrators, employers and policy makers. The key word of that sentence is appropriate. A standardized achievement test of which only 45%

of the items are taught in a school program is not appropriate. A test with 95% recall objectives is not appropriate for a course that espouses to encourage the applications of science and critical thinking. Assessment that depends totally on paper-pencil testing is not appropriate for a course that stresses laboratory work and scientific experimentation.

Unfortunately, these problems exist in far too many schools and states. During May 1989 the New York State Education Department scheduled the first administration of the Elementary Science Program Evaluation Test (ESPET) to fourth graders in all New York schools. The test includes a paper-pencil achievement test, a manipulative skills test as well as student attitude and program environment surveys.

6. *Importance Assigned to Science.* At the elementary and high school levels, science is not perceived as a "basic" subject. It should be! Every student should experience some appropriate science instruction each day, from pre-kindergarten through graduation. Again, the key word is "appropriate." One could argue for the need of science programs to stress application over academics. Obviously, for some students, a theoretical, academic science preparation is appropriate. However, the vast majority of U.S. citizens will be consumers of science, not producers.

A science program for all high school students could focus on "science literacy." This program could be built around a small number of key concepts such as: balance, cycles, energy, force, equilibrium, feedback, system, limits, matter, structure and continuity. These concepts should be experienced through activities in the life, earth, and physical science fields. While excess theoretical and mathematical sophistiction is not needed, these activities will present valid science with its practical applications, limitations and implications. The best science teachers in a school need to be involved with the planning, development, and teaching of these courses for all students. Those students interested in pursuing science and technological fields could additionally enroll in more specialized, academically oriented science and mathematics courses.

7. *Complacency.* Many American high school students have part-time jobs, eliminating large parts of the week as possible studying time. Many high school students do not enroll in elective science or mathematics courses because they are known to be rigorous. Students want to get high grade point averages and class ranks to enter the college of their choice and tend not to enroll in science courses.

Therefore, another hypothesis to explain the low U.S. performance is "complacency." Obviously what is thought of as "complacency" is manifested in few elective science and math courses, little homework, much television watching, and part-time work. In many countries, the perceived "job" of young people is that of a student. Families even support extra-mural study and review programs to help their youngsters to succeed in school.

Reports of "grade inflation" at high schools and colleges reinforce the perception of declining standards and the need to graduate or pass through various educational "hoops." U.S. industries have been charged with reinforcing this mentality by not hiring and differentially placing young people by their level of academic performance and the rigor of their program of study. U.S. parents have also been accused of asking their children "Did you pass?", instead of "What did you learn?"

A number of people and groups have suggested that these current education problems will be solved if we simply lengthen the school day or year, increase course and graduation requirements or strengthen teacher preparation. Each of these clearly is worth discussing and perhaps implementing, but only if viewed in concert with other changes. Changing one component in a very complex enterprise is not a realistic solution. We need to develop an overall strategy that addresses curriculum reform, modes of assessment, teacher preparation, parental support and other related areas.

Rather than pointing fingers, all segments of our society need to demand rigorous academic achievement from all levels of American schools and work with and support the professionals who are chosen to educate our nation's most important national resource—our young people.

REFERENCES

1. Comber, L.C. and J.P. Keever. 1973. *Science Education in Nineteen Countries.* John Wiley and Sons, New York, NY.
2. Rosier, Malcolm. 1987. The Second International Science Study. *Comparative Education Review.* 31(1):106-128.
3. *Science Achievement in Seventeen Countries.* 1988. Pergamon Press, Oxford, England.
4. Jacobson, W.J. and R.L. Doran. 1988. *Science Achievement in the United States and Sixteen Countries: A Report to the Public.* Teachers College, Columbia University, New York, NY.
5. Mullis, I.V. and L.B. Jenkins. 1988. *The Science Report Card. Elements of Risk and Recovery.* Educational Testing Service, Princeton, NJ.
6. LaPointe, A., N.A. Mead and G.W. Phillips. 1989. *A World of Differences.* Educational Testing Service, Princeton, NJ.

Science Education in the United States: Issues, Crises and Priorities. Edited by S. K. Majumdar, L. M. Rosenfeld, P. A. Rubba, E. W. Miller and R. F. Schmalz. ©1991, The Pennsylvania Academy of Science.

Chapter Five

PROFILING AMERICAN STUDENTS STRENGTHS AND WEAKNESS IN SCIENCE ACHIEVEMENT

ARCHIE E. LA POINTE

Executive Director
National Assessment of Educational Progress
CN 6710
Princeton, NJ 08541

Each June, only $7\frac{1}{2}$ percent of our three million high school seniors leave their secondary school experience with an ability to integrate scientific information with other knowledge and skills they have learned in school or through life experience. This means that over 92 percent of our high school graduates are not able to use scientific knowledge from chemistry or physics, for example, to infer relationships or draw conclusions. More specifically, these students were unable to answer questions about the possible dangers of DNA research, the effects of genetic characteristics of parents on children, and the relationship of an object's density to its buoyancy in water.

When one considers for a moment, the issues discussed in most daily newspapers; health, nuclear power, environment, etc., it is difficult to imagine how these young 18-year old voters will be able to cope with the important questions on their local ballots. Their misunderstandings could lead to foolish decisions with implications for the quality of their own lives and perhaps for the lives of all of us.

If one takes a larger view and thinks in terms of global issues, international economic competition, and technological leadership, it quickly becomes apparent that 7.5 percent of 3,000,000 students each year will only yield a pool of 210,000 high school graduates adequately prepared to elect university programs in the sciences, medicine, and engineering. Not all of them are so inclined, so it is not at all surprising that we later find that one half of our graduate students are citizens of foreign countries.

These findings from a 1986 assessment of science knowledge and skills of the nation's 17-year-olds (National Assessment of Educational Progress, 1988—See Appendix-1) generated a flurry of rhetoric, newspaper headlines, and professional debate. The most distressing news of all was that our high school seniors were performing significantly less well in science than they did in 1970, sixteen years earlier. While we may have been shocked by these results, we should not have been surprised. Consider these well-known facts:

- There is no science curriculum in most elementary schools. This means there is little time devoted to its study, there is often no textbook that organizes a course of study, and most elementary teachers are poorly trained in the subject.
- American children don't think much of the sciences. Indeed, many of them bring strong negative biases *against* science to school. This is a very different set of attitudes from those of 15 to 20 years ago. Today's scientists are often viewed by today's children as the causes of problems (bombs, pollution, nondegradable waste, etc.) rather than problem solvers who will help address human issues of hunger and the environment.
- High school students do not elect science courses. Ninety percent study Biology, only 40 percent sign up for Chemistry, and fewer than 20 percent choose to study Physics. A few years ago, all three subjects were required of most college-bound high schoolers.
- Experts who have examined our science curriculum at the high-school level have labeled it a "layer cake" with each science subject treated as a distinct, separate entity. There is strong feeling among the scientific community that these specialties should be merged and treated as interrelated parts of a single discipline; that students should become aware of the "themes of science", should be helped to develop "scientific ways" of looking at their world and should understand the history of science.
- American schools at all levels have too few well-trained teachers in the sciences. This is partly a function of some of the problems listed above concerning inadequate school experiences as well as the broadening array of opportunities presented to successful science majors.
  The obvious, near-in solution seems to be the training of larger numbers of scientists and the retraining of existing teachers.

In 1983, the National Science Board's Commission on Precollege Education in Mathematics, Science, and Technology described the implications of neglecting science education:

Alarming numbers of young Americans are ill-equipped to work in, contribute to, profit from and enjoy our increasingly technological society. Far too many emerge from the nation's elementary and secondary schools with an inadequate grounding in mathematics, science and technology. As a result, they lack sufficient knowledge to acquire the training, skills and understanding that are needed today and will be even more critically needed in the 21st century.

APPENDIX 1**

Ages 9, 13, and 17
National Trends in Average
Science Proficiency, 1969-70 to 1986†

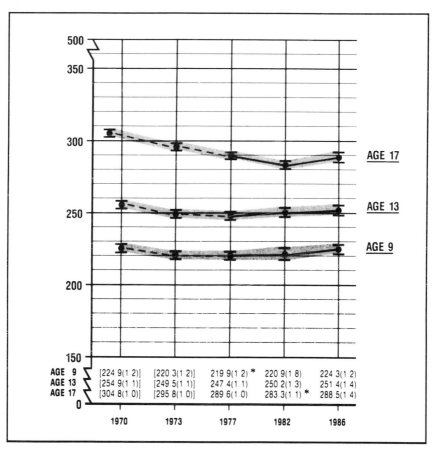

	1970	1973	1977	1982	1986
AGE 9	[224 9(1 2)]	[220 3(1 2)]	219 9(1 2) *	220 9(1 8)	224 3(1 2)
AGE 13	[254 9(1 1)]	[249 5(1 1)]	247 4(1 1)	250 2(1 3)	251 4(1 4)
AGE 17	[304 8(1 0)]	[295 8(1 0)]	289 6(1 0)	283 3(1 1) *	288 5(1 4)

[— — —] Extrapolations based on previous NAEP analyses.

 * Statistically significant difference from 1986 at the .05 level.
Jackknifed standard errors are presented in parentheses.

†Note: While 9- and 13-year-olds were assessed in the spring of 1970, 17-year-olds were assessed in the spring of 1969.

THE NATION'S
REPORT
CARD

 95% CONFIDENCE INTERVAL

**From Educational Testing Service. Science Report Card. Princeton, NJ, June 1988.

Since this statement was made, as many as 100 national reports have been issued calling for greater rigor in science education and suggesting numerous reforms. The nation has responded by updating standards for school science programs, strengthening teacher preparation, and implementing a wide variety of research efforts to deepen our understanding of science teaching and learning. Despite these efforts, average science proficiency across the grades remains distressingly low.

NAEP findings are reinforced by results from the Second International Science Assessment, which revealed that students from the United States - particularly students completing high school—are among the lowest achievers of all participating countries.

- At grade 5, the U.S. ranked in the middle in science achievement
- At grade 9, U.S. students ranked next to last.
- In the upper grades of secondary school, "advanced science students" in the U.S. ranked last in Biology and performed behind students from most countries in Chemistry and Physics.

Given evidence from both the NAEP and international results that our students' deficits increase across the grades, projections for the future do not appear to be bright. The further a student's progress in school, the greater the discrepancies in their performance relative both to students in other countries and to expectations within this country. Because elementary science instruction tends to be weak, many students—especially those in less affluent schools—are inadequately prepared for middle-school science. The failure they experience in middle school may convince these young people that they are incapable of learning science, thus contributing to the low enrollments observed in high-school science courses. Unless conditions in the nation's schools change radically, it is unlikely that today's 9- and 13-year-olds will perform much better as the 17-year-olds of tomorrow.

The results in mathematics as revealed by NAEP are not much more encouraging. (See Appendix-2). In 1986 only 6 percent of our 17-year-olds could "solve multi-step problems and use basic algebra." While it may be argued that science is not *essential* for success in American life, it is difficult to imagine a full and productive life without a grasp of quantitative reasoning skills and the ability to address problems in systematic ways. Only one half (51%) of these same high school graduates can compute with decimals, fractions and percents, recognize geometric figures, and solve simple equations. Even a world filled with calculators demands the abilty to recognize problem components and familiarity with techniques for organizing data to reach solutions.

The list of problems identified by experts studying our school programs is not surprising:

- There is a heavy stress at all levels on arithmetic—the "basics" of mathematics. In fact, this emphasis has resulted in dramatic improvements in *all* students' achievement in the basic functions.

APPENDIX 2**

**National Trends in Average
Mathematics Proficiency for
9-, 13-, and 17-Year-Olds: 1973-1986**

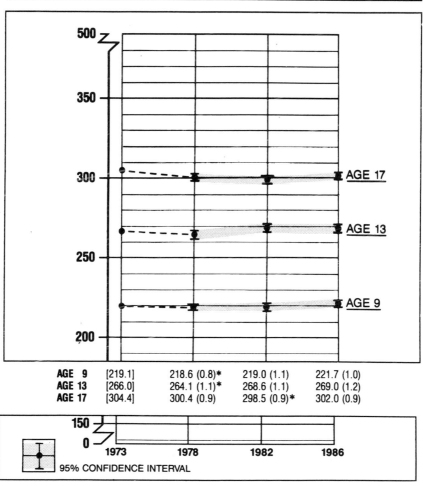

AGE 9	[219.1]	218.6 (0.8)*	219.0 (1.1)	221.7 (1.0)
AGE 13	[266.0]	264.1 (1.1)*	268.6 (1.1)	269.0 (1.2)
AGE 17	[304.4]	300.4 (0.9)	298.5 (0.9)*	302.0 (0.9)

[— — —] Extrapolated from previous NAEP analyses.
 * Statistically significant difference from 1986 at the .05 level.
Jackknifed standard errors are presented in parentheses.

**THE NATION'S
REPORT
CARD** naep

**From Educational Testing Service. Mathematics Report Card. Princeton, NJ, June 1988.

- We seem to devote little class time to teaching problem solving, the so-called "higher-order" skills of mathematics. This becomes obvious as we look at student performance on NAEP and other tests.
- Very little time, less than 10 percent, is devoted to children working in small groups to solve problems—the normal way for people to address most questions. Mathematicians have consistently advised this small-group work as an effective instructional technique in which students learn from each other and are able to develop techniques for using the talents, skills, and experience of companions/colleagues.
- Very few students elect to take many advanced high-school mathematics courses. This situation is changing rapidly as more states and school districts expand course requirements and reinforce higher standards of content rigor and student performance.
- At all school levels, teachers' preparation and comfort level with the subject matter leave much to be desired. As with the sciences, university graduates well trained in mathematics usually have several attractive options available to them upon graduation.

These findings seem to define the issues facing mathematics and science education in the United States of America. While the data are taken from assessments in 1986, both subjects were reassessed in 1990 and there is little reason to expect noticeable improvement in performance although major important changes are being implemented in school practice. While it can be argued whether these results represent a "crisis" or not, it does serve some purpose to consider them in that light since our society tends to react only to crises. Indeed the media as well as the mathematics and science communities treated both reports as reflecting intolerable situations. There was much rhetoric and editorializing about what should/must be done to improve things. Indeed, these data justified President Bush's formulation of his fourth national goal, to be first in mathematics and science by the year 2000.

Most people familiar with school environment, find this goal the most challenging of all. Indeed, a solid majority would describe it as impossible. Others see "the grasp beyond our reach" as the only way we'll have any hope of *dramatic* improvement. And, clearly, that's what is needed.

Setting Some Priorities

As we compare our students' achievement levels with those of other countries, certain observations come clear:
- Our 9-, 13-, and 17-year-olds could do much better in both mathematics and science.
- The importance, the value placed on mathematics and science as respected, important disciplines in our homes, schools, and society are much different from the prestige accorded these two subjects in other cultures.

– The rigor of course content, the amount of class time devoted to its learning, the amount of homework assigned for its mastery are all much less than those of other countries.

As we look within our own schools and examine what our own teachers and students tell NAEP about their experiences, we observe that both the content and structure of our school science curricula are generally incongruent with the ideals of the scientific enterprise. By neglecting the kinds of instructional activities that make purposeful connections between the study and practice of science, we fail to help students understand the true spirit of science.

Some Recommendations

Science:

In the ideal classroom, students would have abundant opportunities to question data as well as experts, to design and conduct real experiements, and to carry their thinking beyond the information given. They would identify their own problems rather than always solving problems presented by test, teachers, or other authoritative sources. Much of their problem-solving might also be in the form of practical experience. Through these experiences, students would come to realize that knowledge in science is tentative and human-made, that doing science involves trial and error as well as systematic approaches to problems, and that science is something they can do themselves.

To provide such instruction, teachers need to be prepared with a keen understanding of the nature of science, rather than just its requisite facts. Like their students, few teachers have had opportunities to conduct real experiments under real conditions; therefore, as a starting point, teacher education should provide opportunities for prospective science teachers to work with students at a variety of grade levels and in a variety of settings. The traditional one-semester methods course required of prospective teachers should give way to two or three semesters of coursework in this area, using video and audio tapes, intense feedback from professionals, and methods instructors who model the types of instruction desired.

Mathematics:

To retain a prominent place in today's technological world, our nation clearly needs to increase the percentage of secondary school students taking advanced mathematics classes. However, care should be taken to implement reforms at all grades, not just at the high-school level. Increasing course requirements at the upper grade levels will ensure that fewer students reject the opportunity to take more mathematics, but it will not address the fact that students in elementary and middle schools also need more challenging curricula.

Only after performance improves dramatically for younger students, particularly those in junior high school, can more high-school students take full advantage of Algebra II and Pre-Calculus or Calculus courses. Throughout the school years, mathematics programs need to focus on developing the higher-level skills and concepts essential to advanced mathematics performance. Success in meeting this challenge will help to define the path of our nation's economic future.

Science Education in the United States: Issues, Crises and Priorities. Edited by S. K. Majumdar, L. M. Rosenfeld, P. A. Rubba, E. W. Miller and R. F. Schmalz. ©1991, The Pennsylvania Academy of Science.

Chapter Six

SCIENCE EDUCATION: STOPPING THE SLIDE

KEITH GEIGER

President
National Education Association
1201 16th Street, N.W.
Washington, D.C. 20036

Undoubtedly, science and technology will determine the role that nations play on the world stage in the 21st century. Economic and political might will be increasingly vested in a nation's ability to be in the vanguard of scientific research and technological development. To meet the demands of world leadership in this new age, America must not simply improve, but revolutionize, its approach to science education.

Scientific literacy has emerged as a central goal of education. And merely grasping the fundamentals of biology, chemistry, physics, and earth and space science will no longer do. To be able to cope with tomorrow's world, all students, no matter their career goals, will need a thorough understanding of the process of scientific inquiry.

There is no time for delay. Immediate action is demanded, if we are to prepare our next generation for a world in which rapid advances in science and technology have become commonplace.

The world of the 21st century will demand scientific and technological knowledge in almost every area of human endeavor. Scientific literacy will be a prerequisite not only for an ever increasing number of occupations, but also for our daily, non-work-related lives. Indeed, as science and technology become continually more central to our civilization, informed participation in public affairs and effective citizenship will be impossible without an understanding of science.

Yet today, fewer than half of our nation's citizens are familiar with a scientific law first stated in the Copernican Theory, namely that the earth revolves around the sun once a year. According to the American Association for the Advancement of Science, a third of our nation's citizens actually believe that radioactive milk can be made safe simply by boiling it.

To eliminate such notions and other popular beliefs so nonscientific as to border on the superstitious will take a complete restructuring of science education. Science must become an integral part of the curriculum from kindergarten on. Inservice training must be accessible for science teachers at all levels. Scientific resource people and state-of-the-art science equipment must be made available to all schools, in poor as well as wealthy school districts.

For real effectiveness, efforts to restructure science education must begin at the elementary level, for it is there that an appreciation for science is first cultivated. It is in those early years that the intriguing world of science first beckons a child—initially as a witness to its miraculous powers, then as a participant in the scientific quest. In the elementary school years we can build on the natural curiosity children have about the world around them. Good elementary science teaching lets children learn the delight of scientific inquiry and discovery.

But it is in elementary school that too many children's difficulties with science begin. To what should be our national dismay, most elementary teachers are not provided the resources and the training they need to effectively meld science into their students' lives. Many elementary teachers, unprepared to teach science, dread this subject area themselves. They present the material in a way that is both boring and difficult, turning young students off before their natural appetite for discovering more about themselves and the scientific world around them is fully whetted. To complicate matters even more, too many elementary science textbooks are hopelessly out of date—and in some schools nonexistent.

Science sequences continue to be a blind spot in many middle and high school curricula. Nearly 30 percent of American high schools today offer no physics classes, and 17 percent offer neither biology nor chemistry. Indeed, a large number of high schools offer no science sequence at all.

But even when instruction is available, something is terribly amiss in the scientific education of our nation's students. This point was recently driven home by a study that compared the scientific education of students in 13 countries. To no one's surprise but everyone's chagrin, U.S. students ranked ninth in advanced physics; America's high school seniors ranked 11th in advanced chemistry and 13th—dead last—in advanced biology.

The National Science Foundation reports that of the 4 million high school sophomores who indicated an interest in the natural sciences and engineering in 1977, just over 200,000 received degrees in those fields in 1984, and only 10,000 are projected to go on to receive their doctorates by 1992.

Today only 7 percent of U.S. college students earn degrees in engineering, compared with some 22 percent in Japan. Estimates are that by the year 2000, America could face a shortfall of as many as 275,000 engineers and 400,000 scientists.

The nation's growing demand for engineers and scientists is complicated by another factor: the demographic changes in our society. Racial minorities will soon make up more than a third of our student population. Black, Hispanic, and Native American students have all too often been led to view science and technology as an elite realm—one they're not encouraged to enter and one in which their prospects for advancement are poor.

Today only 2 percent of America's scientists and engineers are black, only 2 percent are Hispanic, and only 15 percent are women. If our nation is to compete on the new scientific frontiers of the 21st century, talented minorities and women must be actively encouraged to pursue careers in science.

We must inspire these students by bringing from the shadows, where they have been hidden too long, the many contributions of minority scientists, especially black scientists, in the fields of agricultural, medical, nuclear, and space science.

But to consider careers in science and technology, minorities and women must have financial support to pursue these careers. Special scholarships targeting scientifically talented students, especially among women and minorities, must be made available—and their existence well publicized. It is especially critical that students know of the existence and conditions of scholarships for scientific study early in their secondary school years so they can obtain the requisite preparation to meet scholarship criteria. If we fail in the endeavor to build the cadre of scientists our nation needs, we will dilute our national power and further decrease our already fragile ability to meet the social needs of our citizens.

But we need not fail. We already have a historical precedent for rapidly upgrading American scientific know-how. In the late 1950s, after the Soviet Union became the first nation to launch a satellite, Congress passed the National Defense Education Act, targeting billions of federal dollars to encourage the expansion and improvement of education programs to meet critical national needs. The NDEA accomplished its immediate mission. Throughout the country, science education was upgraded.

However, priority attention to the sciences didn't last. By 1969, once the United States had put a human being on the moon, national interest in science began to wane. Many colleges dropped or drastically reduced their science admission requirements, and in turn began diluting their own science curricula and graduation requirements.

Unfortunately, this disinterest in science on the part of our nation's education institutions seemed to accelerate over the next two decades. By the late 1980s the United States faced a serious shortage of math and science teachers. In many school districts, science laboratories, equipment, textbooks, and even curriculum had become obsolete.

Recently, however, some states have moved to stop the slide in science education. Responding to the clearly growing shortage of scientists and the lack of funding for quality science education, these states have opened "super schools"—high schools emphasizing science, math, and technology. With millions of dollars in state-of-the-art equipment, these special schools offer high-level science and technology courses taught by carefully selected instructors to students identified as having high science aptitudes.

Although a step in the right direction, these super schools are not nearly enough. Other, more far-reaching efforts are needed if we are not to give away the right of today's students to compete in the technology-driven world of tomorrow, and if we are not to drastically erode our national wealth.

Solutions to several problems must be found simultaneously. We must reignite interest in the study of science. We must acknowledge the fact that students at all levels are routinely being "turned off" to science.

Asked why they're so turned off, many students blame science teaching methods that are mechanical, dull, and bland. Others say studying science is a passive experience. Still others point to science teachers who lecture too much and try to cover too much material, handing out facts rather than getting students actively involved in the learning process.

Another impediment to quality science education is the age-old 50-minute class structure. Fifty minutes just isn't enough time for students to set up and conduct a meaningful experiment or to get any real, hands-on experience in science.

The discouragement of all but an elite—and largely white male—student body from the advanced study of science is a national tragedy. Better science teaching in America's elementary and secondary schools could stop the early loss of scientific interest among many "average" students who, with stick-to-it-iveness, might actually become better candidates for scientific careers than many of the students who display an early aptitude for science.

If there is a key to developing teaching methods that take into account the broad learning needs of students in science, it is innovation based on research in science education. For that reason, the NEA is encouraged by the project on Scope, Sequence, and Coordination of Secondary School Science, initiated by the National Science Teachers Association.

This major effort to reform and restructure science teaching at the secondary level calls for the elimination of tracking of students, recommends that all students study science every year for six years, and advocates carefully sequenced, well coordinated instruction in physics, chemistry, biology, and earth/space sciences, taking into account the way students actually learn.

Opposing the traditional method of teaching each science in separate year-long disciplines, the NSTA plan calls for spacing the study of each of the sciences over several years. According to NSTA, research on the "spacing effect" indicates that students can learn and retain new material better if they study it at spaced intervals rather than all at once. With the interval method, they revisit a concept or idea at successively higher levels of abstraction.

Another study, "Science for All Americans," the first report of the American Association for the Advancement of Science "Project 2061," was designed to define scientific literacy. Its recommendations are presented in the form of basic learning goals for all American children. A fundamental premise of Project 2061 is that schools do not need to teach more and more, but rather to teach less so that it can be taught better.

The Project 2061 recommendations call for a common core of learning limited to the ideas and skills that have the greatest scientific and educational significance and that take into account the history, economics, and politics of change.

The NEA also applauds many of the ideas advanced in "Tomorrows's Schools: Principles for the Design of Professional Development Schools," a report by a coalition of nearly 100 leading universities dedicated to improving teacher education. This report calls for a national network of professional development schools that would function in much the same way that medical schools use hospitals for training doctors. University professors and education students would work with teachers and administrators in public elementary and high schools to design curricula, conduct research projects, train new teachers, and rejuvenate veteran instructors.

The coalition report emphasizes the need for teaching methods that go beyond rote memorization and promote critical thinking. The professional development schools, as envisioned, would help teachers help students of all ethnic and economic backgrounds to succeed. Other components of the coalition strategy include lifelong learning for teachers as well as students, and more flexible organizational structures involving parents, social service agencies, and local communities.

Professional development schools would be ordinary public schools that would take on extraordinary qualities. Of crucial interest is the coalition's recognition that the success of its plan depends on locating these schools in the poorest communities (where many training hospitals are located).

However, as essential as they are, innovative methods of science education by themselves will not be enough. Science teachers are becoming an endangered species. Many are being lured from their classrooms by the vastly higher salaries offered in corporate America.

Complicating matters further is the fact that many of our nation's science teachers will soon reach retirement age. The average age of high school science teachers in general is now in the mid-to upper-forties; average age for chemistry and physics teachers is 53 or 54. Unless students now entering college elect to pursue careers teaching science, there will be no replacements when the science teachers now in our public schools retire.

A massive commitment from the federal government, corporate America, local communities, and boards of education is essential to revolutionizing science education in our public schools. This commitment must include recruiting and training new science teachers and upgrading the knowledge of science teachers now in our classrooms.

It must also include encouraging and supporting the education of minority scientists to the Ph.D level, with assurance of career advancement to the utmost of their abilities.

State and federal legislation to enhance science education must be passed and implemented. Scholarships and forgiveness loans for future teachers must be available. Summer seminars and reverse sabbaticals should be established to give teachers now in the classroom opportunities to return to school or to spend time in science- and technology-based businesses where they can brush up on their skills.

In a time of tight school district budgets, business should continue and expand the practice of donating equipment essential to improving the quality of science education.

Today, our entire culture bears the imprint of science and technology that only a few truly understand. We seem to be rushing into a scientific dark age where an uninformed majority must follow the dictates of an informed scientific few.

We cannot allow world science to move forward and leave Americans behind, ill-equipped to understand and deal with the environmental, ethical, and international issues raised by scientific advance. It took 700 years for scientific knowledge to move from the theories of Ptolemy to the findings of Einstein, but in the last 50 years alone, scientists have split the atom, engineered flight into space, and broken the genetic code.

This new knowledge has created questions undreamed of by past generations. They run the gamut, from questions about the artificial generation of the human embryo, to genetic engineering, to the use of pesticides, to pollution control and defense policies. Many of these questions must be solved by the generation of students now in our schools. The answers will affect the economic, social, and ethical fabric of our nation.

Every citizen must be prepared to enter the debate. To abrogate this responsibility by failing to nurture scientific curiosity and inquiry in our students, by failing to develop a scientifically literate citizenry, is to imperil our nation's basic philosophy and our system of government by participatory democracy.

Science Education in the United States: Issues, Crises and Priorities. Edited by S. K. Majumdar, L. M. Rosenfeld, P. A. Rubba, E. W. Miller and R. F. Schmalz. ©1991, The Pennsylvania Academy of Science.

Chapter Seven

SCHOOL SCIENCE EDUCATION IN THE U.S.: A HISTORICAL OVERVIEW

EUGENE L. CHIAPPETTA

Professor of Education
Department of Curriculum & Instruction
University of Houston
Houston, TX 72204-5872

INTRODUCTION

Each period in the history of science education leaves its mark on the profession. These impressions are evident in the writings of the period and the curriculum materials produced. Each period brings with it a renewed interest and desire to improve what takes place in the classroom. The intent of those making recommendations is often to emphasize neglected goals in order to produce courses that better serve students and society. Although the recommendations of committees and individuals appear to emphasize balance in the curriculum, they often either omit or de-emphasize important aspects of science teaching. This historical overview of school science education is an attempt to identify important aspects of the discipline that must be included in programs in order to improve the scientific literacy of the nation's youth in the last decade of the 20th century.

If one examines the recommendations and goals of science education over the past 200 years, he/she will observe there are many recurrent themes such as science for all students, updating the content, the nature of science, and the inclusion of technology in the curriculum. These themes are made prominent by the social, economic, and political forces of a given period. For those who desired to better understand science education, an examination of the themes can result in useful insights into the profession. In addition, we must pay attention to the forces in society that push for educational reform approximately every 15 years (Cuban 1990).

The social, political, economic, and cultural forces that drove the reform movement in science education during the 1950s and 60s are much different than those driving the reform movement today (Fensham 1988). The cold war period between the U.S. and Russia, and the launching of Sputnik placed national security and the need for a scientific work force high on the nation's political agenda. The government had to prepare for the Star Wars era with scientists and engineers who could meet the challenge of protecting the country against Soviet aggression. The 1970s brought disillusionment with science and attention on science education was greatly reduced while the country was trying to get over the Vietnam conflict. In the 1980s science education became the focus of a great deal of criticism. Now the nation realizes that it must educate its citizens to become scientifically and technologically literate in order to be a productive society. The emphasis is on science for everyone. Minorities and women are encouraged to participate in careers in science and engineering. The reform movement, this time, is being fueled by the success of the Japanese to diminish the U.S.'s economic leadership in manufacturing and technology.

The purpose for this chapter is to present five periods over the past 200 years that reflect specific trends in school science education. The trends that dominated these periods will be discussed to better understand their impact on science education. Of special importance are the national curriculum projects produced during the last thirty years, for they reflect what some science educators believe should take place in science classrooms. These projects can serve to remind us of the importance for considering the discipline, the student, the teacher, the school, and society as we participate in the reformation that is underway in science education.

THE PERIOD BETWEEN 1800 AND 1910

During the early part of the 1800s very little took place regarding science education. The nation was primarily an agricultural society and the cities were just beginning to take form. Children spent more time working on the farm than attending school. What ever took place in the elementary school under the name of science was didactic in nature, requiring students to memorize facts to support religious doctrines (Underhill 1941).

Between 1860 and 1880 there was a movement to promote "object teaching" which was based upon the teaching of Pestalozzi, a Swiss educator. Object teaching was an attempt to make instruction more concrete and student centered. As opposed to lecturing to children, the intent was to give them real objects with which to experiment and make observations. The object teaching movement was sought to develop student thinking and to de-emphasize the memorization of facts.

Around 1890 the nature study movement began in New York State by Liberty Hyde Bailey, who was a professor at Cornell University. Bailey observed the migration of people into the cities and was concerned that children would grow up knowing very little about nature and the environment. He wanted to stimulate interest in living things and for students to study the natural environment. Bailey promoted his ideas through the Cornell School Leaflets. The early editions of the leaflets centered on the study of birds, flowers, insects, and trees.

Science teaching in the secondary schools began in academies around the late 1700s and early 1800s. Academies were private schools that prepared students of means for college and professional life. During their early period, religious instruction dominated, but gradually other subjects were added to the curriculum. Astronomy, physics, chemistry, botany, geology, geography, navigation, and surveying were some of the subjects included in science.

In the mid 1800s high schools came into being. In 1872 the Kalamazoo Decision to support high schools through a tax opened up high school education to many. This tax bill was important because many people were migrating to the city and the industrialization of the country was dependent upon a work force that possessed the ability to read and communicate. The industrialization of cities created a need for practical science courses that included technology. In the sciences, these courses included zoology, botany, astronomy, surveying, mensuration, mechanics, engineering, geology, and mineralogy. They emphasized the practical arts and citizenship (Lacey 1966).

Science in the high schools was strongly influenced by college science teaching during the last two decades of the nineteenth century. Under the influence of colleges, science courses became a formal part of the secondary curriculum with the lecture and some laboratory work as the dominant teaching modes. However, high school science courses that modeled college and university science teaching were viewed as too specialized and only appropriate for those going on to college, which excluded most students. "When the Committee of Ten (1893) set out to standardize high school curricula it emphasized that secondary schools in the United States did not exist for the sole purpose of preparing boys and girls for college. This report, along with others, promoted articulation between elementary and secondary school programs. It also caused colleges to have less influence over high school science offerings" (Collette and Chiappetta 1989; p. 8).

This early period before 1910 is instructive to study because it shows the beginning of science education in this country. The very fact that science became part of the curriculum indicates that it was perceived to be an important aspect of schooling. As primitive as the schools and curricula might seem, they illustrate some of the same trends and raise some of the same issues that are under discussion today. For example, in the elementary school, the object teaching lessons and nature study movement stressed the importance for having children study real objects and de-emphasized the memorization of facts. These very ideas are in the science education literature today, urging elementary school teachers to focus on concrete learning experiences rather than basing these experiences on textbook reading and completing work sheets. Similarly, the value of the lecture as a strategy for teaching science in the upper grades has been seriously challenged.

THE PERIOD BETWEEN 1910 TO 1955

During the first half of the century, the U.S. grew rapidly and became a modern nation. The country also went through a severe depression and took part in two world wars. The modernization and growth of the country influenced the structure of public schooling whereby the 6-3-3 organizational pattern of grades elementary through high school was put in place. Science in the elementary school was influenced by the work of Gerald Craig. The elementary curriculum that Craig developed centered on generalizations and concepts from the various science disciplines. This influence resulted in readers that taught children science content but was lacking in inquiry and hands-on activities.

The junior high schools flourished during this period of population growth. General science became a ninth grade subject at first and later it was taught in grades seven and eight. Biology was generally taught in 10th grade, chemistry in 11th grade and physics in 12th grade. General science and biology stressed the practical aspects of science and environment, and were more popular courses than chemistry or physics. In fact, advanced general science was proposed for grades 11 and 12.

During the periods after the two world wars, national committees attempted to influence science teaching and promote a general education for all students. These ideas are reflected in the *Thirty-First Yearbook, Part 1* and *Forty-Sixth Yearbook Part 1* of the National Society for the Study of Education in 1932 and 1947 respectively. The recommended objectives for science teaching in the 46th Yearbook are: 1) functional information of facts, 2) functional concepts, 3) functional understanding of principles, 4) instrumental skills, 5) attitudes, 7) appreciations, and 8) interests. These objectives stress a working knowledge and appreciation of science and its applications in society. Although stated in somewhat broader terms, these ideas are similar to the list of thirteen descriptors for a scientifically and technologically literate person put forth by the National Science Teachers Association in 1982.

The influences of the times are reflected in the courses of study suggested and implemented in the schools. The technologies of World War II such as radio, photography, and aviation were often included in general science and physics. The advances made in the sciences during this period of time were not known to the public and therefore did not influence public school science education. The making of the atomic bomb advanced physical science many years ahead in a very short period of time (Rhodes 1986). The Manhattan Project brought together many of the most capable scientists whose work on the atom created new knowledge in the areas of physics and chemistry. Much of this new knowledge did not reach high school or college chemistry and physics textbooks until 10 or 15 years after the end of World War II.

THE PERIOD BETWEEN 1955 AND 1970

After World War II the country was rebuilding its economy and the population began to grow. The number of students increased, and therefore science course enrollment increased in the elementary and secondary schools. At the same time it was realized that few people were going into science and mathematics. This caused scientists and mathematicians to examine secondary school science courses even before Sputnik was launched in 1957. The examination of textbooks and teaching practices brought about a great deal of criticism of the science taught. "The courses, they claimed, lacked rigor, were dogmatically taught, were content-oriented, lacked conceptual unity, were outdated, and had little bearing on what was really happening in the scientific disciplines" (Collette and Chiappetta 1989; pp. 11-12).

The concerns of scientists, mathematicians, and politicians brought about a massive reform in science and mathematics education never before witnessed in public education. Many millions of dollars were spent on the development of curriculum projects that were unique to education at all levels from elementary school through senior high school. Programs were offered across the nation at colleges and universities to update science teachers' content knowledge. Some of these teacher training programs included familiarization with the new curriculum projects, while other programs were offered specifically to train teachers to implement a given curriculum project. Unfortunately, most of the money spent on the reform movement in science education was on the development of new programs, as opposed to training teachers to understand and use the programs. As a result, only a small percentage of the teachers who used these innovative materials had adequate knowledge and understanding of the philosophy behind the programs or the instructional approaches recommended by the developers.

A distinguishing characteristic of the nationally recognized curriculum projects was the emphasis upon the nature of science and teaching science through inquiry. The scientists who developed the programs wanted students to learn that science was a dynamic enterprise where knowledge is constructed through observation and the examination of data. Consequently, laboratory work was a central activity in these courses. Some of the courses consisted entirely of laboratory work, while others placed laboratory work before classroom discussion and textbook reading. The textbooks written for these programs illustrate how scientists discovered certain principles and laws. They de-emphasized memorization of facts and vocabulary, often pointing out to the reader that this is what is known and how it was arrived at. A description of the nationally recognized programs of the 1960s can serve to highlight important curriculum elements that should be re-examined before making crucial decisions about the direction of the reform movement currently taking place in science education.

Elementary School Science Programs

Of the many elementary school science programs developed during the 1960s, three received considerable attention then and continue to be used today, though not as widely. They are the Elementary School Study (ESS), Science A Process Approach (SAPA), and the Science Curriculum Improvement Study (SCIS). These programs all stress hands-on activities and discovery learning. They had no textbook for students to read and placed the teacher in the role of guiding student learning. The influence of these programs can be observed in commercial textbook programs available today.

ESS provides students in grades K through 8 with opportunities to study relationships among objects and events with some of the most innovative materials ever produced for hands-on learning. This highly student-centered approach has the least amount of teacher direction, although it comes with an excellent teacher guide for each unit. The units provide considerable freedom for students to "mess around" in science, pursue individual interests, test out ideas, and discover principles. The ESS units such as "Kitchen Physics," "Small Things," "Bones," "Behavior of Meal Worms," and "Peas and Particles" have been used in many universities and colleges to train elementary school teachers to promote discovery learning. Perhaps "Batteries and Bulbs" is the best known of all, having been used by many science educators to illustrate the discovery approach and introduce current electricity.

SAPA is unique in that it stresses the processes of science and de-emphasizes the content of science. This program is organized into a highly structured sequence of lessons appropriate for students in grades K through 6. The basic science process skills such as ordering, observing, classifying, measuring, and inferring are emphasized in the lessons in the primary grades. The integrated science process skills such as controlling variables, hypothesizing, interpreting data, and experimenting are stressed in the upper elementary grades. The lesson booklets for the teacher present the instructional objectives, rationale, materials, activities, assessment measures, and questions for the student.

The SCIS program attempts to develop scientific literacy through a more balanced approach to elementary school science instruction for grades K through 6. This innovative program emphasizes knowledge, investigation, and curiosity in the development of student appreciation for science and competence in this enterprise (Renner, Stafford & Ragan 1973). The curriculum content centers around four major conceptual schemes—matter, energy, organisms, and ecosystems. The instruction strategy recommended for the program is called the Learning Cycle. This teaching strategy stresses the importance of giving children experiences to explore phenomena so that they can form concepts. That first phase is followed by the opportunity for the teacher to help pupils invent ideas to explain the phenomena under investigation. The newly formed concepts are then reinforced through their application in other situations.

Junior High School Science Programs

Two junior high school curriculum projects that were developed during the late 1950s and early 1960s stand out. They reflect the inquiry orientation which was prevalent among many of the nationally recognized curriculum projects. The first of these programs was the Earth Science Curriculum Project (ESCP), which began in 1958 under the direction of the American Geological Institute. ESCP was developed for ninth grade students and the textbook is still available today (*Investigating the Earth;* Mathews, et al. 1981). The laboratory investigations in the textbook convey to the learner that science is ever changing and there are many unanswered questions. Nevertheless, it is important that one appreciates this ambiguity and participates in the discovery of answers and new information regarding the Earth. In fact, at the end of each textbook chapter there is a section called "Unsolved Problems." The textbook chapters include laboratory activities for students to carry out that illustrate the principles and phenomena under discussion. The program and the textbook de-emphasize the memorization of facts and vocabulary terms. The textbook develops ideas for the reader and often uses historical accounts to add meaning, such as the vignette titled "Winds Cause Currents at the Ocean's Surface:"

> When Benjamin Franklin was Deputy Postmaster for the northern American companies, it came to his attention that mail ships took two weeks longer than whaling ships to make the voyage from England to America. He asked his cousin Timothy Folger, a whaling captain from Nantucket, Massachusetts, to explain the extra speed of the whaling ships. Captain Folger replied that the whalers knew of a place in the ocean where the water flowed like a river. He went on to say that whaling captains... (Mathews, et al. 1981; p. 130).

Another nationally recognized junior high school science program that received considerable attention is the Introductory Physical Science (IPS) project. IPS is a laboratory based course whose major goal is to help students conceptualize the atomic model of matter. This guided discovery process is directed through a student textbook that consists mainly of laboratory exercises. The textbook is designed so that the exercises in each chapter build upon one another, continually stressing the idea that the particle model of matter is useful when attempting to explain experimental results and the nature of matter. The reader can get a sense for this approach from a segment of the text where radioactivity in a cloud chamber is discussed:

> If there are more particles more tracks will be left; but the individual tracks are not affected by the number of particles. This abstract description of the radioactive process in terms of particles is an example of a "model" or theory. This particle model clearly accounts for the increase in the blackening of a photographic plate with increasing exposure time. (Haber-Shaim, et al. 1972; p. 137).

Senior High School Science Programs

Many science courses of study were developed for the high school biology, chemistry, and physics. Two and in some cases several courses were developed for each of the discipline areas. Noteworthy among these efforts were the many biology courses produced by the Biological Sciences Curriculum Study (BSCS), one of the most ambitious curriculum reform groups. This organization produced three major high school biology courses, a laboratory block series, courses for the educable mentally handicapped, film loops, readers, and many other instructional materials.

BSCS produced three textbook-based courses for high school biology known as the Yellow, Green, and Blue Versions. The Yellow Version was similar in many respects to traditional courses. However, its treatment of levels of organization was much more comprehensive, ranging from sub-cellular to organismic interactions with major emphasis placed on the cell. The Green Version was ecological in its approach to the study of life. The Blue Version presents biology on the molecular level. Both the Green and Blue Versions are still published.

Teaching science as inquiry was a major theme of BSCS courses and one that seems to have had a major influence on school science education. The inquiry model is elucidated when BSCS's high school biology textbooks are contrasted with the conventional textbooks of that time:

> Conventional high school texts were organized around a series of unqualified, positive statements. That kind of exposition—known as rhetoric of conclusions—has the advantage of simplicity and economy of space, but it gives the student the impression that science consists of unalterable, fixed truths, and that all is known.
>
> In addition, the rhetoric of conclusions approach fails to show that scientists are human and capable of error. Much of scientific enquiry has dealt with the correction of error...The aim of the enquiry approach is to show the conclusions of science within the context in which they arise and are tested. Thus, the student can see the problems posed and the experiments performed, the data found, and the interpretation that con-

verts data into scientific knowledge.

Another approach is to replace the rhetoric of conclusions with a narrative of enquiry. Thus, instead of telling the student the present state of genetic knowledge, the story is developed step by step through a description of the experiments performed, the data obtained, and the interpretations made of them. Still another way of emphasizing enquiry is to present laboratory work as a series of investigation activities into problems for which the text does not provide answers. (Myer 1970; p. 13).

The Chemical Education Material Study program (CHEM Study) deviated markedly from the traditional high school chemistry courses when it was produced during the 1960s. The explicit intent of the project was to encourage high school students to study chemistry and to pursue it as a career. The developers made every effort to help teachers as well as students understand how the knowledge base of chemistry is formed and to learn about the latest advancements in the field. A large variety of instructional materials was developed for this innovative chemistry course, which included: a textbook, a laboratory manual, a teacher's guide, tests, programmed sequences in mathematics, 26 motion pictures for the classroom, 17 motion pictures for teacher training, and film loops. The program recommended the inductive approach to the study of chemistry whereby the student conducted a laboratory exercise first, followed by the discussion of results and textbook reading (Merrill and Ridgway 1969).

Noteworthy among the nationally recognized physics projects was Harvard Project Physics (HPP). This high school physics course was different from the Physical Sciences Study Committee (PSSC) physics course developed in the 1950s and the traditional physics courses in existence. The aim of the developers was to produce a humanistically oriented course of study that would stimulate more interest in physics.

The Project Physics course includes a large variety of instructional materials, including a textbook, handbook, resource books, readers, programmed instructional booklets, film loops, transparencies, films, and laboratory equipment, that were designed to personalize the study of physics. The textbook focuses on six conceptual themes that form its units. Each unit begins with a prologue that provides historical, cultural or philosophical background, and an epilogue that summarizes and adds relevance. The *Project Physics Reader* is a collection of articles and book passages on physics that pertain to everything from historical events in science to the relationship between science and art. This interesting collection of writings was meant for browsing and intellectual arousal. The *Project Physics Handbook* is more than a compilation of laboratory exercises. The *Handbook* contains background information and suggestions on how to record data, explanations of concepts, photographs on how to set up apparatus, procedural hints, etc. This wonderful guide to laboratory work contains a large number of laboratory "experiments" so that students can choose among many investigations.

discipline base. The movement introduced technology and society into school science, and the interaction among science, technology and society; it is referred to as Science, Technology, and Society (STS).

Within the STS movement, some science educators are stressing social issues as a central theme. For example, Hofstein and Yager (1982) believe that social issues should be used as organizers for the science curricula. An issues based curriculum would include content that is relevant to the lives of students and eliminate content that is of little value. Brinkerhoff (1986) recommends frequent use of social issue vignettes to stimulate values clarification and awareness of problems that society must address.

Ramsey (1989) stresses the importance of social issues in the curriculum and their vital role in preparing students to be good citizens. He believes that science education within the traditional discipline based framework trivializes the intent of an appropriate education for all students. Ramsey lists the following outcomes for science programs in order to meet the general education needs of students: (1) the ability to identify science-related social issues; (2) the ability to analyze issues and identify the "players" (i.e., the individuals and/or groups involved in issues) as well as their beliefs and values; (3) the ability to use scientific problem solving skills to investigate these issues in order to identify the facts surrounding them and their social, economic, political, legal, and ecological ramifications; (4) the ability to evaluate issues and determine the most effective means of resolving them; (5) the ability to use a decision-making model to develop an action plan that can be implemented to resolve or help the issues, and (6) the ability to execute the action plan if it is consistent with the student's personal values (Ramsey 1989; p. 41).

Not all science educators promote the STS concept as enthusiastically as the seemingly many proponents of this new movement. For example, Kromhout and Good (1983) pointed out the importance of coherent structure in science and the inherent lack of structure in current social issues. They are fearful that an issues based curriculum might open up science education to "manipulation and perversion by social activists into a vehicle for reinforcing one-sided or narrow perceptions of social, political, intellectual and religious ideas" (p. 650). Good et al. (1985) stress the importance of the nature of science as the basis of science courses. Bybee (1987) suggests that the STS theme should be only a part of science education. He points out that science education should enhance the personal development of students and contribute to their citizenship. Although Rubba (1987) supports the STS movement, he cautions science educators, emphasizing that they must go beyond issue awareness and help students develop the capabilities to investigate issues using problem solving and inquiry and take action. (Rubba and Wiesenmayer 1988).

As the decade of the 1980s came to a close several nationally recognized projects and programs appeared that reflect the STS movement. Project 2061 is a long range undertaking designed to completely revamp science education by the year 2061 when Haley's comet returns. Sponsored by the American Association for the Advancement of Science, the intent of the project is to design programs of science that teach less but develop more understanding for what is taught (American Association for the Advancement of Science 1989). Its four goals are: (1) develop new curricular models; (2) improve the teaching of science, mathematics, and technology; (3) develop a realistic understanding of what it will take to achieve significant and lasting reform nationally, and (4) initiate collaborative action on many fronts.

Developed by the American Chemical Society (1988) with funds from NSF, Chemistry in the Community (ChemCom) is a high school chemistry course that attempts to improve scientific literacy by illustrating the impact of chemistry on society. This STS oriented course uses principles of chemistry to help students think intelligently about current issues that involve science and technology. The focus of each ChemCom unit is the community. The textbook includes readings, laboratory exercises, discussion questions, problem solving activities, and decision making activities.

Chemical Education for Public Understanding Program (CEPUP) is under development at the University of California at Berkeley (1990) through grants from NSF and private foundations. This innovative program provides instructional materials for middle school science that focuses upon chemicals and their interaction with people and the environment. The materials use scientific principles, processes, and evidence for public decision making. The program focuses on such topics as toxicity, threshold limits, parts per million, household solutions, and ground water. Although many modules have been completed, many are still to be developed and published.

DISCUSSION

The past two centuries have formed the discipline of science education in the U.S. The curriculum projects, courses of study, and articles have contributed significantly to this discipline. As a result the profession possesses a rich history of programs and literature with many themes that stand out as important. For example, during the first fifty years of the 20th century, the practical aspects of science were emphasized. An examination of the textbooks of that period shows that aviation, radio, photography, the telephone and many other technologies of the time were deemed important. General science, which was taught in the junior high school, in particular focused on the practical aspects of science. This course had some similarities to the STS programs that are gaining popularity today. However, science and societal issues were not included. Also, during the first half of this century the nature of science and the organization of science content into carefully thought out conceptual schemes were not of great concern to educators and textbook writers.

During the Golden Era of science education, between 1955 and 1970, the most drastic reforms occurred. This period brought to science educators and science teachers a new view of teaching science and an ambitious update of course content. The inductive, hands-on approach, which was part of most curriculum projects in grades K-12, was revolutionary for the time. The projects stressed teaching science through inquiry and helping students form concepts and laws, while de-emphasizing the memorization of facts and vocabulary terms. However useful these projects appeared, they were not teacher proof. Consequently, many of the programs were not used in a manner in which their developers intended. The teacher training necessary to properly implement the programs was never carried out to a great extent.

The programs of the 1950s and 1960s had some serious omissions. They left out technology and the practical applications of science. The laboratory work associated with most programs centered on pure science-type investigations and equipment. One can question how could it have been possible for students to relate what they studied in the laboratory with what took place in their daily lives. Science courses must always be relevant to the lives of the learner—a central theme of the STS movement that is gaining in popularity.

The next generation of curriculum projects that were implemented in the 1970s attempted to address some of the weaknesses of the first generation but had their own shortcomings. Programs like ISCS and ISIS were highly individualized courses of study with extensive teacher training materials. Unfortunately, many school systems never adapted to self-paced instruction, flexible scheduling, and a highly hands-on approach characterized by these programs, because many teachers found them difficult to implement and manage day-in and day-out. Furthermore, governmental funding for science teacher training and implementation of curriculum projects was practically eliminated.

The STS trend that gained impetus in the 1980s and is likely to continue into the 1990s may have answers to some of the problems of science education. The programs and recommendations by committees emphasize the importance of relevancy in the curriculum and the interaction among science, technology, and society. Project 2061 and reviews of research (Linn 1987) suggest that science teachers cover less material and provide greater depth of understanding of major topics. Amongst all of the excitement and change that is bound to take place in science education, teaching students about the scientific enterprise and involving them in activities that can help them to appreciate this discipline—the nature of science—may be neglected.

The programs and recommendations for science education must include a proper balance among the major themes of scientific literacy. They should stress science as a way of thinking, science as a way of investigating, science as a body of knowledge, and the interaction among science, technology, and society. Central to the science education of all students is that science is a way to view the world we live in; it is a human activity that helps to satisfy our curiosity and world we live in; it is a human activity that helps to satisfy our curiosity and to better understand phenomena. Science is a way to study nature using intuition and reasoning in search of evidence either to support hunches or replace them with better ideas.

Science instruction must reflect the many methods of science, always using those that best match the intellectual development of a given group of the students. Subject matter knowledge should be constructed from inquiry activities that center around major principles and conceptual schemes. Skillfully guided learning experiences must be designed around the major concepts of science as well as those that relate to the lives of students without compromising the integrity of scientific knowledge. The study of science and technology-related societal issues can serve to reinforce the methods and thinking related to the scientific enterprise, making science a more meaningful discipline that is studied within a contemporary context familiar to students and relevant to their lives.

REFERENCES

American Association for the Advancement of Science. 1989. *Science for all Americans.* Washington, DC.

American Chemical Society. 1988. *Chemistry in the community.* Kendall/Hunt, Dubuque, IO.

Brinkerhoff, R.F. 1986. *Values in school science: some practical materials and suggestions.* Philips Exeter Academy, Exeter, NH.

Bybee, R.W. 1987. Science education and the science-technology-society (STS) theme. Science Education, 71:667-684.

Collette, A.T. and E.L. Chiappetta. 1989. *Science instruction in the middle and secondary schools.* Merrill, Columbus, OH.

Cuban, C. 1990. Reforming again, again, and again. *Educational Researcher,* 9(1):2-13.

Fensham, P.J. 1988. Familiar but different: some dilemmas and new directions in science education. *In:* Peter J. Fensham (eds.). *Development and Dilemmas in Science Education.* The Falmer Press, London, England.

Good, R., J.D. Herron, A.E. Lawson and J.W. Renner. 1985. The domain of science education. *Science Education,* 69:139.

Haber-Schaim, U., J.B. Cross, G.L. Abegg, J.H. Dodge and J.A. Walter. 1972. *Introductory physical science.* Prentice-Hall, Toronto, Ontario.

Helgeson, S.L., P.E. Blosser and R.W. Howe. 1977. *The status of pre-college science, mathematics, and social science education: 1955-75. Vol. 1: science education.* ERIC Center for Science and Mathematics Education, The Ohio State University, Columbus, OH.

Harms, N.C. and R.E. Yager. 1982. *What research says to the science teacher.* National Science Teachers Association, Washington, D.C.

Hofstein, A. and R. Yager. 1982. Social issues as organizers for science education in the 80s. *School Science and Mathematics,* 82(7):539-547.

Hurd, P.D. 1969. *New directions in teaching secondary school science.* Rand McNally, Chicago, IL.

Krumhout, R. and R. Good. 1983. Beware of societal issues as organizers for science education. *School Science and Mathematics,* 83:647-650.

Lacey, A. 1966. *Guide to science teaching in the secondary school.* Wadsworth, Belmont, CA.

Mathews, W.H., C.J. Roy, R.E. Stevenson, M.F. Harris, D.T. Hesser and W.A. Dexter. 1981. *Investigating the earth.* Houghton Mifflin, Dallas, TX.

Mayer, W.V. 1978. The BSCS and its influence on biological education. *In:* William V. Mayer (ed.). *BSCS: Biology Teachers' Handbook.* John Wiley and Sons, New York, NY.

Merrill, R.J. and D.W. Ridgway. 1969. *The CHEM Study story.* W.A. Freeman, San Francisco, CA.

National Commission on Excellence in Education. 1983. A nation at risk: the imperative for education reform. U.S. Government Printing Office Stock No. 065-000-001772, Washington, D.C.

National Society for the Study of Education. 1932. *The Thirty-first yearbook of the national society for the study of education, part 1:* a program for teaching science. University of Chicago Press, Chicago, IL.

National Society for the Study of Education. 1947. *Forty-sixth yearbook, part 1: science education in American schools.* University of Chicago Press, Chicago, IL.

Ramsey, J.M. 1989. A curricular framework for community-based STS issue instruction. *Education and Urban Society,* 22:40-53.

Renner, J.W., D.G. Stafford and W.B. Ragan. 1973. *Teraching science in the elementary school.* Harper & Row, New York, NY.

Rhodes, R. 1986. *The making of the atomic bomb.* Simon and Schuster, New York, NY.

Rubba, P. 1987. Perspectives on science-technology-society instruction. *School Science and Mathematics,* 87:181-189.

Rubba, P. and R. Wiesenmayer. 1988. Goals and competencies for precollege STS education: recommendations based upon recent literature in environmental education. *Journal of Environmental Education,* 19(4):38-44.

Stake, R.E. and J.A. Easley. 1978. *Case studies in science education.* University of Illinois Center for Instructional Research and Curriculum Evaluation, Urbana, IL.

Underhill, O.E. 1941. *The origins and development of elementary schools.* Scott Foresman, New York, NY.

University of California, Berkeley. 1990. *Chemical understanding for public understanding program.* Addison-Wesley, New York, NY.

Weiss, I.R. 1978. *Report of the 1977 national survey of science, mathematics, and social studies education.* Center for Educational Research and Evaluation Park, Research Triangle Park, NC.

Science Education in the United States: Issues, Crises and Priorities. Edited by S. K. Majumdar, L. M. Rosenfeld, P. A. Rubba, E. W. Miller and R. F. Schmalz. ©1991, The Pennsylvania Academy of Science.

Chapter Eight

SCIENCE FOR MIDDLE AND JUNIOR HIGH SCHOOL STUDENTS

STEVEN J. RAKOW[1] and JAMES P. BARUFALDI[2]

[1]Associate Professor of Science Education
University of Houston
Clear Lake
Houston, TX 77058
and
[2]Professor of Education
Science Education Center
University of Texas at Austin
Austin, TX 78712

INTRODUCTION

Science education at the middle and junior high school levels has suffered from an identity crisis. Long neglected, the education of early adolescent learners has vacillated, in orientation and philosophy, between an extension of elementary school to a junior version of high school. Many of the issues and crises facing science in the middle grades are related to this attempt to define the appropriate role for science education as it seeks to meet the unique developmental needs of the early adolescent student.

WHAT IS MIDDLE LEVEL EDUCATION?

Middle level education (middle and junior high school) faces many of the same challenges as the middle child in a family. Questions such as "Where do I fit in?" and "How do I get my fair share?" are common.

This identity crisis has characterized much of the thinking about middle level education to the extent that the issues affecting the education of early adolescents have often been ignored rather than to face these thorny questions of articulation. Yet the middle level years are among the most critical in a student's life. Students at this age are going through tremendous physical, social, and emotional changes. With these changes comes a new sense of independence. Middle level students begin to make decisions about their education and their future careers. For this reason, the exposure that they receive in their science classes can have a life-long impact.

A differentiated educational experience for early adolescents is not a new idea. Junior high schools for grades 7, 8, and 9 sprung up in the decade between 1910 and 1920. In 1940, Gruhn and Douglas (cited in Hurd, et. al.)[1], in a now classic description, identified six functions of the junior high school:

1. **Integration** of previously acquired and newly learned basic skills, attitudes, and understandings into coordinated effective behavior
2. **Exploration** of specialized individual interests, aptitudes, and abilities
3. **Guidance** in making intelligent decisions regarding present and future educational and vocational decisions
4. **Provision** of differentiated educational facilities and learning opportunities suited to the varying backgrounds, interests, aptitudes, abilities, and needs of students
5. **Socialization** of early adolescents for participation in a complex democratic society
6. **Articulation** from preadolescent education to an educational program suited to needs of adolescents.

Throughout the next decades, the junior high movement came under criticism for a number of reasons including inadequate training of teachers and an inability to meet the needs of early adolescent learners. In the late 1950s, a new organizational plan, the middle school, emerged. Middle schools are distinct from junior high schools in a number of ways. Typically, middle schools encompass some combination of grades five through eight. These schools take more of a child-centered approach to curriculum than the content-centered focus typically found in junior high schools. There is an emphasis on interdisciplinary instruction with groups of teachers from different disciplines meeting and planning together on a regular basis. Although distinct in nature from elementary schools, the middle school attempts to use many of the approaches common to the elementary school in providing an educational experience that meets the unique needs of the heterogenous population of early adolescent students.

Recently, the Center for Research on Elementary and Middle Schools of the Johns Hopkins University conducted a national survey of schools that enroll seventh grades.[2] They found that 81.3 percent of the nation's seventh graders were enrolled in one of three types of schools: a sixth to eighth grade middle school (39.3%), a seventh or eighth grade school (24.6%) and a seventh through ninth grade junior high school (17.4%). Thus, although the middle school is a relatively recent educational structure, it has gained widespread popularity for the education of early adolescent students.

WHAT IS UNIQUE ABOUT MIDDLE LEVEL EDUCATION?

Children in the age range of ten to fourteen are going through a period of rapid change. Physically, they are developing at a rate faster than at any other time in their lives with the exception of the first three years of life. Often their rate of physical development does not match their rate of emotional and social development leading to tremendous concerns about how they are perceived by others.

This physical development is marked by increases in height, body breadth and depth, lung capacity, heart size, and muscular strength. Because body growth is often faster than muscular development, poor coordination can result. Individual growth patterns, however, are not simultaneous according to age level. Hence, some thirteen-year-olds may be very tall and well developed while other classmates still retain much more child-like characteristics. This concern over "will I ever grow up" becomes a major preoccupation for many children of this age.

As adolescents develop, they identify less with the authority of their parents and teachers and more strongly with the pressures of their peer group. This is the root of the term "adolescent rebellion." Although the cause of many a parent's grey hair, this behavior is an indication of a developing level of moral maturity which allows the child to recognize that there are a variety of alternative solutions to problems. Hence, decision-making becomes difficult. The students are less likely to accept teacher or parental directives without question. Though they have difficulty making up their minds, they do not want others to do it for them. Thus, they see their peers as an important and non-authoritarian aid in making decisions, and their dependence increasingly shifts from that of their parents and teachers to that of their friends.

Much of what we know about the cognitive and reasoning skills of adolescents comes from the work of Jean Piaget. According to Piaget, individuals progress through four stages of cognitive development. Each stage is characterized by specific patterns of reasoning that individuals are able to apply to problem-solving situations.

Adolescents typically span two of these developmental stages, the concrete operational stage and the formal operational stage. In the concrete operational stage, one's thinking is related to concrete objects. Students at this age need the opportunity to see and feel real objects (manipulatives). Math and science are both especially appropriate subject fields for providing early adolescent students with the opportunity to learn by means of real objects.

Some early adolescents may be in the formal operational stage. These students are able to think in terms of abstract ideas and concepts. They have the ability to apply probabilistic and proportional reasoning to their problem solving, to consider all possible combinations of two or more objects or events, and to apply theoretical reasoning to the identification and control of multiple variables.

Psychologists recognize that students in the early adolescent years may exhibit characteristics of both the concrete operational stage and formal operational stage. These individuals, termed by Eichhorn[3] "transescents," may demonstrate formal operational skills when faced with tasks for which they have some previous experiences, but demonstrate concrete operational skills when faced with novel situations. The key for teachers of early adolescents is to recognize that the typical classroom will contain children who exhibit concrete operational abilities, as well as those who exhibit characteristics of both stages.

The heterogeneous nature of the middle level classroom poses serious challenges to the teachers of early adolescents. Yet relatively few middle or junior high school science teachers have specific training for teaching science to early adolescents. The National Science Teachers Association has recently issued guidelines for the preparation of middle and junior high school science teachers which recommend a minimum of three courses in each of the major science areas (life, physical and earth/space sciences) as well as a science methods course appropriate for early adolescent students. Fewer than one quarter of those teaching junior high science meet that guideline.[4] This is of particular concern for teachers of earth and space science. Twenty-two percent of these teachers reported that they had never taken an earth or space science course and another 30 percent had taken only one or two related courses.

In addition to general inadequacies in their content preparation, most teachers who teach science at the middle level were prepared to teach science at either the elementary or senior high school level. Very few states offer middle grade certification. Exemplary programs do exist such as the Middle School Science Teacher Preparation Program at the University of Georgia and the Iowa Plan. The overall crucial issue is the unfortunate dichotomy between the sweeping reformation movement at the national level and the lack of articulation between these reform efforts and institutions and state agencies involved in teacher preparation and certification.

The most commonly taught courses at the middle and junior high school level are general science, life science, physical science and earth/space science. In a 1985-86 survey, Weiss[5] found that a majority of seventh to ninth grade science classes were taught using the textbooks of two publishers, Merrill (37%) and Holt, Rhinehart, Winston (16%). Thus to the extent that the textbook drives the curriculum there appears to be a de facto national curriculum at this level in a majority of the nation's schools.

Lecturing appears to be the most common form of instruction with 83 percent of junior high science teachers reporting that a lecture was a part of their most recent lesson. Hands-on activities were part of the most recent lesson of only 43 percent of the teachers although more than three-quarters of the teachers agreed that laboratory-based science classes are more effective. Perhaps a reason for this discrepancy is inadequate facilities, a factor perceived by one-quarter of the junior high teachers as being a serious problem affecting science instruction in their school.[6]

Thus, the picture that emerges of middle level science education is one of varying philosophies, with teachers attempting to meet the need of a widely diverse population of students, often with little or no specific preparation to meet the needs of these students and frequently handicapped by inadequate resources. A bleak picture, perhaps, but one that presents interesting challenges and opportunities for middle level science educators.

CRITICAL ISSUES FACING MIDDLE SCHOOL SCIENCE CURRICULUM

The imperative need to restructure or reform science curriculum in the middle years has been well documented. Since the 1950's, the perception that middle level science is nothing more than a "clone," albeit somewhat modified, of what is offered to students at the high school level, has permeated much of the literature dealing with the middle school years.

It was noted that the current concept of middle schools began to develop in the late 1950's as a result of a growing general dissatisfaction with the structure and curriculum of the junior high schools. Critics of the junior high school began to question the tendency to copy high school programs at the junior high level and believed that subject-centered curriculum, departmentalization, competitive athletics and the variety of social activities were not appropriate for the early adolescent; others perceived middle schools as the "stepchild" of junior and senior high schools. Middle schools frequently received out-dated equipment and books from high schools. Faculties often consisted of surplus elementary-prepared teachers or secondary-prepared teachers that tended to remain in a "holding pattern" until "a more desirable position" could be located teaching high school biology, chemistry, or physics.

The emerging popularity of the middle school movement during this period initiated the first generation of curriculum reform in science through the conceptualization of the *Human Sciences Program* (HSP), developed by the Biological Sciences Curriculum Study (BSCS) and funded by the National Science Foundation (NSF). Historically, this program served as a "landmark developmental effort" during this transitional time of curriculum reformation. The purpose of HSP was to design, develop, implement, and evaluate science curriculum for middle school science students. Characteristics of the early adolescent, physical, social, cognitive, and psychological, were also of prime importance and served as the organizational elements throughout the development process. HSP differed from most other curriculum projects in that it viewed four fields of knowledge as essential in designing curriculum for this unique population: natural and social science disciplines, characteristics of the learner, and major societal trends. An extensive needs assessment was conducted in addition to a synthesis process in forming a conceptual framework for the curriculum.[7]

Many have noted that HSP truly reflected the essence of science for the middle school student and was responsive to concerns of previous development efforts funded by NSF. The primary goal of HSP was to facilitate the transition of students from the lower levels of cognitive and affective development to higher levels (formal). The subject matter focused upon student-centered problems through multidisciplinary or interdisciplinary activities. Modules were designed to accommodate various classroom management organizations. The instructional activities were highly stimulating, acknowledging different learning styles.

In terms of overall adoption of this innovative program by school districts during the 1970s, many would concur that the impact of HSP at the middle school level was minimal. Yet, in terms of HSP serving as the "benchmark" in curriculum development efforts for middle school science, the program was an enormous success. It not only acknowledged the complexity of problems and issues confronting middle school years but also responded to those unique characteristics of the early adolescent and to the plight of science during those critical years charactered by intense social, political, and economic crises.

Hurd[8] recommended a major transformation in the present middle school science curriculum and believed that such a transformation might result in a curriculum that is interdisciplinary, one that is selected for general usefulness in life, one that emphasizes decision-making and includes values, and one in which instructional strategies are consistent with the goals of instruction and the nature of the learner. It appears that the potential for this transformation is on the horizon for middle school science.

THE NEED TO IMPROVE SCIENCE CURRICULUM FOR THE MIDDLE SCHOOL YEARS

Today, the recognition has surfaced again that the middle grade schools—junior high, intermediate, and middle schools—have been unrecognized, understaffed, unprepared, and undereducated. This has resulted in the need to improve educational programs for the early adolescent. It is the belief of many educators that the middle grade schools "are potentially society's most powerful force to recapture millions of youth adrift, and help every young person thrive during early adolescence".[9] Barufaldi[10] has noted that "Within the next 20 years or so...minority ethnic groups will become the majority in many states and cities. The school drop out rate among these populations is dangerously high (Texas 40%). Across the nation, approximately 2,000 students drop out of school each day. You must realize that students who drop out don't drop off the face of the earth! Besides the tremendous and unacceptable loss in human potential, these drop-outs cost taxpayers 75 billion dollars a year in lost revenues and increased welfare payments, not to mention 35 billion dollars per year in business and industry spent on retooling efforts and on remediation programs" (p. 16).

The need for a new science curriculum has become a major focus in science education. It has been noted that, "A volatile mismatch exists between the organization and curriculum of middle school grades and the intellectual and emotional needs of the young adolescents. Caught in the vortex of changing demands, the engagement of many youth in learning diminishes, and their rates of alienation, substance abuse, absenteeism, and dropping out of school begin to rise"[11]

Science educators are facing many crucial issues dealing with science curriculum for the middle school. The need to rethink programs at the middle school level was the focus of a position paper published by the National Science Teachers Association (NSTA) which stated that, "The primary function of science education at the middle and junior high level is to provide students the opportunity to explore science in their lives and to become comfortable and personally involved in it. Certainly science curriculum at this level should reflect society's goals for scientific and technological literacy and emphasize the role of science for personal, social, and career use, as well as prepare students academically"[12] The statement also included recommendations for curriculum which focused on all disciplines with frequent interdisciplinary opportunities, the development of written and oral communication skills, the use of process skills, the implementation of social issues that involve decision making, and the inclusion of information about careers. This position statement has served as a "rallying point" for middle school science since it differentiated middle school curriculum from both junior and senior high school programs.

FORCES AFFECTING DEVELOPMENT OF SCIENCE CURRICULUM FOR THE MIDDLE SCHOOL YEARS

Three major forces are affecting the success of development efforts in science curriculum for the middle school student; NSF funding of four middle school science programs, the 1989 American Association for the Advancement of Science (AAAS) report, and the *NSTA Project of Scope, Sequence, and Coordination of Secondary School Science*. The following section will briefly discuss each contributing factor and related issues and concerns.

Intervention of the National Science Foundation

NSF is currently funding four major programs in middle level science. These are being developed by the BSCS, Education Development Center, Education Systems Corporation, and the Florida State University. NSF supports the notion that materials developed should integrate science with other subjects using the hands-on approach, establish a coherent pattern of science topics, capitalize on the interests of students, use research findings in teaching and learning and identify standards of student achievement.

The BSCS program, *Science and Technology: Investigating Human Dimensions,* is rather typical of those recently funded and will integrate life, earth, and physical science in the context of issues that are meaningful to middle school students. The curriculum will be designed to incorporate the middle school philosophy with current educational innovations such as cooperative learning, an instructional model, and technology.[13]

During the late 1980's NSF also supported the development of innovative elementary school science programs. These programs are frequently referred to as the "triad programs" because of their involvement with a publisher, a university, and a school district. Publishers and developers are currently designing, writing, and field testing curricula for the 1990s.[14]

Partnerships have also been formed during the conceptualization and development of the recently funded middle school science programs. In addition, all programs intend to incorporate a strong, research-based instructional model, subscribe to cooperative learning opportunities, integrate science with other subjects, focus on decision-making skills, present the application of science to everyday life experiences and include career information. According to the publication, *The National Center for Improving Science Education: Summaries of Reports,*[15] the curriculum and instruction frameworks for the NSF funded projects for the middle years correlate well with the proposed frameworks for the elementary years. This correlation is especially noteworthy since it may lead toward a well articulated and integrated scope and sequence. Historically, the lack of integration of goals, content, and skills in science between grade levels has been a recurring, unresolved problem. It appears that this problem has been addressed appropriately.

Curriculum development is somewhat more sophisticated today than what transpired during the 1960's because of greater understanding of the developmental process, instructional and design strategies, assessment methodology, program evaluation and implementation strategies. Since the first generation of science curriculum reform in the early 1960s much research has been conducted addressing how middle level students can best learn skills and conceptual knowledge. As the result of this research we are now beginning to understand how best to teach science to early adolescents. Yet, if developers identify "all that is known" about learning and teaching in the middle level environment and fail to integrate these understandings within our social fabric, the success of these programs will be placed in great jeopardy. Science curriculum for this unique age group of students cannot (and should not) be separated or isolated from the community nor from the real problems of the world. The isolation of "what and how we teach" will only intensify our overwhelmingly complex social, political, and economic problems. These problems are part of the real dilemma facing curriculum developers and focus on a major challenge that was adroitly stated in the publication, *Educating Americans for the 21st Century*,[16] "The educational system must provide opportunity and high standards of excellence for all students—wherever they live, whatever their race, gender, or economic condition, whatever their immigration status or whatever language is spoken at home by their parents, and whatever their career goals." Hopefully, these new NSF funded projects have been sensitive to our diversity and to complex societal issues and problems such as drug abuse, crime, health, erosion of the family unit, violence, racism and sexism that form our social fabric and will develop appropriate instructions, strategies, and delivery systems to enhance the success of science materials.

Middle school science programs must also reflect the dramatic changes in demographics in the United States, and the low level of educational productivity and limited academic success among our children. We cannot remove ourselves from the "hard data"; the nation's report card in science education is quite depressing. New science programs for the 21st century and beyond must acknowledge these issues, concerns and challenges and use them as the driving forces for all curriculum development. If not, one then must question if much progress will be made or if these new endeavors in curriculum stimulated by federal intervention become part of the many "royal journeys" that we have taken in the name of progress.

Project 2061: Science For All Americans

Project 2061: Science For All Americans,[17] is a publication about scientific literacy and presents information on scientific knowledge, skills, and attitudes that all students should acquire, kindergarten through high school. During the first phase of the project recommendations were made concerning the nature of science, the nature of mathematics, the nature of technology, the physical setting, the living environment, the human organism, the human society, the designed world, the mathematical world, historical perspectives, common themes, and habits of the mind. The report also suggests a number of teaching strategies such as engaging students actively, de-emphasizing the memorization of technical vocabulary, using the team approach, and providing a historical perspective. Currently the project is in its second phase and teams of scientists and educators are attempting to translate the aforementioned recommendations into several curriculum models. These models will become the templates or blueprints to reform science education—curriculum, materials, technology, policy—in the United States. The last phase will use the resources generated and disseminate them to move the nation towards scientific literacy. The project will probably continue for a decade or longer.

One might want to assume that the newly funded middle level science projects are incorporating the information gained from Project 2061 into their conceptual frameworks and philosophical and psychological underpinnings. Yet, it appears that development efforts at the middle level are "out of sync" with recommendations and anticipated outcomes of Project 2061. A strong possibility exists that the products for both the new middle level science programs and the Project 2061 blueprints for reformation will converge in the educational community approximately within the same time period. Hopefully, the models for curriculum restructuring will be able to accommodate the new middle level science programs and be supportive and compatible with each other. If not, school districts will experience much confusion and difficulty in implementing the innovative ideas for reformation purposes.

NSTA Project of Scope, Sequence, and Coordination of Secondary School Science

The report, *Essential Changes in Secondary Science: Scope, Sequence and Coordination,*[18] endorsed by NSTA, outlines science curriculum reform aimed at improving science education of grade levels 7 - 12. (One immediately observes the lack of differentiation between the middle level and high school years.) Basically the report supports the idea of "spaced learning", with fewer science concepts spread over a longer period of time, in a coordinated way and emphasizing practical applications. Students would study current topics from biology, earth science, chemistry, and physics over multiple years across grades 7 - 12, possibly devoting up to seven hours of class time per week to science. Program development, supported by NSF, is currently occurring and will be implemented in pilot schools during the 1990-91 school year beginning in grade 7.

It is quite probable that the proposed structure, 7—12, will not provide opportunities to address the special needs of the early adolescent learner. With this inherent omission the middle school may again go unrecognized and unrepresented. Also, the proposed restructuring has similar problems to Project 2061 with synchronization. Differences in planning, developing, and implementing the product from NSTA's Scope and Sequence will send mixed signals to school districts who are encouraging and anticipating change in middle level science.

NSF interventions, new program development for middle school science, Project 2061 and the NSTA Scope and Sequence are major factors that will have a tremendous impact, both directly and indirectly, on the science education of early adolescents. NSF funding during the 1990s is welcome throughout the education community. It is rather unfortunate that NSF was not sufficiently farsighted to provide the necessary leadership that would have provided for a more coordinated effort. The initial funding of innovative elementary school science programs followed by science programs for the middle schools appeared to be reasonable; nevertheless, the funding of Project 2061 and the NSTA Scope and Sequence, emphasizing goals, organizational themes, and guidelines should have occurred before major program development. The lack of communication and creation of meaningful linkages among the recently funded programs may lead to the demise of this renewed interest in middle level science. What should have been cooperative professional endeavors are in danger of becoming unnecessarily competitive.

ADDITIONAL POINTS TO PONDER

One must not forget the role of the teacher during this reformation process; the success of these noble efforts is dependent upon the nature of the teaching force. Hopefully the NSF funded projects have also recognized characteristics of today's teaching force during the development of materials. Numerous reports have documented indicators of inadequacies in the teaching force.[19] Characteristics of our nation's precollege teaching force indicate that standardized test scores of teacher candidates declined throughout the 1970s. Many surveys have also shown that academically talented people remain underrepresented among those who pursue teaching as a career. The proportion of minorities who taught science in 1986 was pathetically low. In addition, the average teacher is now 52 years of age and more than 50% of them will retire by the year 2000.

Teachers, like most people, favor progress but tend to resist change. Barufaldi[20] has questioned the receptivity of this unique cadre of teachers toward these innovations and has noted that, "The process of change is a process, not an event. The process of implementation is a very time-consuming endeavor requiring patience, support, and perseverance, as well as an organized follow-up plan" (p. 10). He further questioned the desire of these teachers to change their philosophy and strategies toward teaching and learning at this point in their careers. Needless to say, these statistics and concerns have an enormous impact on the new reformation movement in science curriculum at the middle level.

The new middle level science programs are contemporary in approach and will reflect a science/technology/society (STS) philosophy. In viewing this philosophy, educators believe that "...they (students) can learn that science is an attempt to construct rational explanations of the natural world, and that technology is an attempt at providing solutions to human problems. They can also see that explanations are tentative, and technological solutions are incomplete and always have side effects. It is well within the reach of these students to explore the interrelationships between and among science, technology, and society, and to develop a knowledge base in the earth, physical, and life sciences".[21]

It has been noted that technology is seldom presented as a topic of study for the middle and junior high schools and current programs give no recognition to the nature and history of technology or information about the personal and social contexts of science and technology. Students at the middle level are aware of "Earth Day"-oriented issues such as pollution, energy, nutrition, environmental awareness, pesticides, and disease, but they have limited opportunities to study topics such as these within an integrated school curriculum.[22] For the STS philosophy to form the underpinnings of restructuring of "what and how we do it" in middle level science, revolutionary changes must occur in curriculum, instruction and assessment. Renewed commitment and support must emerge from teacher preparation institutions, state and national professional associations, and from state, local, and national governments.

The change and commitment necessary for restructuring will come to fruition only if our profession and its members make them part of their personal agenda. A personal agenda implies that we must instill in our profession the notion of ownership and encourage this feeling among students, parents, and the community during this new generation of curriculum development in middle and junior high school science. Ownership reflects a special relationship, a relationship that forms a bond or connection between individuals, groups, and systems.[23] Only then will middle level programs be effective by maximizing the science learning of these young adults.

REFERENCES

1. Hurd, P.D., J.T. Robinson, M.C. McConnell and N.M. Ross, Jr. 1981. *The Status of Middle School and Junior High School Science, Volume 1 Summary Report.* Center for Educational Research and Evaluation, Louisville, CO. p. 3.

2. Epstein, J.L. 1990. What matters in the Middle Grades—grade span or practices? *Kappan* 71:438-444.

3. Eichhorn, D.H. 1966. *The Middle School.* Center for Applied Research in Education, New York, NY.

4. Weiss, I.R., B.N. Nelson, S.E. Boyd and S.B. Hudson. 1989. *Science and Mathematics Education Briefing Book.* Horizon Research, Inc., Chapel Hill, NC.

5. Weiss, I.R. 1987. *Report of the 1985-86 National Survey of Science and Mathematics Education.* Research Triangle Institute, Research Triangle Park, NC.

6. *Ibid.*

7. Biological Sciences Curriculum Study. 1978. An update: The Human Sciences Program. *BSCS Journal,* 1(2):1-5.

8. Hurd, et al. *op. cit.*

9. Carnegie Council on Adolescent Development. 1989. *Turning Points, Preparing American Youth for the 21st Century.* Carnegie Council, Washington, D.C., p. 8.

10. Barufaldi, J.P. 1990. *The Tunnel at the End of the Light in Science Education.* Paper presented at the United States-Japan Seminar on Science Education (Honolulu, HI).

11. Carnegie Council, *op. cit.,* pp. 8-9.

12. Brunkhorst, B.F. and M.J. Padilla. 1986. Science education for middle and junior high students. *Science and Children,* 24(3):62-63.

13. Biological Sciences Curriculum Study. March, 1990. "Middle school science." *BSCS The Natural Selection.* BSCS, Colorado Springs, CO.

14. Champagne, A.B., B.E. Lovitts and B.L. Galinger. (Eds.). 1989. *This Year in School Science 1989: Scientific Literacy.* AAAS, Washington, D.C.

15. National Center for Improving Science Education. 1990. *Summaries of Reports "The Middle Years."* The National Center for Improving Science Education, Andover, MA.

16. National Science Board Commission on Precollege Education in Mathematics, Science, and Technology. 1983. *Educating Americans for the 21st Century.* National Science Foundation, Washington, D.C.

17. American Association For the Advancement of Science. 1989. *Science for All Americans.* AAAS, Washington, D.C.

18. Aldridge, B.G. 1989. *Essential Changes in Secondary School Science: Scope, Sequence and Coordination.* Paper written for the National Science Teachers Association, Washington, D.C.

19. Barufaldi, *op. cit.*

20. *Ibid.*

21. National Center, *op. cit.,* pp. 5-6.

22. Rakow, S.J. 1989. Middle/junior high science speaks out. *Science Scope,* 12(7):24-30.

23. Barufaldi, *op. cit.*

Science Education in the United States: Issues, Crises and Priorities. Edited by S. K. Majumdar, L. M. Rosenfeld, P. A. Rubba, E. W. Miller and R. F. Schmalz. ©1991, The Pennsylvania Academy of Science.

Chapter Nine

QUALITY HIGH SCHOOL SCIENCE

RUSSELL H. YEANY

Professor of Science Education
School of Teacher Education
University of Georgia
Athens, GA 30602

INTRODUCTION

Attempting to describe quality high school science is a risky venture. One can be somewhat inductive and assume that well defined quality components sum to a quality whole. Or, one can be more deductive and assume that an identifiable quality whole is made up of identifiable quality components. Most of us have been around long enough to realize that either approach is subject to failure some percent of the time. We have seen many programs with the right pieces but lacking in quality because of intangibles. Conversely, we have seen many programs that exist with weak components but possess quality because of intangibles.

Nevertheless, on the basis of twenty plus years of observing, researching and helping to develop effective school science programs, thinking related to several critical ingredients of quality high school science will be presented in this chapter. These ingredients are: the curriculum, with special emphasis on the selection of content and thought skills; the instruction, with attention drawn to the need for expanding verbal dialogue and critical analysis of laboratory activities; the testing, with little hope provided for change; and, special knowledge required of the high school science teacher.

THE SCIENCE CURRICULUM

The high school science curriculum is a place to promote the development of content knowledge, thinking skills, some technical skills and an understanding of the nature of science. School science needs to present and represent science as a special human endeavor that has many facets, and as a powerful force that has influenced both the substance of our knowledge and understandings and the way that knowledge and understanding is generated.

Curriculum Content

The content of the high school curriculum needs to reflect: (1) the state of our scientific knowledge; (2) a relevance to our past and future; and, (3) relationships and importance of science to other areas.

In order for the content to reflect the state of our scientific knowledge, it must be current and accurate. It can be argued that any knowledge that is accurate is current. However, the way that it is presented can affect the currency dimension. For example, if the study of light does not include mention of lasers and fiber optics, the curriculum is not current. Or, if a genetics unit begins with Mendel and ends with Watson & Crick or even protein synthesis with no mention of recombinant DNA work and its far reaching possibilities and ramifications, the curriculum is not current. The major problem confronted by high school curriculum developers is that there is simply too much current and accurate scientific knowledge for school science to deal with at a truly meaningful level. The curriculum must be focused on selected content.

How does one select? What should students know? What will they need to know? These are extremely important questions being addressed currently by national science curriculum projects such as *Project 2061* from the American Association for the Advancement of Science (AAAS). One product of these efforts resulted from an attempt by the AAAS to articulate an answer to the question: What science should all graduating high school seniors know? Their answer, constructed over a three year period by scientists and the AAAS staff, has been published as the book: *Science for All Americans.* This volume attempts to capture and communicate the broad conceptual schemes that represent minimal scientific literacy as we move into the 21st century. Persons engaged in the difficult task of selecting content for high school science texts or curriculum materials can be well served by consulting this work. They categorize the content of a 21st century science curriculum as follows:

The Nature of Science
   The Scientific World View
   Scientific Inquiry
   The Scientific Enterprise

The Nature of Technology
   Science and Technology
   Principles of Technology
   Technology and Society

The Physical Setting
   The Universe
   The Earth

Forces That Shape the Earth
The Structure of Matter
Transformations of Energy
The Motion of Things
The Forces of Nature

The Living Environment
Diversity of Life
Heredity
Cells
Interdependence of Life
Flow of Matter and Energy
Evolution of Life

The Human Organism
Human Identity
Life Cycle
Basic Functions
Learning
Physical Health
Mental Health

Human Society
Cultural Effects on Behavior
Group Organization and Behavior
Social Change
Social Trade-Offs
Forms of Political and Economic Organization
Social Conflict
Worldwide Social Systems

The Designed World
The Human Presence
Agriculture
Materials
Manufacturing
Energy Sources
Energy Use
Communication
Information Processing
Health Technology

Historical Perspectives

Displacing the Earth From the Center of the Universe
Uniting the Heavens and Earth
Uniting Matter and Energy, Time and Space
Extending Time
Setting the Earth's Surface in Motion
Understanding Fire
Splitting the Atom
Explaining the Diversity of Life
Discovering Germs
Harnessing Power

Common Themes
Systems
Models
Constancy
Patterns of Change
Evolution
Scale

Habits of Mind
Values and Attitudes
Skills

Selection of curricular content also should be based on relevancy. While determining accuracy and currency is largely an objective process, the determination of relevancy is somewhat subjective. Many logical arguments for relevancy can be mounted for any list of science knowledge one thinks high schoolers should acquire. For example, a particular bit of information may not have singular importance but it may be the key to understanding a highly relevant concept of greater importance. Or, content may be argued as relevant to link the past with the present. Often we need to study the history of a concept and the seminal thinking which generated an understanding of a scientific phenomenon in order to gain that understanding ourselves. Quite often the "intuitive knowledge" of a student will mirror the thinking of early scientists. For example, a true grasp of a sun centered solar system may need to be preceded by discussions of an earth centered theory. Or, students could be motivated toward deeper insights in inheritance while learning that respected scientists such as Lamarck developed misconceptions similar to naive views held by themselves.

The most difficult relevancy question regards future relevance. Will the content learned be adequate in the future for which the student is preparing? Of course we can never be sure. However, we can be sure that scientific knowledge will be an important ingredient in both shaping the future and helping individuals function in a changing world. Two tests for future relevance can be used. First, will the knowledge per se have future utility; and second, will it be prerequisite as a building block for knowledge not yet defined. Who has the crystal ball, or how much does it really matter? We could be so driven by the future utility and prerequisited knowledge arguments that every scientific fact, concept, and theory could be defended as part of the high school curriculum. As stated earlier, that is simply not possible to do and still expect meaningful learning to occur. However, there is a solution. That is to recognize that: first, even an argument for direct future utility will have to account for gross changes in the context of the application and probably much broader understandings of the principles themselves; and, second, knowledge by its very nature is always a building block for new knowledge. Therefore, the test for future relevance should lie in the perspective and level of presentation or expected learning. This mandates that the content of the curriculum be focused more on the conceptual level with a great reduction in the amount of facts and detail. It also mandates that students be helped to gain the perspectives that, since our scientific understanding is changing, the learning of science is a life-long process. A major goal of high school science is not to prepare scientists but to prepare life-long science learners.

Selection of content for the high school science curriculum needs to be done in a way that reinforces the relationship among the sciences and the relationship and importance of science to other areas. Currently, the science curriculum at all levels, elementary through college, is much more distinct and segregated than the practice of science. The high school science curriculum is particularly susceptible to this charge. Much attention needs to be focused on selecting content and modifying the configuration of that content to allow students to appreciate the highly interrelated and overlapping activities that cross over the traditional boundaries of biology, chemistry, physics, and the various earth sciences. The present configuration not only leads to a distorted and segmented view of the scientific world but also insures that most persons are never involved in any indepth learning of physics, chemistry, or earth science concepts. This has become at least a partially accepted norm that can be broken by selecting content that shows linkages among the sciences or by actually integrating the content across the traditional courses. For example, the esoteric task of memorizing all the bones and muscles groups could be supplemented (or replaced) with a study of the mechanics of the two systems, including mathematical calculations of the mechanical advantage of different arrangements and amount of energy required and work accomplished through certain actions.

In regard to selecting content to show how science is important in other areas of human endeavor, several areas are receiving much deserved attention. These are the interfaces among science, technology and societal concerns or actions— STS. Most scientists would agree that scientific knowledge is neutral, but one's frustrations with societal problems related to issues such as pollution, resource exploitation or global warming allows one to view science as the culprit. Specific curricular examples need to show where the applications of scientific knowledge can be either positive or negative depending on the motivation of those involved. And, in fact, for many of the science and technology-related societal issues we face, a resolution supported by scientific knowledge is available.

Cognitive Processes of Science

One of the most important impacts that the study of science can have on an individual is the influence on thought processes. After all, the basic characteristic of science that distinguishes it from other human endeavors is that science is a unique way of "knowing", a way of "finding out." All scientific knowledge and the resultant content in the school curriculum is a result of these thought processes. The high school curriculum has not been as attentive to this critical dimension of science as has middle and elementary school science through selected materials. This is unfortunate because the scientific thinking skills are generalizable and lasting; whereas, the factual or conceptual knowledge that we hold paramount in the high school curriculum is both specific and tentative. Of course, the conceptual knowledge should be learned, but it should be done in the framework of and provide a vehicle for learning scientific thinking skills. High school students do not need to discover all of their knowledge, but they do need to acquire skills in the identification of variables, formulation and testing of hypothesis, designing of experiments, collection and communication of data and drawing conclusions on the basis of rational processes. These themes should pervade the high school science curriculum and give a special perspective to learning science content. Unfortunately, they are too often reserved for special projects, special students, or special chapters on the "scientific method."

SCIENCE INSTRUCTION

Currently, much of the instruction associated with high school science is verbal, mostly lecture or discussion with questions. There is nothing inherently wrong with lecture-based instruction; but, its extensive use leaves little room for students to have direct experience with science or scientific thinking. More time needs to be spent on activities that allow the learners to engage in science as an active participant rather than a passive recipient. In addition, the active participant experiences must be examined to insure that the learner is having a rich experience and not simply engaging in low level, redundant manipulative activities.

Verbal Science Instruction

Many metaphors have been used to describe the act of teaching and learning. But the prevailing image one constructs when viewing a high school science class is that of recording on magnetic audio tape, some playback to test the system, more recording, and then extensive playback on exam day. Sadly, students and teachers have become so conditioned to interacting in this manner that few question the practice. Yes, the human brain is capable of receiving, storing, and retrieving upon proper stimuli, large amounts of information. But more important, every high school student is capable of much more. They can construct new and richer meaning. They can engage in inductive and deductive reasoning. They can evaluate for consistency. They can modify the old in relation to the new. They can identify problems and construct solutions. They can compare and contrast their own and other's meanings. In order to facilitate this type of science learning, verbal instruction must go beyond lecture and simple questions related to recall. The science teacher and the learners must engage in a dialogue that allows the learner to actively and critically process the information while allowing the teacher to closely assess the amount and level of assimilation occurring.

Non-Verbal Science Instruction

Laboratory activities usually come to mind with the mention of non-verbal science instruction. Lab-based exercises are a traditional part of high school science and have helped set it apart from the other subjects in the school curriculum. To the casual observer, high school science lab exercises may appear to differ only on the basis of content, materials, and specific activities. Certainly, these are key components that characterize a lab; however, a more fundamental difference exists when different science labs are compared. These differences emerge when one examines how much of the activity was predetermined and structured by the teacher or curriculum. To simplify the discussion, let's assume that each science lab has a problem, a method, and a solution. It is not difficult to uncover cases where all three of these are being prescribed by the teacher or the curriculum materials. In these cases, little or no science is occurring. In fact, the situation shows little variation from following recipes in the home economics kitchen or painting by numbers in a child's playroom. To truly reflect science, a lab activity would not consist of prescribed methods, and certainly not predetermined solutions. There should be room in a quality high school science curriculum for students to determine the nature of the problem as well as the method, and to seek their own unique solution.

There is another difference in lab activities that bears analysis and in most cases cries for significant modification. That is the relative amount of time a student spends in setting up the experiment and collecting the data when compared to the amount of time spent in critical analysis, interpretation and communication of the data, hypotheses testing and refinement, and conclusion drawing. All too often, the redundant low level task of data collection completely obscures the real or higher order purposes of the activity. Both the teacher and the students feel justified to hurry through the completion of an activity after a tedious, time-consuming ordeal of collecting a set of data. Often, it would lead to more meaningful learning if a teacher would provide the data and spend a much longer and more productive period of time in data interpretation. That is the role of the scientist. Students can spend far too much time in the role of lab technician if the teacher or curriculum materials are not sensitive to this important difference.

There has been a great call for more "hands-on" science in the school curriculum over the past decades. In fact, some states have legislated minimum amounts of time to be spent in the science laboratory. A careful and systematic analysis of all student activities needs to be carried out to reduce the amount of mindless, redundant manipulation of materials and objects in favor of richer activities that exist in a progression beyond those being carried out.

TESTING

For the most part, pencil and paper tests serve as the primary means of assessing the quality and quantity of learning affected by the high school science curriculum. Both students and teachers have been incultured to this form of testing, most of us have. We criticize and complain, attempt to modify, collect data to support or attack; but seldom do we abandon the practice. Can the pencil and paper test be improved? Are there alternatives?

Pencil and Paper Tests

The typical form of the pencil and paper science test uses multiple-choice format, i.e., a correct response is selected from among several alternatives. The refinement of this testing format was spurred by the desire or need to carry out mass testing and eventually to facilitate machine scoring and data entry for computer analysis. Arguments that led to further refinement were based on objectivity of scoring and reliability of students responses, both of which are enhanced by this type of assessment.

The wide-spread and sometimes exclusive use of this testing format has led many to be concerned about its limitations. One main concern is the reliance engendered in the student on stimulated recall. The correct answer is there among the choices. It need not be generated by the students. "Test-wise" students soon learn how to make best advantage of the situation and may not have the functional knowledge attributed to them by the test score.

Another concern is that items that measure higher order thinking skills, such as analyzing, synthesizing, comparing and contrasting, or interpreting, are difficult (or impossible) to measure with multiple choice items. Therefore, the tests,

especially those constructed by busy high school teachers, tend to dwell on factual low level learning. Another concern related to multiple choice testing is
the guessing factor. On the average, a guess will be correct a number of times
in proportion to the number of choices selected from. For example, if a student
randomly picks (guesses) from among four choices, a twenty-five percent probability of correctness exists. On the other hand, if the choices have been narrowed through limited but insufficient knowledge of the science concept to two
choices, the probability of successfully guessing the correct answer is fifty percent, always worth the try. In order to combat this confounding tendency for
guessing, several practices have been devised. The most common used in standard test scoring is to assume that the more items one misses the more items one
probably got right by guessing and mathematically adjust the score accordingly. This can be appropriate for large groups of data but could be a serious
injustice for an individual learner. Another technique used in National Assessment of Educational Progress is to simply allow the student to select "I don't
know" as a response. But the students know they are not being graded on these
items.

An effective technique that may not eliminate guessing but greatly reduces
the chances of a totally correct answer due to guessing is the double response
format. Each item requires the selection of two correct responses. One for the
correct answer to the question; the other for the reason that the answer is correct. Figure 1 contains an example.

The Pendulum's Length

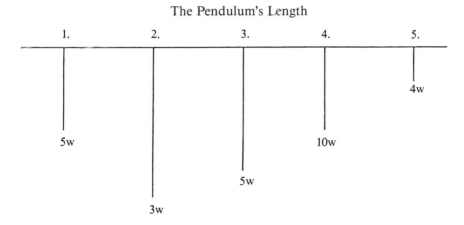

Suppose you wanted to do an experiment to find out if changing the length
of a pendulum changed the amount of time it takes to swing back and forth.
Which pendulums would you use for the experiment?

a. 1 and 4   d. 2 and 5
b. 2 and 4   e. all
c. 1 and 3

Reasons

1. The longest pendulum should be tested against the shortest pendulum.
2. All pendulums need to be tested against one another.
3. As the length is increased the number of washers should be decreased.
4. The pendulums should be the same length but the number of washers should be different.
5. The pendulums should be different lengths but the number of washers should be the same.

FIGURE 1. A Double Response Item.

Of course there are other pencil and paper testing formats that don't have the problems spoken to above. Test items can require the construction of responses rather than a simple selection. They include short answer completion or essay type responses. Both of these provide greater insight into a student's understanding of a particular concept and, in the typical high school situation, are used in conjunction with the multiple-choice format often in the same test.

Alternative Testing

Performance based testing is gaining popularity for assessing science skills and knowledge. In Israel, for example, high school students are tested on their ability to successfully complete specific science related tasks. Whole days are set aside for such testing at the national level. In our own country, the effects of curriculum change have been assessed by measuring the ability of students to perform science related tasks and verbally respond to queries that seek to expose their thinking skill level or understanding of a concept. Many research studies employ interview testing to help discover and understand the types of science misconceptions held by students. Results from performance based testing often reveal that the concern about the weakness of pencil and paper tests is valid. Some students hold many misconceptions not identified by the tests while other students tested exhibit an understanding of the concepts not predicted by their scores on pencil and paper tests.

Unfortunately, this will result in little change in high school science tests. Performance based testing requires time. High school teachers have little time to devote to the practice of alternative testing. Classes meet in structured time blocks with structured groups. Little or no paraprofessional help is available at the high school level. Mass education means mass testing, and mass testing usually means multiple-choice pencil and paper with a limited number of student constructed responses.

TEACHER KNOWLEDGE REQUIREMENTS

The knowledge base required to be a successful high school science teacher is extensive and varied. Little needs to be said about the need for a strong academic background. No less than a bachelors level in a science discipline is required for successful planning, teaching, and development of student potential. Additionally, knowledge of general classroom pedagogy and curriculum material, knowledge of the learner and learning theory, along with knowledge of schools is essential.

In addition to these, a newly conceptualized category of teacher knowledge has been shown to be highly indicative of successful teaching. This knowledge has been labeled: content pedagogical knowledge. Science teachers who exhibit this type of knowledge have the ability to: (1) develop helpful analogies for the students; (2) enlist a meaningful metaphor; (3) provide relevant examples; (4) construct demonstrations, and so forth. All are valid and effective linkages of content and pedagogy. After all, this is where teaching and learning occur.

SUMMARY

A quality high school science program will be based on a curriculum with content carefully selected to represent the state of scientific knowledge, insure relevance, and to show relationships among the sciences and other disciplines. Special emphasis will be placed on the development of thought processes. The instruction will be varied to provide a true science experience for the student. Meaningful dialogue will occur between the teacher and the students. Laboratory activities will be frequent with special attention to insure that data analysis and interpretation takes precedence over data collection. Testing should be conducted in such a way as to diagnose misconceptions and encourage meaningful learning on the part of the students. Finally, the key ingredient for a quality program is a teacher with a strong background in science content and classroom pedagogy and the special knowledge to effectively link the two.

SUGGESTED READINGS

Yeany, R.H., M. A. LaRussa, T.M. Kokoski and M.E. Hale. 1989. *A Comparison of Performance-Based Versus Pencil and Paper Measures of Science Process and Reasoning Skills as Influenced by Gender and Reading Abilities.* Department of Science Education, University of Georgia

Gilbert, J.K. and D.M. Watts. 1983. Concepts, misconceptions and alternatives conceptions: Changing perspectives in science education. *Studies in Science Education,* 10:61-98.

Treagust, D.F. 1988. Development and use of diagnostic tests to evaluate students' misconceptions in science. *International Journal of Science Education,* 10:159-170.

Funk, H.J., R.L. Fiel, J.R. Okey, H.H. Jaus and C.S. Sprague. 1985. *Learning Science Process Skills.* Kendall/Hunt Publishing Company.

Driver, R. and M. Keynes, (Eng.). 1983. *The Pupil as Scientist.* The Open University Press.

Science for All Americans. 1989. American Association for the Advancement of Science.

Science Education in the United States: Issues, Crises and Priorities. Edited by S. K. Majumdar, L. M. Rosenfeld, P. A. Rubba, E. W. Miller and R. F. Schmalz. ©1991, The Pennsylvania Academy of Science.

Chapter Ten

WHY IS SCIENCE BASIC IN ELEMENTARY SCHOOL?

JOHN R. STAVER

Professor of Science Education
Center for Science Education
Kansas State University
Manhattan, Kansas 66506

INTRODUCTION

The historical purpose of public education in the United States is to prepare all youngsters to function in society as literate, productive citizens. We live today, and our children will live tomorrow, in a sophisticated, technological, democratic, and pluralistic society. Moreover, science influences every segment of our lives. Thus, the goal of K - 12 science should be to prepare all high school graduates for productive roles in such a society. The educational community has coined the term "scientific literacy" to represent this goal.

Scientific literacy, however, is a complex, multidimensional concept. Moreover, certain segments of scientific literacy should be emphasized at different points in K - 12 education. The central point of this chapter is to present an answer to the title question, "Why is science basic in elementary school?" The answer is based on two conditions and contains three parts. Elementary school science is basic if science is taught:

1. with a heavy emphasis on concrete activities, and
2. utilizing a guided inquiry model.

If the above conditions are met, then elementary science is basic because science:

1. develops fundamental learning skills,
2. stems
3. provides the foundation upon which scientific literacy is constructed.

FUNDAMENTAL LEARNING SKILLS

The central core of science is a disciplined activity of coming to know the natural world by means of observation, measurement, analysis, and experimentation. The products of basic science, known as theories, are themselves tentative, held up for refutation by the scientific community, and revised and replaced by improved theories that better explain and predict the natural world. The products of applied science, which include technology, are similarly tentative.

The first fundamental contribution that elementary school science can make is to teach youngsters how to do science. Teaching elementary children how to do science involves teaching science process skills (e.g. observing, predicting, hypothesizing, experimenting) within the context of hands-on activities and through a guided discovery format. The benefits of activity-based, materials-oriented elementary science instruction, carried out through a guided discovery format, are very clear in the research literature. A few studies are discussed to emphasize the point.

Renner, Stafford, Coffia, Kellogg, and Weber (1973) investigated several fundamental issues in a series of studies. Renner and his coworkers compared the performance of elementary students who studied science through the Science Curriculum Improvement Study (SCIS) with counterparts who studied in textbook based science programs. They found that: 1) the SCIS program was superior in fostering the development of science process skills compared to a textbook science program; and 2) the SCIS Material Objects unit was superior in aiding the acquisition of Piagetian conservation reasoning compared to a textbook science program. In addition, the SCIS Material Objects unit was found to yield larger gains on five of six subtests of the Metropolitan Reading Readiness Test compared to a commercial reading readiness program. SCIS students recorded higher gains in Word Meaning, Listening, Matching, Alphabet, and Numbers subtests. The control group out gained SCIS students on the Copying subtest.

More recently, Shymansky, Kyle, and Alport (1982, 1983) quantitatively synthesized data from a body of research that compared activity-based, materials-oriented elementary science curricula such as the Elementary Science Study (ESS), SCIS, and Science: A Process Approach (SAPA) with traditional textbook science programs. Meta-analysis was the statistical procedure used to synthesize the data. The results showed that students who studied science via ESS, SCIS, or SAPA performed better across all performance criteria than students who studied science through traditional textbook programs. When performance criteria were separated, the superiority of students in ESS, SCIS, and SAPA classrooms was exhibited with respect to science process skill development, science content achievement, attitudes, problem solving capability, Piagetian reasoning level, and related skills (reading, arithmetic computation, and communication).

The data were based on both standardized and special tests, but regardless of the test, students in ESS, SCIS, and SAPA classrooms outperformed their counterparts in textbook based classrooms. In an independent meta-analysis, Bredderman (1985) corroborated Shymansky, Kyle, and O'Brien's (1982, 1983) findings, although he found somewhat smaller effect sizes in favor of students in ESS, SCIS, and SAPA classrooms.

In 1990, Shymansky, Hedges, and Woodworth reanalyzed the data collected earlier by Shymansky, Kyle, and O'Brien using refined statistical procedures for meta-analysis and found that earlier conclusions regarding the effects of ESS, SCIS, and SAPA were generally supported. The effect size across all performances measures at the primary level (K - 3) in favor of ESS, SCIS, and SAPA students increased. When performance was grouped as in the original meta-analysis, the effect size for development of science process skills in favor of ESS, SCIS, and SAPA primary level (K - 3) students also increased. The effect size for science content achievement decreased slightly but remained significant.

At the intermediate elementary level (4 - 6) only the attitude effect remained significant. The authors state, "...we suspect that the architects of the new curricula would probably be very pleased with this pattern of performance - students developed their process skills and interest in science at the elementary grade level and then increased their achievement and continued their process skill development in later grades" (p. 139).

The research discussed above focused on American students. However, a substantial part of current national concern about the quality of science education stems from international performance comparisons which show American students lagging behind. There exists, however, data based on international comparisons that indicate the benefits of activity-based, materials-oriented science. Staver and Small (1990) examined the science process skill development of three large cohorts of eighth grade students. The smallest cohort contained over 800 students, and the largest group numbered over 1,400 youngsters.

One cohort was from North Carolina; these students had studied science in elementary school via traditional textbook programs that emphasize reading about science with few, if any, activities. The second cohort was from Japan. The Japanese elementary science program is heavily materials-oriented and activity-based, having been developed by the Japanese Ministry of Education based upon Japanese curriculum developers' examinations and adaptations of American activity-based science curricula such as ESS, SCIS, and SAPA. Th third cohort was from Schaumburg, Illinois and included only students who had attended school continuously in District 54 since grade 2.

The Schaumburg District 54 elementary science program is exclusively activity-based and materials-oriented. It is a dynamic, gradually changing product of twenty years of development under the coordination of Mr. Larry Small, who is the district's Science and Health Curriculum Coordinator. Moreover, the Schaumburg program is well known throughout the nation as an exemplar, having been identified in a 1983 search for excellence by the National Science Teachers Association as one of 12 exemplary science programs across the nation (Penick & Johnson, 1983).

Analysis of data (using p < .01) from the Test of Integrated Process Skills II, TIPS II (Burns, Okey, & Wise, 1985) revealed that the Japanese cohort out performed the North Carolina cohort on the total score as well as three scales (i.e., Identify Variables, Operational Definitions, Graph and Interpret Data). No significant differences existed for two scales (i.e., Identify and State Hypotheses, Designing Investigations). The Schaumburg cohort out performed the North Carolina cohort on all scales and the total score. No significant differences were found between the Schaumburg and Japanese cohorts on four of five TIPS II scales. On the Identify Variables scale and on the total score, 2 point significant differences existed in favor of the Japanese cohort. Analysis of the total score results revealed clearly that the 2-point difference was due to the difference on the Identify Variables scale.

Prior to the study Mr. Small anticipated that the Schaumburg students might not do well on the Identify Variables scale because identifying independent/ manipulated and controlled variables in experiments was not emphasized by faculty. Mr. Small's prediction was born out in the data and appeared to explain the lone difference in the Japanese and Schaumburg scores.

The research described above exemplifies an emerging stockpile of evidence suggesting that science process skills are fundamental learning skills not only in science but also across the curriculum. For example, mathematics skills include classifying and sequencing; among skills essential to social studies are collecting data and interpreting graphs. Moreover, the report of Renner and his co-workers establishes a connection between science process skills, Piagetian conservation reasoning, and reading readiness.

Mechling and Oliver (1983) cite an appropriate ancient proverb: "Give me a fish and I eat for a day. Teach me to fish and I eat for a lifetime" (p. vii). Learners with well developed science process skills are learners who are skillful at learning. Moreover, the evidence indicates clearly that science process skill development is best fostered in an activity-based, materials-oriented elementary science program.

CONSTRUCTIVIST EPISTEMOLOGY

Learning is acquiring knowlege by means of a constuctive process in which the learner interacts with the external world, mentally considers data, and renders it meaningful in terms his or her prior knowledge. Thus, learning is always an interpretive activity characterized by individual constructions (Tobin, 1989). This epistemology, called constructivism, does not deny that knowledge exists in the external world; however, constructivism does deny that a learner can ever acquire direct knowledge of the external world. Rather, learners must construct such knowledge interpretively, sensing phenomena of the external world. Constructivism implies that the mind is not a tabula rasa and that knowledge is not simply transferred, recorded and memorized. To know requires the learner to receive, interpret, and relate incoming information to already existing knowledge. Moreover, knowlege is dialectical, interactional, and the tentative and relative result of a learner's own mental activity.

Within constructivist epistemology, students' prior knowledge is recognized as worthwhile and important, even though it may not be consistent with accepted knowledge. Learning experiences take into account the existing knowledge possesed by learners and provide opportunities for learners to construct new knowledge by fitting it into, revising, or replacing prior knowlege frameworks. Teachers, therefore, are not active transmitters of knowledge, and students are not passive receivers of knowledge. Rather, the teacher's role is to facilitate students' knowledge construction by designing learning experiences and providing resources that students can use to constuct knowlege.

"Teaching for conceptual change" is one outgrowth of constructivist epistemology. Such teaching emphasizes process skills, higher order thinking in concept learning, and discovery experiences. Interactive teaching methods that utilize teacher questioning and cooperative student learning are emphasized. Finally, a non-threatening classroom climate must be fostered, a climate in which students feel free to discuss and exchange ideas, to analyze and interpret data, to contribute and know that such contributions are valued.

One example of constructivism and the need for conceptual change is illustrated by my eldest daughter, Amanda, who is now in third grade. Two years ago as a first grader, Amanda, studied the seasons as part of science class. She constructed a simple theory to explain why days become longer in summer and shorter in winter. According to Amanda, the earth spins slower in summer and faster in winter. When she shared this with her younger sister, mother, and I at supper one evening, I decided to probe further:

Dad: How do you know that days get longer in summer and shorter in winter?

Amanda: Because in summer the sun is up before I get up and is going down when I go to bed. In winter the sun comes up when I get up and goes down before I go to bed.

Dad: Why do days get longer?

Amanda: Because it takes the earth longer to go around once.

Dad: Why do days get shorter?

Amanda: Because the earth goes around quicker.

Dad: Does the earth's spin suddenly speed up in summer and suddenly slow down in winter?

Amanda: No, it speeds up and slows down like the days do.

Dad: How do you know the earth spins? Mrs. Ireland said so (Mrs. Ireland was Amanda's first grade teacher).

Though naive by our scientific understanding of the seasons, Amanda's theory is quite an achievement for a then six year old. She accepted that the earth spins on the authority of her teacher. However, she observed that days become longer as summer approaches and shorter as winter approaches. She constructed a direct cause-effect relation to explain what she observed. Simultaneously, the theory qualifies as an alternative conception because two years later she still uses the theory to explain the shortening and lengthening of days.

More importantly, how should I, or her teachers, help Amanda to reach a more scientifically accurate understanding of why the length of a day changes? More specifically, what learning experiences should Amanda have that will provide her with evidence and opportunities to construct a more scientifically correct model? Following conceptual change theory and practice, her teachers and her parents must provide patient guidance and allow Amanda to examine her own experiences, confront the inconsistencies in her thoery about the seasons, and find a path toward a more scientifically appropriate understanding.

Any discussion of constructivism and its implications for teaching science is incomplete without a credit to Piagetian thoery. Although the logical formalisms of Piagetian theory have been cast aside in favor of modern cognitive science, which includes conceptual change teaching, teachers and researchers should remember that the epistemological basis of Piagetian theory was constructivism. Moreover, the wealth of data collected by Piaget, his co-workers, and other scholars clearly illustrate that elementary age, high school, and adult learners do not think qualitatively alike. Whereas adult learners are capable of abstract deductive reasoning, elementary youngsters rely primarily on inductive reasoning, which functions best when grounded in concrete materials and experiences. New theories in cognitive science must explain and predict such phenomena.

Constuctivist epistemology and developmental capability represent two factors that undergird the effectiveness of guided discovery instructional models such as the Learning Cycle, which was invented by Robert Karplus for the Science Curriculum Improvement Study elementary science program. Readers who desire a comprehensive discussion of the thoeretical foundations and empirical research surrounding the Learning Cycle are referred to *A Theory of Instruction* by Lawson, Abraham, and Renner (1989). The studies cited and discussed by Lawson, Abraham, and Renner yield conclusions that are quite consistent with research presented above. Moreover the Learning Cycle has been implemented effectively in junior high school, secondary school, and college level science curricula.

The bulk of research shows that students in SCIS programs exhibit more favorable attitudes as well as science content achievement. Lawson, Abraham, and Renner (1989) state, "the learning cycle approach appears to have considerable promise in areas of encouraging positive attitudes toward science and science instruction, developing better content achievement by students, and improving general thinking skills. It has showed superiority over other approaches, especially those that involve reading and demonstration-lecture activities" (p. 69).

Already mentioned above as an important part of conceptual change teaching, cooperative student learning (Johnson and Johnson, 1987) is actually a distinct concept with its own research base and foundation in constructivist epistemology. Cooperative learning systems are characterized by positive interdependence of learners within a cooperative group, accountability for learning by individuals as well as the group, heterogeneity within the group, and the development of leadership, partnership, and social skills. Positive interdependence is created by structuring goals and tasks such that students are part of a team endeavor in which team performance and individual performance are mutually important. Each individual is accountable for the performance of all team members; thus, no student can coast on others' work. Cooperative learning groups are heterogeneous to take advantage of differing individual characteristics, styles, and backgrounds. Each individual has leadership responsibilities for a specific role within the group, and members must work in partnership to complete assigned tasks. Students develop social skills by working together as they learn.

Learning outcomes that are promoted by cooperative learning systems include higher achievement, retention, intrinsic motivation, and self esteem; more positive peer relations, attitudes towards learning, psychological adjustment, and health; and greater use of high level reasoning strategies, critical reasoning, social support, collaborative skills, and attitudes necessary for working effectively with others.

The influence of cooperative learning systems and guided discovery instructional models can be seen by their presence in recently funded NSF curriculum projects. The Biological Sciences Curriculum Study has recently completed national field testing of its new elementary curriculum, entitled *Science for Life and Living: Integrating Science, Technology, and Health*. Major features of this program include a cooperative learning system and a five-stage instructional model (engage, explore, explain, elaborate, evaluate) which is an outgrowth of the original SCIS Learning Cycle. The *Full Option Science System*, FOSS, is a 3 - 6 science program developed by staff at the Lawrence hall of Science. FOSS features what its developers call a collaborative learning system. Moreover, the original three stage SCIS Learning Cycle (explore, invent, discover) has been renamed (exploration, concept invention, concept application) and carried forward in this curriculum.

Principles of constructivist epistemology are clearly evident in guided discovery science instructional models, particularly in the Learning Cycle. Empirical research has demonstrated the effectiveness of the Learning Cycle and cooperative learning to the extent that curriculum developers are now making guided inquiry and cooperative learning systems integral parts of new elementary school science curricula. In sum, the evidence indicates clearly that science process skill development and content achievement are fostered better not only in an activity-based, materials-oriented elementary science program, but also a program that is inquiry-rooted.

A FOUNDATION FOR SCIENTIFIC LITERACY

Whereas scientific literacy is the phrase used to describe the goal of K - 12 science education, defining what scientific literacy means is another task. Champagne and Lovitts (1989) argue that scientific literacy is a concept in search of a definition. Moreover, the myriad of national reports on reform in education provide no help. Authors of the reports concur that scientific literacy is important to the nation, but they fail to set forth a definition that enables observation, assessment, and evaluation.

Attempts to develop an operational definition of scientific literacy represent only failures at possibly an impossible task. Perhaps a workable approach to defining scientific literacy can be developed by analogy. The discussion regarding scientific literacy parallels the debate about the use of performance objectives to guide instruction. Readers will recall that goals containing verbs such as "know" or "understand" are vague, as these verbs carry different meanings. The logic of performance objectives is inductive in nature. A teacher develops a list of objectives that include conditions, specific action verbs, and descriptions of acceptable performance. If, after suitable teaching, learners can attain acceptable performance as described in the objectives, then the teacher may infer that the learners know or understand. The validity of the inference depends heavily on the quality of the performance objectives. If one examines performance objectives according to Bloom's Taxonomy, one looks for objectives that require performance at the higher levels, such as application, analysis, synthesis, and evaluation. Inferences as to knowing and understanding based on performance objective reflective of the entire taxonomy are more valid than inferences based on performance objectives requiring only memory.

It may prove useful to approach scientific literacy in an analogous manner. Consider first the concept as delineated by Showalter (1974a), who proposed seven dimensions of scientific literacy. The scientifically literate person:
1. understands the nature of scientific knowledge;
2. accurately applies appropriate science concepts, principles, laws, and theories in interacting with his universe;
3. uses processes of science in solving problems, making decisions, and furthering his own understanding of the universe;
4. interacts with the various aspects of his universe in a way that is consistent with the value that underlie science;
5. understands and appreciates the joint enterprise of science and technology and the interrelationships of these with each other and with other aspects of society;
6. has developed a richer, more satisfying, and more exciting view of the universe as a result of his science education and continues to extend this education throughout his life;

7. has developed numerous manipulative skills associated with science and technology. (p.2)

Thus, understanding and applying the nature of science, concepts, processes, value, science-technology-society relationships, interests, and skills constitute scientific literacy according to Showalter (1974 a,b).

Recently, Collins (1989) set forth a list of minimal opportunities that elementary science curricula should provide for children. The list below is adapted from Collins' list in that it is a list of performance statements. Upon matriculation into grade 7, learners should be able to:

1. observe and describe natural events;
2. pose questions about natural events;
3. explain natural events using scientific terms accurately;
4. recognize and describe the concepts that support scientific terms, including the role of humans and the tentative nature of concepts;
5. exhibit the skills required to describe, explain, and predict natural phenomena;
6. design experiments that test predictions about natural events.
7. exhibit the skills required to work with other students to produce scientific descriptions, explanations, and predictions;
8. describe and explain the nature of scientific knowledge, including its complexity and tentativeness, who constructs it, and the purposes of its use;
9. follow and discuss a debate about a science related topic in the media.

It is not a coincidence that the majority of the performance statements focus on processes. If elementary youngsters are able to accomplish the kinds of performance tasks stated above, then I am willing to infer that such students have developed a broad foundation for attaining scientific literacy.

IN CONCLUSION

To repeat a point made in the introduction, science is at its core a disciplined activity of constructing knowlege about the natural world. Activity-based, materials-oriented, inquiry-rooted science that emphasizes process skills, is conceptually rich, relates science and technology to individual needs, discusses the connections between societal issues and science and technology, and provides career awareness experiences seems to provide the best foundation for all students in their early school years as our educational system prepares them to take their places as useful, productive citizens in society. Such science programs are a fundamental part of elementary school curriculum.

REFERENCES

Bredderman, T. 1985. Laboratory programs for elementary school science: A meta-analysis of effects on learning. *Science Education*, 69 (4), 577-591.

126   Science Education in the United States: Issues, Crises and Priorities.

Burns, J.C., Okey, J.R., & Wise, K.C. 1985. Development of an integrated process skills test: TIPS II. *Journal of Research in Science Teaching*, 22(2), 169-177.

Champagne, A.B. & Lovitts, B.E. 1989. Scientific literacy: A concept in search of a definition. In A.B. Champagne, B.E. Lovitts, and B.J. Calinger (eds.) *Scientific Literacy* (pp. 1-14). Washington, D.C.: American Association for the Advancement of Science.

Collins, A. 1989. Elementary school science curricula that have potential to promote scientific literacy (and how to recognize one when you see one). In A.B. Champagne, B.E. Lovitts, and B.J. Calinger (eds.) *Scientific Literacy* (pp. 129-156). Washington, D.C.: American Association for the Advancement of Science.

Johnson, D.W. & Johnson, R.T. 1987. *Learning Together and Alone: Cooperation, Competition, and Individualization.* Englewood Cliffs, NJ: Prentice Hall.

Lawson, A.E., Abraham, M.R., & Renner, J.W. 1989. A theory of instruction. *Monographs of the National Association for Research in Science Teaching.* Number 1.

Mechling, K.R. & Oliver, D.L. 1983. *Science Teaches Basic Skills.* Washington, D.C.: National Science Teachers Association.

Penick, J. & Johnson, R. 1983. Excellence in teaching elementary science. In J. Penick & R. Yager (eds.), *Focus on Excellence: Elementary Science.* Washington, D.C. National Science Teachers Association, Vol. 1, 2.

Renner J.W., Stafford, D.G., Coffia, W.J., Kellogg, D.H., & Weber, M.C. 1973. An evaluation of the science curriculum improvement study. *School Science and Mathematics, 73*(4), 291-318.

Showalter, V. (1974 a, Spring). Program objectives and scientific literacy. *Prism II, 2*(3), 1-3, 6-7.

Showalter, V. (1974 b, Summer). Program objectives and scientific literacy. *Prism II, 2*(4), 1-3, 6-8.

Shymansky, J.A., Hedges, L.V., & Woodworth, G. 1990. A reassessment of the effects of inquiry-based science curricula of the 60's on student performance. *Journal of Research in Science Teaching, 27*(2), 127-144.

Shymansky, J.A., Kyle, W.C., & Airport, J.M. 1982. How effective were the hands-on science programs of yesterday? *Science and Children, 20*(3), 14-15.

Shymansky, J.A., Kyle, W.C., & Airport, J.M. 1983. The effects of new science curricula on student performance. *Journal of Research in Science Teaching, 20*(5), 387-404.

Staver, J.R. & Small, L. 1990. Toward a clearer representation of the crisis in science education. *Journal of Research in Science Teaching, 27*(1), 79-89.

Tobin, K.G. 1989. Learning in science classrooms. In *Curriculum Development for the Year 2000* (pp. 25-38). Colorado Springs, CO; BSCS.

Science Education in the United States: Issues, Crises and Priorities. Edited by S. K. Majumdar, L. M. Rosenfeld, P. A. Rubba, E. W. Miller and R. F. Schmalz. ©1991, The Pennsylvania Academy of Science.

Chapter Eleven

IMPROVED OPPORTUNITIES FOR HIGH SCHOOL SCIENCE EDUCATION

COLLEEN G. FIEGEL[1] and FREDERICK W. FISHER[2]

[1]Science Department
L.W. Higgins High School
7201 Lapalco Blvd.
Marrero, Louisiana 70072
and
[2]Chair, Science Department
Abington Senior High School
900 Highland Avenue
Abington, Pennsylvania 19001

INTRODUCTION

Science education in the high school classroom has improved significantly during the last decade. Heretofore, the primary mode of instruction was large group lecture, with little emphasis on lab experiences. Often, the science instructor was at a loss to find innovative ways in which to demonstrate the interrelatedness of science and society. For students to appreciate the relevance of science to their everyday lives, they must, in some way, feel connected to the large body of scientific knowledge. The advent of educational technology has opened new avenues of opportunities which can only help to propel the science educator into the 21st century. This chapter will focus on several innovative programs which are currently being implemented in many science classrooms across the country.

IMPROVED OPPORTUNITIES IN BIOLOGY EDUCATION: COLLEEN G. FIEGEL

One of the greatest problems facing teachers today is how to gain the attention of students and keep them motivated. Textbooks have long been used as the primary tool for instruction. However, modern technology has really challenged teachers to go above and beyond traditional teaching techniques.

sequence. The "interactive" feature is provided by a computer. The learner responds by typing on a keyboard, touching the screen, or manipulating objects attached to the computer. A variety of levels of interaction can be achieved, ranging from linear video to learner-directed sequencing of instruction. Fully interactive video features embedded questions, response feedback, and branching within the lesson. Depending on the response of the learner, the computer can branch to another section of the program and provide remedial instruction rather than continuing on in the same lesson. This feature allows for individualized, self-paced instruction. Most students, including low achievers, learn more and retain longer when lessons are presented at a reasonably fast pace because interest and participation is stimulated and because more content is covered (Wyne, Stuck, White & Coop, 1986).

This technology can be used by individuals for such things as tutorials, drill and practice exercises, or simulations. It can be used by small groups for cooperative problem solving or projects. Using this technology as a group visual display device also makes it appropriate for large group instruction. (See Table 2 for uses of videodisc technology.)

The most effective biology teachers use a combination of individual, small group, and large group instruction. Textbooks and lectures traditionally favor students who are able to independently build conceptual models from information which is read or heard. Videodisc presentations, when used to complement effective teaching strategies, enable more students to develop a clearer understanding of the terminology and concepts necessary for success in biology.

Biotechnology

A relatively new and exciting field of research can now be brought into the high school biology classroom: biotechnology. Biotechnology, the use of biological processes to develop new products, is one of the most important developments of the 20th century. It is a combination of genetics, molecular biology, and biochemistry. Students need to be aware of the impact this new research can have on their lives and the lives of future generations. These new products can be useful in agriculture, medicine, and industry. By altering a cell's

TABLE 2
Advantages of Classroom Use of Videodisc Technology

1. —enables students to problem solve, reason, think critically, learn cooperatively
2. —encourages peer teaching and a renewed enthusiasm for subject matter
3. —allows for individualization and personalized instruction
4. —allows for active participation
5. —can supply a large number of photographically realistic images at relatively low cost
6. —with computer integration permits development of lab procedures, experiments, and biological processes

genetic makeup, we now have the ability to make crops more resistant to insects and disease, make large quantities of pharmaceuticals that were once scarce and expensive, and develop bacteria that can help in the breakdown of PCB's, dioxins, oil slicks, and many slow degrading insecticides and herbicides.

Today, students in high school biology can use plasmids to transform bacteria, restriction enzymes to dissect DNA, gel electrophoresis to analyze restriction fragments, and ligation to place separated fragments into plasmid vectors for placement into different organisms.

Many teachers are already wondering how they will add a new topic to an already overcrowded curriculum. James Luken (1987) has suggested ways of incorporating various biotechnology topics into the existing curriculum. Recombinant DNA and gene splicing can be included after teaching DNA structure, protein biosynthesis and molecular genetics; cell culturing techniques can be taught after cell biology. Monoclonal antibodies fit perfectly into the unit on the immune system and regulation and testing of altered organisms can be incorporated into the ecology unit.

Biology education today should include basic biotechnology as well as applied and ethical aspects of genetic engineering. This technology is already having an enormous impact on agriculture, human health care, and industry. As we increase our ablity to manipulate genetic material, complex ethical issues for individuals and society may arise. Students must be given factual information in order for them to make informed decisions about these issues. In analyzing the ethics of a particular procedure, students must think critically and weigh the benefits and risks of various ethical dilemmas.

WHERE DO WE CURRENTLY STAND?

This certainly is an exciting time for biology educators! Current educational technology offers possible solutions to some of the issues facing biology educators today. While the preceding section has focused primarily on the issue of student motivation and interest and the use of technological aids, we would be remiss not to mention several other areas.

The biology curriculum in the past was primarily fact or skill driven. Emphasis was placed on the acquisition of specific and required facts, not necessarily on the scientific processes and concepts involved. Biology had become a subject for which students simply memorized facts in order to pass the course. Teachers are now challenged to find creative ways to manipulate the written curriculum in order to emphasize scientific processes and concepts.

In this era of growing concern over humane treatment of animals, the issue of dissection in the classroom is thrown into controversy. Teachers must balance the needs for providing hands-on experience for students against the ethical issues of "animal rights". Working with local and state school boards and the local scientific community to establish policy and clearly defined guidelines becomes an important responsibility of the biology educator.

Another major problem which is faced by many local school districts is the lack of adequate funding for biology classrooms. All of the technological aids previously discussed are costly expenditures. While local school districts may realize the value of such science equipment, many are prevented from ordering these items because of the cost factor. Therefore, it is encumbent upon local districts to find ways to either (1) reestablish priorities in order to allow for such expenditures or (2) find additional sources of revenues to provide funding. It may even be necessary for individual teachers to submit grant applications for single school sites.

One thread that runs constant throughout the field of biology education is the importance of science teachers' networks. Across the country, teachers are working together to form strong networks to provide not only support for each other, but also to help link teachers with the many resources available through today's technology.

IMPROVED OPPORTUNITIES IN CHEMISTRY AND ONE SCHOOL'S APPROACH TO SCIENCE EDUCATION: FREDERICK W. FISHER

WHAT'S HOT IN CHEMISTRY? CHEM COM!

One of the most recent and exciting developments in the teaching of chemistry is the Chem Com curriculum (Chemistry in the Community). Initially developed and written in 1982-83 by a group of high school and college chemistry teachers, the program was co-funded by the American Chemical Society and the National Science Foundation. Currently, the Kendall/Hunt Publishing Co. is contributing an additional $50,000 per year toward teacher training efforts.

In the past, traditional college prep chemistry courses have always centered on chemical theory rather than relevant applications to societal issues. Perhaps as a direct result of this, ninety percent of students taking a traditional chemistry course in high school end their chemistry study at that point. If such a trend is allowed to continue, tomorrow's citizens will be unable to cope with environmental issues by using their science knowledge to solve emerging problems. It is imperative that future political leaders and legislators (local to national) understand the scientific ramifications of decisions they make. Chem Com is an attempt to make that all-important connection between theory and application. Its approach addresses the very concern raised by the American Association for the Advancement of Science's Project 2061 regarding science literacy among the general populace.

In a Chem Com classroom, the focus is on science-based community issues to be solved. As students need chemical knowledge to apply to those issues, such knowledge is presented. It is important to note that a majority of the concepts and skills contained in a traditional chemistry course are also taught in a Chem Com course, the only exceptions being quantum mechanics and equilibrium constant calculations. It should also be noted that Chem Com is a rigorous college preparatory course, not suited to slow learners, and that it is an appropriate first course for students considering future college-level chemistry courses.

The societal issues around which Chem Com is centered include: supplying clean water; conserving chemical resources; using petroleum as both a fuel and a chemical feedstock; feeding the world; nuclear energy production, isotope utilization, and waste disposal; air quality; personal health; and the contemporary role of chemical industries in our society. Laboratory experiments, using standard high school equipment, remain an integral part of a Chem Com course.

Preliminary studies seem to indicate that Chem Com does succeed. My prediction is that a significant number of students will begin to become excited about science, and, as a result, opt for science-related careers.

ONE SCHOOL'S APPROACH TO SCIENCE EDUCATION
ABINGTON HIGH SCHOOL-MONTGOMERY COUNTY'S M.I.T.
OVERALL VIEW, GRADES 1-12

In the 1988-89 school year, 13.5% of Abington's senior class were enrolled in an Advanced Placement (A.P.) science course. These students had already taken chemistry, physics, and biology. Forty-three percent of the 1989 graduating class had taken physics, 66% had taken chemistry, and 100% had taken biology, Fig. 1.

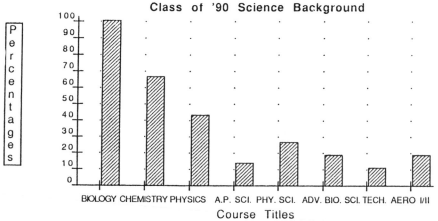

FIGURE 1.

Such statistics may not seem impressive for an exclusive private preparatory school, but Abington draws its students from a community diverse in both socio-economic and racial background. The community's economics span the gap from the poverty level to households with incomes well in excess of $100,000. Racial make-up consists of 16.4% Black, 3.3% Asian, 1% Hispanic and 79.3% White, Fig. 2.

Obviously this is not an "exclusive" community, but the science program is highly successful. Why? Two identifiable reasons are an experienced, dedicated, well-educated staff and a 1-12 curriculum which allows students to excel in science.

STAFF: 92% master's degree or higher in their field

First and foremost, Abington science teachers are highly skilled in their subject matter. Past practices have emphasized hiring the best qualified professionals available and paying excellent salaries for their services. Any teacher not having a master's degree was/is encouraged to enroll in a degree program. Money is an incentive. Attracting highly educated, skilled teachers has been possible because of competitive salaries. A teacher with a master's plus 30 graduate credits will earn $63,978 in 1991-92, with additional compensation for those earning a doctorate.

HIGH SCHOOL

Even though the state of Pennsylvania has recently mandated that all graduating students complete three years of science, the curriculum at Abington

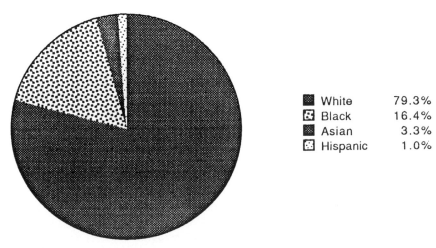

■ White	79.3%
▨ Black	16.4%
▨ Asian	3.3%
▨ Hispanic	1.0%

FIGURE 2.

had already evolved into quite a diversified program aimed at meeting the needs of all interest and ability levels. An enrollment of 1300 students enables Abington to offer an appropriate variety of courses in which all students have the opportunity to succeed.

Abington offers 14 different science courses for its 10th through 12th graders. Those students talented in science may choose one of the three advanced placement (A.P.) courses available in chemistry, biology, or physics. Because these talented students have been afforded the opportunity to accelerate their coursework since the 7th grade, they have already completed one year of biology, chemistry, and physics by the end of their junior year. This leaves the senior year open for A.P. courses and helps to keep Abington's A.P. program thriving. Out of the 500 school districts in PA, one school, Abington, accounted for over 2% of students taking the A.P. Chemistry exam in 1989. In fact, no Montgomery County, PA public school comes close to equalling Abington's number of students currently enrolled in A.P. courses. Class sizes here and in all science courses are limited to a maximum of 24 students.

The strength of the A.P. courses is threefold:
1. well qualified, popular instructors;
2. rigorous curricula;
3. appropriate lab equipment, supplies, and facilities

For other college bound students, the science department offers biology, lab-based physics (including PSSC), and chemistry, as well as a lab-based Chem Study course. It is anticipated that Chem Com will be introduced here in the very near future.

Physics lab work at Abington occupies 40 to 50% of class time and labs are a potpourri of the best experiments from a variety of sources. The text used is College Physics by Raymond Serway and Jerry Faughn, chosen because, in our estimation, all of the high school texts had been watered down to the point of being inappropriate.

Additional science courses are also available for those interested in science or those attempting to meet science graduation requirements. Students may choose from science technology, physical science ($\frac{1}{2}$ yr. chemistry, $\frac{1}{2}$ yr. physics), 2nd year biology, 1st or 2nd year aerospace (which includes flight time in an on-site flight simulator), ecology, and field biology. It should be noted that in the latter two courses students are actively involved in field studies. In addition, it has been found that many students opting for the physical science course have been successful in moving on to a later full year course in chemistry or physics. The aerospace course includes the private pilot ground test administered by the FAA at the end of the Aero II course. Although optional, this test was taken by 14 students during the '89-90 school year.

Abington's science technology course is geared to the ideas presented in the American Association for the Advancement of Science's Project 2061. Rather than memorizing a multitude of facts and figures, students develop an appreciation and understanding of science-related problems and "miracles" occurring in society. Students are also asked to consider possible moral and ethical dilemmas posed by such advances in technology. These include topics such as:

1. Hi-Tech Babies (in vitro fertilization, day-after pills, etc.);
2. Nuclear Power - Pros & Cons;
3. Nuclear War: The Effects (nuclear winter, fire storms, etc.);
4. Energy - Sources and Conservation;
5. Water - Sources, Conservation, and Contamination;
6. The Greenhouse Effects - Causes and Cures;
7. Acid Rain;
8. Conservation - Reusing, Recycling, and Reducing;
9. Hi-Tech Medicine - Transplants, Replacements, and Genetic Engineering.

The sci-tech curriculum was written and developed by Abington staff members during a summer, paid curriculum project. The course is continually being updated and relies heavily on current events presented through magazines, newspapers, and the TV/VCR.

With such diverse course offerings from which to choose, Abington has made great strides toward ensuring that all students are scientifically literate upon graduation.

But a successful program is more than statistics. Abington's science staff shares in a number of crucially important attitudes about young people and science.

1) Students need to be actively involved. To this end, all chemistry, physics, and biology courses rely on appropriate lab experiences.

2) Students need to be successful. Teachers at Abington strive continually to recruit and screen students for future courses which best fit their talents and interests.

3) Students need teachers who have high expectations for them. Abington's staff believes that no college-bound student is well-rounded without a year of physics. As a result, 43% of the class of 1989-90 graduated with such a course at a time when many high schools struggle to get enough students to offer physics. The staff also encourages students to stretch their aspirations by taking A.P. exams and entering academic competitions. The result of this staff encouragement is a significant number of students not only taking A.P. exams, but scoring three or above. In addition, a number of students receive awards and honors each year in science competitions. During the '89-90 academic year, Abington's high school science students have achieved:

1st place, Delaware Valley Science Bowl Competition (Philadelphia, PA);
1st place, Northeastern U.S. Science Bowl Competition (Valley Forge, PA);
4th place, National Science Bowl Competition (San Diego, CA);

1st place, Montgomery County, Pa Junior Academy of Science Competition (Penn State);

2nd place, State PJAS Competition;

1st place, Energy in the Environment Photo contest (sponsored by the Energy Education Advisory Council);

2 finalists, Philadelphia Science Competition (sponsored by the Philadelphia Science Council);

2 Honorable Mentions, 1990 Mars Settlement Illustration Contest. A total of 10 honorable mentions were awarded nationally (sponsored by NASA).

JUNIOR HIGH—THE SCHOOL'S FOUNDATION

Abington's junior high school science program is flexible and attempts to discover exceptionally talented science students and challenge them, while providing significant interest and complete science background for all students. To this end, a three-track program has evolved based upon I.Q. tests, and standardized math and reading scores. Input from parents, teachers, counselors, and administrators is also considered.

The program calls for accelerated students to proceed from the elementary grades directly to physical science, earth and space science, and finally biology or honors biology in 9th grade. Regular students take a year of life science first before continuing with physical science and finally studying earth and space science in 9th grade. (Fig. 3)

	LIFE SCI.	PHYS. SCI.	EARTH & SPACE	BIOL.	HONORS BIOL.
ACC.		7	8		9
TRACK		7	8	9	
REG.TRACK	7	8	9		

FIGURE 3.

It should be emphasized that while the system is sequential, it is not rigid. Recognizing the adolescent's penchant for changes in attitude, work habits, and goals, the staff is continually attempting to match students with appropriate courses. And again, Abington's science staff proves to be a sterling group, with 70% having a master's degree in their field.

ELEMENTARY PROGRAM—ABINGTON'S SCIENCE RENAISSANCE

In 1985, a survey of elementary teachers in Abington found that science was the subject most often avoided in classrooms, grades 1 through 6. Despite the few "shining stars" in science, quite a few teachers did not feel comfortable with the subject matter. The survey also found that science supplies were often hard to find and/or order and the text was outdated. Abington then set about to correct the situation. Upon perusal of a variety of elementary science curricula and texts, it was concluded that rather than the curriculum being of primary importance, it was the approach taken by staff that determined the success or failure of a program.

With that in mind, the district selected volunteers among staff members to specialize in science. They contracted Dr. Ray Rose, science professor from Beaver College, to act as paid consultant to elementary staff volunteers. Over the next two years, Dr. Rose, whose children attend the Abington schools, gave every elementary science teacher 45-60 hours of "support instruction." (figure 4) During this year before the new curriculum was to be taught, teachers themselves, using a half day released time each month, experienced hands-on completion of all experiments while updating their own personal science knowledge.

FORMULA FOR ELEMENTARY SCIENCE SUCCESS

1985-86	1986-87	1987-88	1988-89
SURVEY OF ABINGTON ELEMENTARY STAFF	SELECTION OF NEW CURRICULUM AND TEXT DECISION TO HIRE CONSULTANT	1 st YEAR FOR IMPLEMENTATION OF NEW CURRICULUM AND TEXT IN GRADES 1-3 CONSULTANT WORKS WITH ELEMENTARY STAFF, GRADES 1-3 AND "SCIENCE SPECIALISTS" GRADES 4-6 45 TO 60 HOURS——→	1st YEAR FOR IMPLEMENTATION OF NEW CURRICULUM AND TEXT IN GRADES 4 AND 6 SUPPORT WORK CONTINUED

FIGURE 4. Formula For Elementary Science Success.

When the new program was introduced in the schools during 1988-89, Dr. Rose continued his support work with staff during half-day monthly sessions, and the immediate result was a "dramatic" increase in class time spent on science.

Unlike many other districts, Abington's commitment to its elementary science program did not stop here. In order to ensure a continuing "rebirth" of science in the primary and intermediate grades, Abington has further taken the following steps:

1. creation of a science supply manager position in each school (at extra pay) whose job is to coordinate the ordering and distributing of science supplies for that school as well as maintaining the science storerooms;

2. the designation of "science specialists" in grades five and six who are responsible for teaching the science curriculum to all students in those two grades;
3. creation of an Elementary Science Coordination Committee comprised of teachers, administrators, and the district Director of Curriculum. This group meets to discuss existing or emerging problems concerning the elementary program.

It is important to note that such a renaissance might seem to be the direct result of a simple, rational plan. However, none of the new enthusiasm or updated knowledge and interest would be possible without Abington's commitment of money. Money is essential for new texts, consultants, support programs, and establishment of ongoing positions. It is encumbent upon society to understand that this money will be returned to them a hundredfold. When future citizens who have benefited from such programs are able to grapple with difficult environmental and technological issues, it will be a direct result of such committment.

A VIEW FROM THE TRENCHES
THE OVERSEAS THREAT—STOP COMPARING SUSHI TO APPLE PIE

As educators, we are continually bombarded by the media with "the overseas threat" . . . America is falling farther behind the Japanese, Koreans, Germans, etc. in science and technology. Whether or not this is true, hard-working teachers are becoming paranoid. Is this "failure" of the American education system our fault? Our curricula, our textbooks, and our teachers seem to be the focus of blame of every white paper and special report on the educational crisis in this country.

Let's take a look at a few salient points. The average Japanese student has a variety of cultural differences from the average American student. While the Japanese student is studying and/or being tutored after classes, the American student is out for one or more sports activities and/or working a part-time job (salaries of which rarely go for educational expenses). As a teacher, I'd be wealthy if I had a dollar for each time students have declined after-school extra help because of their job or sports practice. There are exceptions, but they are just that . . . exceptions. The average student seems to come to class as an unwilling participant, and until this attitude improves, we as a nation are at risk.

In a survey I performed during the 1988-89 school year, I inquired of science honors students earning a "C" or below why they thought they were performing so poorly. Almost all students felt that their teachers were highly capable and that their textbooks and course content were appropriate. On the other hand, 76% of those students said they were getting C's because they weren't working to their capabilities. After school jobs, extra-curricular activities were considered more important. These were conscious decisions made by these students and their families. These were honor students, those whom we expect to be in the forefront of American science and technology in the near future.

INDUSTRY ≠ SCHOOLS

Another way the public tends to view our present "crisis" is to look at the "industry" model to which schools are forever being compared unfavorably. "If we in industry continued to turn out as poor a product as the schools do, we'd be out of business" goes the argument. But wait! Industry controls the raw material coming in so that they may better control the product going out. The schools cannot. We must, and gladly do, accept all the raw material, but are expected to produce an equally acceptable product from all of it. The fact is that children are not products. They are each different and bring with them a variety of wonderful talents and abilities along with a plethora of physical, psychological, social, and family problems. Industry seeks to homogenize its raw material into a marketable product. Education seeks to make of each student a knowledgeable, ethical, responsible citizen who, in his/her way, will contribute to the future of American society.

HAVE AMERICANS TAKEN A "HYPOCRITIC" OATH?

Simply put, schools continue to need money to complete their mission to society. Education is America's largest business, yet is woefully underfunded in most areas. It is an absolute myth that there are untold dollars "wasted" in our schools. Money is needed . . . to attract quality teachers, to purchase up-to-date equipment and labs, to insure small, safe class size, to finance the continual update of teachers' professional knowledge, and to support the rejuvenation of outdated curricula. Such needs exist in every school and America needs to put its money where its mouth is. As a society, we say we value education; we pay athletes and entertainers well, but not teachers. We say we can't afford school bond issues; we buy the latest computer games and videos for our children. We say that teaching is a noble, important profession; we advise our best and brightest Ivy-League caliber children not to become teachers because there's not enough status or money involved. After all, isn't it foolish to pay $80,000 to $100,000 for a college education leading to a job that pays $20,000? In many states this salary never goes above $35,000 or $40,000 per year.

GLOOM AND DOOM?

All is not lost. If I were convinced that it is, I wouldn't remain in education.

American schools can and must improve. Abington has instituted excellent programs, pays its staff well, and holds everyone to the highest standards. As such, Abington is one of those select districts that will remain an excellent example of what schools can accomplish when a community and its educational staff work hand-in-hand to achieve mutual goals. Money is not the entire answer . . . only part of it. As a society, we must 1) Stop "dumping" all of our ills on the schools, our most visible target. We must, instead, be positive about the good things our schools accomplish. 2) Begin to encourage our children to think of their school work as the most important part of their lives, perhaps even at the expense of after-school athletics or jobs. 3) Stop comparing our system to Japan or to a cookie factory. We are not Asia and children are not cookies. 4) Begin to encourage our very brightest, most capable sons and daughters to choose teaching as a profession. Make them proud that they did by giving them the status they deserve. After all, they will "touch the future" because "they teach."

CONCLUSION, MORE QUESTIONS THAN ANSWERS

In conclusion, improved opportunities for students in biology, chemistry, and indeed, in all the sciences are available in some areas today. In select, quality districts, the sciences have leaped beyond the traditional classroom into the labs and fields; they've sped past their traditional textbook to the use of computers, video aids, and high-tech instrumentation.

However, such opportunities are much the exception rather than the rule. Because inequities abound in local tax bases, such inequities also abound in the opportunities afforded their students. How can a student who was taught physics by a football coach (who was assigned the class to fill out his schedule) be expected to compete with others taught by highly qualified science teachers with advanced degrees in their fields? How can local boards in poor districts attract and hold such highly qualified staff by offering meager salaries? And perhaps the most important questions of all . . . assuming that all funding became adequate and equal, how do we convey to our children the sense of urgency we feel about the importance of the sciences and their generation's responsibility to keep the United States in the forefront of science exploration and technology into the twenty-first century?

ACKNOWLEDGEMENTS: BIOLOGY

I would like to thank the following people for their assistant and support: Kathleen M. Kilgore, Administrator, Orleans Parish Public Schools, New

Orleans, LA; Sue Ellen Lyons, Science Department Chairperson, Holy Cross High School, New Orleans, LA; Carolyn Van Norman, Science Department Chairperson, L.W. Higgins High School, Marrero, LA.

C.G.F

REFERENCES

Lounge, J.P. and Walker, J.E. (1988). An Education school prepares for the future. *Electronic Learning,* 8, 30-32.
Whitten, R.H. (1988). Invertebrates on Video, *Carolina Tips,* 51: 9, 33-36.
Bunderson, C.V., Baillo, B., & Olsen, J.B. (1984) Instructional effectiveness of intelligent videodisc in biology. *Machine Mediated Learning, 1,* 176-215.
Wyne, M.D., Stuck, G.B., White, I.P. & Coop, R.H. (1986). *Carolina teaching performance assessment system.* School of Education, Group for the Study of Effective Teaching.
Luken, J. (1987). Does biotechnology have a place in introductory biology? *The American Biology Teacher, 49*:6, 351-354.

ACKNOWLEDGEMENTS: CHEMISTRY

In completing this manuscript, I received invaluable assistance from a number of Abington School District Colleagues. My thanks go to Dr Nancy Allbaugh and Gene Niclolo for proofreading. Chuck Ginter and Ruth Fechter for computer assistance, and Kathryn Christiana for input regarding the elementary program.

Dr. Joseph Sckmuckler, Temple University, graciously provided specific information regarding Chem Com.

Special thanks go to Kathleen M. Snider of the Council Rock School District for her untiring efforts in assisting with the editing of the manuscript.

REFERENCES: CHEMISTRY

1. Chem Comments, American Chemical Society, Vol. 3, No. 1, March, 1990, page 4.
2. American Chemical Society, 1988. A complete high school student text for a chemistry in the community curriculum. *Chem Com.*

Science Education in the United States: Issues, Crises and Priorities. Edited by S. K. Majumdar, L. M. Rosenfeld, P. A. Rubba, E. W. Miller and R. F. Schmalz. ©1991, The Pennsylvania Academy of Science.

Chapter Twelve

THE ROLE OF ATTITUDES IN SCHOOL SCIENCE INSTRUCTION

ROBERT L. SHRIGLEY

Professor of Education
College of Education
Department of Curriculum and Instruction
168 Chambers Building
The Pennsylvania State University
University Park, PA 16802

THE ROLE OF ATTITUDES IN SCHOOL SCIENCE INSTRUCTION

Attitudes are required to explain much human behavior. A sophomore from a religious home supports creationism but she earns the top score in a course on biological evolution. A physicist pickets the opening of a nuclear energy plant. With the same information available to all, prochoice and prolife marchers duel with stinging slogans at the same public rally. Anguish was my lot when visiting my childhood farm home overturned for its coal seam 100 feet down, but hope wells up within me when I view cows grazing in kneedeep in green pasture on reclaimed coal strippings down the road a ways. All involve impassioned feelings that influence important decisions in life. Allports'[1] pithy quote is as valid today as a generation ago: "The term itself may be indispensible, but what it stands for is." Attitudes filter and umpire our perceptions especially when confronted with scenarios in life not previously encountered. Attitudes ". . . draw lines about and segregate an otherwise chaotic environment; they are our methods for finding our way about an ambigous universe," reports Allport.[2]

This report documents the mission of science attitudes in the school and the community. More specifically, appraised are the relationships of science attitudes to student performance, student election of science courses and science careers, and a citizenry favorably disposed toward science and technology.

THE AFFECTIVE DOMAIN OF LEARNING

The interest of educators in the affective domain began in the 1950s when Benjamin Bloom and his colleagues[3] set out to design a taxanomy that would help teachers systematically assess instruction. Three domains of human learning evolved from their yearly conferences: the cognitive, psychomotor and affective domains.

Long considered the gatekeeper to learning, cognition is the better developed domain. Here Bloom's taxonomy helps teachers assess student learning of factual information in the sciences. Arranged hierarchially, the levels of cognitive objectives are knowledge, comprehension, application, analysis, synthesis, and evaluation. Employing the taxonomy science teachers tutor students through several levels of learning beginning with simple recall of science facts, the application of facts and concepts to the natural environment, and finally evaluation, where students are challenged to make judgments—often value-laden choices. The psychomotor domain appraises motor skills. Adjusting a microscope, bending glass tubing over a flame or measuring accurately with a metric scale would be examples in a science class.

The affective domain, the thrust of this report, assesses beliefs, interests, opinions, and attitudes in many curricula, including science. In their survey of 1855 middle and high school students, Lazarowitz and his colleagues[4] disclosed that almost 43 percent of their reasons for liking science were affective in nature, e.g., students like to see things live and grow. The emotional dimension of human learning, affection seems central to the learning act. Affection fine-tunes both hemispheres of the brain so that they function as a single entity when processing scientific information.[5] Science attitudes are central to the affective domain of learning.

SCIENCE ATTITUDES: WHAT ARE THEY?

Attitudes that affect science instruction are manifested on four fronts (see Figure 1): 1) those likes and dislikes of students toward science-related objects and ideas, e.g., acid rain and evolution; 2) science anxiety, that fear of science and all it stands for; 3) the historical fear-trust cycles of science rendered by scientific invention and innovation; and, 4) scientific attitudes, those beliefs about the nature of science modeled by scientists.

Likes and dislikes of science

Here science attitudes are defined as those favorable or unfavorable feelings students, teachers, and the general population have toward such objects as the metric system, prepared specimens, the periodic table, and or any of a hundred ideas or science-related objects.

Science anxiety

Science-anxious students see nothing good about science. They conjure up many questionable perceptions: its perceived difficulty, its formulas, obstensibly eccentric teachers, exploding beakers, and the allegedly weird profile of the few who prize science.

The Fear-Trust Cycle

Positive and negative science attitudes operate daily within school communities. But globally, a fear-trust cycle has pulsated throughout history, especially with the advent of the industrial revolution. Chronicled by historians, those valley of fear and hilltops of trust add another attitude dimension that permeate science instruction at the local level. Meeting in West Berlin in 1979, thirty scholars probed and analyzed those historical fear-trust cycles of science. This section draws upon their findings.[6]

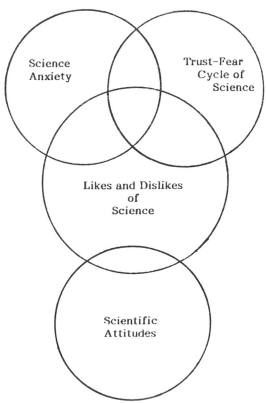

FIGURE 1. Science Attitudes

Best documented are the valleys. As early as 1479, an improved loom in Danzig forced weavers to come to grips with a more efficient model but not before the inventor was threatened with drowning. The madder plant, once a major source of dye, was reduced to a botanical curiosity late in the 19th century when dye-making became a cheap by-product of Britain's coal tar industry. Wiped out was the livelihood of madder farmers around the Mediterranean.

Inventions early in our century provided a high for science. However, in the 1930s the trust in science faltered when it could do little to calm economic woes. But in the 1940s and 1950s science was the ally to human happiness. Science was credited for our victory over the Axis Empire, but it sired the atom bomb. At the same time we promptly embraced penicillin, hybrid corn, and nylon stockings. NASA's successful space program of the 1950s and 1960s enjoyed a global blessing.

As the 1990s open up before us science bumps along in a valley glimpsing sporadically over the rim at the successes in Silicon Valley, headway in genetic engineering, and the like. The thought of malfunctioning nuclear energy plants terrorize even the strong-hearted. Industrial pollution in our midwest returns as acidic raindrops rendering lakes high in the Adirondacks crystal clear but lifeless. Our hot, summer drought in 1988 handily correlates with all the talk about global warming. The Challenger misfires before our eyes. Conquering cancer comes in small steps and the cost of a gallon of gasoline seldom dips below a dollar.

Contemporary failures and successes in science propel social issues through the classroom door where a teacher's opportunity to influence student attitudes abound.

The scientific attitude

Scientific attitudes are peripheral to the standard definition of attitudes. More cognitive than affective, the scientific attitude is a phrase coined by science educators that concerns the nature of science, and philosophical beliefs that often direct the behavior of scientists. Wolfinger[7] lists seven beliefs valued by men and women of science: objectivity, curiosity, open-mindedness, willingness to suspend judgment, skepticism, and a positive approach to failure. Exemplar science teachers model these attributes in science instruction expecting students to imitate them.

Information Drives Attitudes

Students learn positive or negative feelings toward science from experiences, both direct and vicarious ones, in and outside the school classroom. If science attitudes are learned, they can also be taught. Attitudes are transient enough to change but enduring enough to be stable once they are learned.

The attitudes are learned may have been best expressed by Baron and his colleagues: "Heroes may be born, but bigots are clearly made. No one would seriously suggest . . . that children spring from the womb with all the complex attitudes they will later show as adults firmly in place."[8] Unlike human intelligence, attitude has escaped the nature-nurture controversy.

In 1960 Katz[9] wrote that attitudes develop as they promote basic human needs. One need is knowledge. Another is the need to express our values. Information helps us make sense out of an otherwise chaotic world, and attitudes serve as social and emotional filters. Our decision-making processes motivate us to seek knowledge, often constrained by time. So we turn to information, both factual and non-factual, in order to cope with life's problems.

Information avails itself in many forms: facts, beliefs, opinions—even misinformation. And the sources of information vary: keen observation, empirically-driven data from the lab, philosophy, logic, court judgments, the *Bible,* people important to us, and the list could go on and on.

Factual information

Schools are in the information business. With knowledge a basic human need and information the source of science attitudes, does the learning of science facts enhance science attitudes?

Human learning is never that simple. Wilson[10] analyzed 43 studies involving thousands of students from grades three through college, and the science attitude-science achievement correlations averaged only 0.16. Wilson's study suggests that science knowledge, at least the type assessed by standard science achievement tests, has only a negligible effect on science attitude. In a recent experimental study, Showers[11] disclosed that learning basic factual information on the atom failed to affect the nuclear energy attitudes of high school science students.

Straight-forward factual knowledge seems to do little to drive science attitudes in the classroom, e.g., defining the three classes of levers, memorizing ROY G BIV, explaining the airfoil, or learning by rote the periodic table of chemical elements. Requiring the enrollment of high school students in more science courses will expose them to more science knowledge, and learning advanced knowledge in biology and the other sciences may open science-related careers later in life, but the data suggest that learning more factual information may do little to enhance science attitudes. When does knowledge influence science attitudes?

Personal information

Attitudes filter affective information, personal knowledge important to our well being. For example, knowledge of the biological growth of the tobacco plant might do little to persuade young students to say "NO"! to smoking.

But data on the debilitating effect of tobacco on the athlete's performance on the court would be more apt to motivate negative attitudes towards smoking. Here knowledge is personal; it has immediate utility in the life of a 17-year-old basketball star. The science teacher or parent who actively seeks and passes on knowledge immediately relevant to a youth will probably influence attitudes. But here information is rather select and it often requires that we deal with the needs and interests of students on an individual basis rather than within a typical classroom setting.

Discrepant information

Sinking ice cubes does not jive with a student's previous encounters with ice. The experience is discrepant and teachers who carefully choose such puzzling investigations motivate even the inattentive student to seek an answer. Students employ rapt attention—much like their scrutiny of a magician's tricks. Such investigations heightens student curiosity and baffle their thinking. Learning here is mildly discomforting as students reject offhand their perception that ice cubes sink in "water." Everyone knows that ice floats in water! Always! Yet this ice cube hovers near the bottom of the liquid-filled glass container. Two perceptions hurdle head-on.[12]

Seeking a new understanding—in this case the liquid perceived as "water" is really alcohol—restores a student's mental equilibrium. Students who get close enough to smell the alcohol breathe a sigh of relief, chuckle, and vow never to be tricked by the sinking cube again! Now that we have their undivided attention, students can be easily challenged to deal with the differing densities of the two liquids. Most students agree that this kind of science is fun.

Young children perceive many new experiences as discrepant. A magnet in the hands of a five-year old or a fifth grader's first gyroscope is greeted with wonder. Or take a simple mystery box. Here a child is encouraged to tilt a sealed box slowly observing and inferring its contents while temporarily deprived of the sense of sight. The science teacher or parent who introduces youngsters to science through such surprise encounters augments positive attitudes toward science instruction. Alfred Friedl's[13] book for elementary teachers describes and illustrates many science investigations that are discrepant in nature.

The most disquieting current finding in educational research is the loss of students' enthusiasm for science as they advance through the American elementary and secondary school.[14] That almost universal wonder very young children express toward their environment must be sustained and nourished by teachers and parents who employ such attitude-enhancing practices as the discrepant event and other principles cited in this report and elsewhere.

The discrepant concept can also affect attitude through classroom discussions or debates where more mature students argue a point of view alien to their current personal opinions. For example, the college student opposed to tough drunk driving legislation is challenged to gather and discuss with others data favoring the statute. Assimulating new information, often ignored up to this point, the student is prone to generate through self-persuasion a new attitude more supportive of tough legislation on drunk driving.

Discrepant information made personal

If dissonance is coupled with personal knowledge, the probability of attitude change rises. For example, if the basketball player mentioned earlier is already a smoker, cognitive dissonance could influence the attitude and behavior when encountered with research findings on tobacco use. Here the athlete could be confronted with two discrepant thoughts:

Thought A: "I smoke."

Thought B: "My teacher (or parent) tells me that smoking will slow me down on the basketball court."[15]

If the athlete accepts Thought B, he/she will experience cognitive discomfort by the two inconsistent thoughts, and he/she will set out to reduce the tension. Our athlete can exercise several options that will restore mental equilibrium. Two are shared here, the first of which represents a positive attitude and behavior change:

1. Change Thought A to "I no longer smoke." Now Thought A and B agree and all is well on the basketball court.
2. Exchange misinformation for Thought B: "I don't inhale when I smoke so my athletic prowess is unaffected." Again, A and B are consistent until the athlete is confronted with and accepts information refuting Thought B.

Our national concern with physical fitness, nutrition, and environmental awareness provides opportunities for teachers and parents to add dissonant information to the thought life of our teenagers. For example, that suntanning accelerates skin wrinkling may serve as the dissonant thought to the youth whose Thought A is, "I love a deep tan."

Vivid Information

Myers[16] portrays the powerful influence of vivid experiences on attitudes. Subtle and telling experiences can promote positive attitudes. How about the subtlety of a safety-conscious teacher who displays a set of safety glasses with one shattered but intact lens? The word is out that the glasses saved a teacher's eye from serious inquiry. No caption is needed. Such anecdotes are vivid; they are affective and they have staying power![17]

Consider the school principal resolved to influence ecological atitudes by challenging the kids to sweep across the playground and the nearby woodland, and collect candy wrappers, crumpled school papers, sun-bleached clothing, and the like. Instead of bagging the litter and carting it off with the kitchen waste, the students stationed the refuse in a long, netlike bag and wound it alongside the school sidewalk for all to see. Here again, no caption nor explanation is needed. A memory of this mass of garbage conjures up graphic images easily retrieved when one is about to toss a candy wrapper to the wind.

Science abounds with animated, lifelike, and colorful encounters, some aesthetic in nature: the symmetry of a leaf; the release of monarchs hatched and cared for by youngsters; or an afternoon rainbow so vivid that traffic slows, and some stop to marvel at its beauty. The Lazarowitz[18] survey reveals that almost 48 percent of 1855 students cited the beauty of science as one of their reasons for liking the subject. Here parents play an important role in building positive science attitudes in their children when stops on a family trip include museums, caverns, planetaria, or a simple lakeside rest stop where a family of mallards can be seen swimming along the water's edge.

Persuasive information

Since the 1950s persuasion has been ardently researched for its effect on adult attitudes. Here the message is often non-factual, i.e., options or beliefs. For example, persuasive message expressing the importance of the scientific literacy of American children can influence positively the science attitudes of young teachers.[19] The credibility and trustworthiness (and even attractiveness) of the persuader positively affect the message. Saying "NO!" to drugs advocated by a celebrity easily recognizable to all TV viewers can be persuasive. Here the instant recognition of the celebrity summons the viewer's attention; and it sustains the retention of the message.

Showers[20] designed a persuasive message for high school students based on the belief that nuclear energy can foster overall benefits to society as a means of generating electricity. Here persuasion enhanced their nuclear energy attitudes in a positive direction even tested several weeks after they viewed the videotaped message.

Some insist that persuasion connotes unethical practices. And persuasion can be manipulative—even diabolical. But as Myers has written, "persuasion is neither inherently good nor bad."[21] The message content classifies persuasion as manipulative or salutary. It must be obvious that the advocacy of a green environment, clear air, and clean water is salutary.

It is manipulative to: 1) persuade females and minority teenagers that they can be successful in science; 2) advise students that success in science does not require the brains of a genius; and 3) advocate that qualified scientists and technicians will foster our nation's competitiveness in the future world market?

Endorsing the legalization of street drugs, proposing the serving of alcoholic beverages at the school prom, or defending rough and unsportsmanlike behavior on the basketball court would be unethical and manipulative. The content of the persuasive message must be consistent with the goals of public education in a free society.

Valued information

Attitudes serve as a major conduit through which we express our values, those moral and ethical standards that help us discern right from wrong. Science content can be neutral but much is value-laden. In science instruction, the values of students conflict and teachers walk a tightrope. Evolution conflicts with creation in biology class and "No smoking" signs are hard to come by in tobacco capitals. As suggested earlier, many values can be openly expressed by students in schools where teachers choose topics compatible with the goals of public education.

For example, Americans seem ready to better value the planet Earth. The twentieth anniversary of Earth Day is celebrated as this report is written. Tomorrow morning I will place our bright, plastic bin near the curb filled with newspapers, properly arranged and metal cans flattened as directed. Recently, two young boys down the street passed out pleas hand-printed on waste computer paper urging ocean fisherman to spare dolphins caught in tuna nets.

Concerned that animals may swallow latex and die when helium-filled balloons burst and litter our eastern waterways, several rural Pennsylvania adolescents accompanied their legislator to the statehouse to lobby for a bill limiting balloon launches. Here, their value for aquatic turtles they had never seen took precedence over fun-filled balloon launches.[22]

Mandated behaviors can also foster new attitudes. That is, new attitudes can follow behavior. Encouraged to keep the schoolyard litter-free, students may become more positive in their attitude merely by following the principal's initial challenge. Young children evolve a taste for favorite family recipes as parents admonish them to ". . . try just one more bite." The process spirals: eating green beans this week, even reluctantly, can foster more positive attitude toward green beans; next week the behavior becomes easier which, in turn, improves the attitude toward beans. An endless chain is Myers' analogy.[23]

A major trend today is a science curriculum committed to the symbiotic relationships between the amenities of science and societal needs. Science-Technology-Society, as it is coined by science educators, delivers the environmental ethic to the classroom door. Encouraging our youth to exercise positive attitudes toward concepts highly valued in a free and democratic society is the heartbeat of public school instruction.

Implicit information from others

Establishing a social identity is another basic human need. Those considered important to our social well-being mediate our attitudes and behavior. In 1954

Allport wrote that we are ". . . influenced by actual, imagined, or implied presence of other human beings."[24] Sherif adds that ". . . my preferences or antagonisms have little consequence apart from my ties with others, apart from my reference persons and groups. . . "[25] The sophomore, mentioned earlier, who as a staunch creationist scored high on an exam covering the theory of evolution may have been more influenced by the values of parents or Sunday School teachers than her biology teacher. She and her parents also may value high grades in science.

The well-prepared science teacher who instructs with enthusiasm within an instructional climate where students err without penalty can implicitly influence others. Liking, identifying with, and choosing a science teacher as a significant other can affect positively that student's attitude toward science or almost any other attitude or behavior modeled by the teacher. Female science teachers can serve as role models for female students. In the Lazarowitz survey 21 percent of the students cited teachers as one of their reasons for liking science, and one-third of the reasons for disliking science were chalked up to poor instructional behaviors, e.g., one who teaches too fast; the teacher who overdoes book work.

Information from direct experience

To John Dewey "learning by doing" was the hallmark of quality instruction. And science teachers have long known that students like hands-on science. Unlike literature, history or even mathematics teaching, science abounds with opportunities for students to directly investigate the natural world. Babies spring from the womb ready to explore.

In the Lazarowitz survey eight of the 18 reasons for liking science alludes to some form of student investigation. Almost 90% liked the outdoors and its relationship to science; other enjoyed experiments, working with their hands, inventing and dissecting things, and solving problems.

Not only does direct experience enhance favorable feelings toward science, such investigations can nurture open-mindedness, objectivity, curiosity, willingness to suspend judgment, etc., those beliefs highly valued by scientists, labeled as scientific attitudes by science educators. Fostering those beliefs exalts important science processes, a component in science curricula subordinate to the learning of science facts and concepts: observing, classifying, measuring, inferring, predicting, and experimenting.

STUDENTS WHO OPT OUT OF SCIENCE

Like it or not, students have little choice but to attend required science courses in the elementary, middle, and secondary school. But electing advanced science courses in high school and college is another matter. Here students control their destiny and many who should stay bid farewell to science. A disproportionate number who exit are females and students representing minority groups. Here a negative attitude begets instant dropouts to science shrinking our pool of human resources for a scientifically-informed citizenry.

Pollack[26] explored the motivations of 1300 high school students for electing advanced science courses. Interest in and attitudes toward science were the major motives for enrollment. Perceived difficulty of science courses and poor scores were the major reasons for sidestepping advanced science courses.

Pollack reports that significant others, especially parents, influence the enrollment of teenagers in science courses. Older brothers and sisters are influential, too. High school and college advisors influence the science attitudes of teenagers. Access to occupational information makes counselors credible in the eyes of parents and students. Also, the attitudes and beliefs of counselors prevail. Advisors who reject the view that science is a masculine subject will advise more females into advanced science courses. The advisor who believes that only science career students should elect advanced chemistry and physics courses frustrate those of us who recognize our national need for scientifically-literate laymen and laywomen.

THE GENDER GAP IN SCIENCE

The low science attitude scores of females is one of the most consistent research findings in science education.[27] With few exceptions, females throughout the Western World score lower than males on science attitude scales.

That males might inherit a more favorable aptitude for the sciences than females is fiercely debated. The gender difference in science is probably cultural; we implicitly teach females that science is a masculine enterprise. Females taught to hold this view dutifully fulfill it.

The gender gap is especially troublesome in our nation's elementary schools where about 80 percent of our teachers are females. Research reveals that female elementary teachers have not escaped the gender difference in science attitudes. The attitudes of female teachers can be enhanced by many of the same interventions already expresssed in this report where our youth are targeted. Persuasion that American children deserve excellent science instruction has a positive force on teacher attitude.

COPING WITH SCIENCE ANXIETY

Here attitude is not a simple preference of biology over chemistry or oat bran over bacon and eggs; it is consuming fear of science and all it stands for. Fear of science is learned and avoiding the feared object in the option chosen by many. Fear not only thwarts the motivation of students enrolled in required science courses, science-anxious students justify opting out of elective courses and boycotting science careers altogether.

Conventional science attitudes are high on the research priority list of science educators. Fear of science is not. However, Mallow[28] describes a counseling procedure that appears to be successful at reducing the fear of science. Here therapy teaches the fearful to cope with negative and often irrational self-thoughts. Many science-anxious thoughts are remedied with reason. Counselors encourage subjects to replace negative self-statements with more positive and rational ones, a belief-shifting exercise known as cognitive restructuring. The fear-reducing process also involves students in science learning skills and relaxation skills.

CONCLUSIONS AND IMPLICATIONS

Attitudes are learned which means that they can be taught, and information is the driving force. But the types of information are restricted. Factual information, simple recall that is commonly assessed by science achievement tests, has only a negligible effect on science attitudes. Personal and immediately relevant information influences science attitudes. Discrepant investigations, those that surprise, baffle, and then delight students, motivate positive attitudes toward classroom science.

Vivid and subtle information, often in the form of anecdotes, grasps the attention of students, facilitates comprehension, and renders lifelike images easily retrieved from memory. Seeking beauty in nature raises our spirits, too. Beliefs and values, in the form of non-factual information, also influence attitudes. Our values serve as conduits for expressing science-related attitudes.

Teachers and parents who involve our youth directly with the natural environment foster positive science attitudes. Also, science investigations school our youth in such important science proceses as observing, classifying, and experimenting, those skills needed to solve everyday problems.

Parents, and almost anyone significant to the social and emotional well-being of our youth, influence for better or worse many science attitudes. They also influence student election of advanced science courses, the precursor to an informed citizenry and a national pool of scientists and technicians.

The less-than-positive science attitude of many females is troubling. This concern swells when it reaches into our elementary schools where most teachers are females. The science attitudes of female teachers can be ameliorated through persuasion.

Poor science teaching and widespread folklore about the scientific enterprise feeds science anxiety, a fear of anything associated with science. Subjective and often irrational beliefs can be annulled by positive self-statements initiated through a counseling process.

Science teachers in our schools and universities must shoulder the major responsibility for instructing our youth in the sciences. However, influencing science attitudes positively requires a community effort. Here the efforts of teachers, parents, siblings, counselors, good friends, and other referents must be orchestrated, much of it outside formal classroom instruction and commonly one-on-one through counseling, modeling, persuading, and valuing.

At the same time, raising up a scientifically-literate citizenry will require that we attend to three major concerns: 1) the tendency for children's science attitudes to grow more negative as they proceed through school; 2) the less-than-positive science attitude of female elementary school teachers; and, 3) the attitudes of those who drop out of science, especially females, minorities, and youth suffering from science anxiety.

REFERENCES

1. Allport, G. W. 1954. The historical background of modern social psychology, In G. Lindzey (Ed.) *Handbook of Social Psychology*. Addison-Wesley, New York, pp. 45.
2. Allport, G. W. 1935. Attitudes. In: C.A. Murchison (Ed.) *Handbook of Social Psychology,* pp. 46. Clark University Press, Worcester, MA., pp. 1195.
3. Bloom, B. S. 1956. *Taxonomy of educational objectives, Handbook I: Cognitive domain.* David McKay, New York, pp. 207.
4. Lazarowitz, R.J., H. Baird and V. Allman. 1985. Reasons why elementary and secondary students in Utah do and do not like science. *School Science and Mathematics, 85:* 663-672.
5. Wolfinger, D. 1984. Teaching science in the elementary school. Little, Brown,
   Boston, pp. 443.
6. Markovits, A. S., and K. W. Deutsch (Eds.) 1980. *Fear of science: Trust in science.* Oelgeschlager, Gunn, and Hain, Cambridge, MA, pp. 263.
7. Wolfinger, *op. cit.*
8. Baron, R. D., and D. Byrne. 1977. *Social psychology.* Allyn and Bacon, Boston, p. 105.
9. Katz, D. 1960. The functional approach to the study of attitudes. *Public Opinion Quarterly, 24:* 163-204.
10. Wilson, V. L. 1985. A meta-analysis of the relationship between science achievement and science attitude. *Journal of Research in Science Teaching, 20:* 839-850.

11. Showers, D. 1986. A study of the effects of informational and persuasive messages on the attitudes of high school students toward the use of nuclear energy for electrical production. The Pennsylvania State University, Doctoral Dissertation.

12. Shrigley, R. L. 1987. Discrepant events: Why they fascinate students. *Science and Children, 25:* 24-25.

13. Friedl, A. E. 1986. *Teaching science to children: An integrated approach.* Random House, New York, pp. 301.

14. Yager, R. E. and S. Yager. 1985. Changes in perceptions of science for third, seventh and eleventh grade students. *Journal of Research in Science Teaching, 22:* 347-358.

15. Festinger, L. 1957. *The theory of cognitive dissonance.* Stanford University Press, Stanford, pp. 291.

16. Myers, D. G. 1983. *Social psychology.* McGraw-Hill, New York, pp. 674.

17. Shrigley, R. L. and T. R. Koballa, Jr. 1988. Ancedotes: What research suggests about their use in the science classroom. *School Science and Mathematics, 89:* 293-298.

18. Lazarowitz, *op. cit.*

19. Shrigley, R. L. 1983. Persuade, mandate, and reward: A paradigm for changing the science attitudes and behaviors of teachers. *School Science and Mathematics. 83:* 204-215.

20. Showers, *op. cit.*

21. Myers, *op. cit.*

22. *Centre Daily Times,* State College, PA. April 5, 1990.

23. Myers, *op. cit.*

24. Allport, *op. cit.*

25. Sherif, C. 1976. *Orientation in social psychology.* Harper and Row, New York, pp. 230.

26. Pollack, N. 1982. The relative importance of selected variables involved in the decision of students to enroll or not enroll in grade ten science. New York University. *Dissertation Abstracts, 43:* 1105-A.

27. Gardner, P. L. 1975. Attitudes to science: A review. *Studies in Science Education, 2:* 1-41.

28. Mallow, J. V. 1981. *Science anxiety: Fear of science and how to overcome it.* Van Nostrand Reinhold, New York, pp. 232.

Science Education in the United States: Issues, Crises and Priorities. Edited by S. K. Majumdar, L. M. Rosenfeld, P. A. Rubba, E. W. Miller and R. F. Schmalz. ©1991, The Pennsylvania Academy of Science.

Chapter Thirteen

TECHNOLOGY IN THE SCIENCE CLASSROOM

ROBERTTA H. BARBA
Associate Professor of Science Education
San Diego State University
San Diego, CA 92182

THE ROLE OF TECHNOLOGY

Movie projectors, cassette tape players, television monitors, microcomputers, and interactive video workstations are examples of audiovisual technologies that have at one time or another entered the science classroom as tools to facilitate the teaching/learning process. Since the first school museum opened in St. Louis in 1905 (Saettler 1968, p. 89), science educators have looked to technology for solutions to pedagogical problems. In 1913, Thomas Edison proclaimed that, "Books will soon be obsolete in our schools . . . . It is possible to teach every branch of human knowledge with the motion picture projector." Edison went on to predict that "our school systems will be radically changed (by motion picture projectors) within ten years" (Saettler, 1968, p. 98).

History of Audiovisual Technology in Science Instruction

Magic lanterns (stereopticons and stereoscopes) became instructional tools in the science classroom during the later half of the nineteenth century (Anderson, 1962). Edison's movie projector became the visual tool of choice for science teachers during the decade of the 1910s (Saettler, 1968). Following World War I motion pictures with accompanying sound became popular instructional tools (Finn, 1972). Educational applications of audiovisual technology multiplied with the advent of World War II. Military applications of instructional technologies included: the use of overhead projectors for large group instruction, slide projectors for teaching aircraft recognition, audio tape instruction for teaching foreign languages, and the use of multi-media flight training simulators (Olsen & Bass, 1982). Following World War II, teachers adopted these same technologies for instructional purposes in the science classroom.

The 1950s and 1960s were the decades of instructional television (Gumpert, 1967). Educational television during this time period was viewed as a means to provide quick, efficient, and inexpensive science instruction (Hezel, 1980, p. 173). The National Defense Education Act: Title VII, which was in effect for one decade, spent more than $40 million dollars on 600 projects (Filep & Schramm, 1970) including instructional television. Private foundations, such as the Ford Foundation, which spent $170 million dollars during this same time period (Blakely, 1979), supported instructional programming as a means of improving instruction nationwide. With minor exceptions, the disappearance of educational television has left science instruction fundamentally unchanged (Carnegie Commission of Educational Television, 1967, pp. 80-81).

Microcomputer Technology

In 1951, Remington Rand introduced the first commercially available computer system, the UNIVAC I (Lockard, Abrams & Many, 1987, p. 11), which was constructed of vacuum tubes. By the middle of the 1950s the vacuum tubes of the UNIVAC had been replaced by transistors. The second generation of computers was faster and more complex than its predecessor. Early in the 1960s, the third generation of computers run by integrated circuits became commercially available. Widespread science classroom usage of microcomputer technology began in the 1970s with the invention of the fourth generation of microcomputers, those based on microchip technology (Leuhrmann, 1980, p. 144).

Software Evolution

Just as there has been a revolution in technology which has allowed the microcomputer to become an affordable instructional tool in the science classroom, there has been an evolution in the software that runs that hardware. The first generation of educational software used in the science classroom consisted of networked programs such as PLATO, which required the computing power of mainframe computers (Denenberg, 1988, p. 313). The first microcomputer software widely used in science classroom was written in BASIC and required the regular science teacher to have a knowledge of BASIC commands (such as "LOAD <FILENAME>" and "RUN <FILENAME>"). By the beginning of the 1980s "Press <RETURN> to continue" became the instructional standard of the second generation of science software (Jonassen, 1988, p. 151). Software in the early 1980s became menu driven and user friendly, but the pedagogical base of that software remained rooted in a mathemagenic, drill and practice model. Indeed, 85% of the software used in public schools in the United States today is drill and practice software (Cohen, 1983).

A third generation of software, software that encourages high levels of user interactivity is beginning to find its way into the science classroom. The term interactivity refers to the users ability to engage in direct and continual two-way communication with the computer (Lockard, Abrams & Many, 1987, p. 144). Third generation science software establishes a transactional triangle among the learner, the microcomputer, and the natural world (see Fig. 1).

Science software ought to allow the learner to explore the natural world and relationships in that natural world, to do science, rather than to "teach about science". As science software is modified and refined, microcomputers will become tools to assist the learner in exploring the natural world.

First and second generation software was based largely on the behavioral principle of connectionism, traditional drill and practice courseware assumes that enough practice ultimately produces correct performance (Jonassen, 1988, p. 151). This mathemagenic model for computerized instruction has failed to achieve desired results in the science classroom because: 1) it fails to accommodate the principles of educational psychology, 2) it results in shallow or low levels of information processing, and 3) microcomputer and interactive video technologies have outgrown programmed instructional models. With the third generation software, microcomputer software is evolving from stimulus-response learning to information processing learning models.

The goal of third generation software in the science classroom ought to be to support the acquisition, retention, and retrieval of information within the individual learner. It should promote and guide active mental processing on the part of the student (Merrill, 1988, p. 72). To do this science software needs to have embedded generative cognitive strategies (Wittrock, 1978) that facilitate the transfer of information from short-term to long-term memory (Brunig,

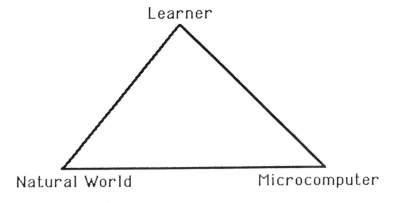

FIGURE 1. The Transactional Triangle.

1983). Third generation science software should allow the learner to actively explore the natural world and to support the activities of the learner.

RESEARCH INTO CAI EFFECTIVENESS

Computer Assisted Instruction (CAI), the use of the microcomputer as an instructional tool (Lockard, Many & Abrams, 1987, p. 392), has been available to the science teacher for more than two decades. During that time, literally hundreds of research studies have been conducted into the effectiveness of this instructional media. Unfortunately, most of the studies into the effectiveness of CAI have been flawed by poor research designs, improper control of variables, and inappropriate use of statistics (Kracjik, Simmons & Lunetta, 1986). Overall, findings from properly controlled studies indicate that CAI makes significant differences in student learning in four areas: achievement, learning retention, learning time, and learner attitude.

Achievement

Perhaps the most comprehensive study of the effectiveness of computer aided instruction was undertaken by Kulik, Bangert, and Williams (1983). The authors used a meta-analysis technique to examine and synthesize 51 studies into CAI involving students in grades 6 to 12. They found that generally students who received CAI scored higher on objective tests than did students who received "traditional" instruction. Results from their meta-analysis indicated that the average student in the control groups scored at the 50th percentile, while the average student treated with CAI scored at the 63rd percentile. The gain for students treated with computerized instruction was .32 of a standard deviation.

Studies conducted by the Educational Testing Service indicate that students who use computerized drill and practice packages for only ten minutes a day score significantly higher on mathematics achievement tests than do students who do not have access to microcomputers (Bracey, 1982). Roblyer (1985) in summarizing the results of CAI studies reports that "computer based instruction achieves consistently higher effects than other instructional treatments in experimental situations".

Science teachers need to integrate this body of research into their every day instructional practice. Information at the lowest levels of Bloom's taxonomy, i.e., knowledge and comprehension, lends itself to electronic learning. Hence, bodies of factual information, such as the symbols for the elements of the periodic table or names of plant and animal structures can be taught more efficiently with computerized drill and practice packages.

Retention

Kulik, Bengert and Williams (1983) also investigated student retention of information in their meta-analysis study. They found that CAI improves retention of learning. They reported that four of five studies investigating retention over a period of two to six months showed greater retention of information for students who used CAI. Roblyer's (1985) survey of CAI research indicates that there were differences in the performance of CAI students and traditionally instructed students, but the differences were not statistically significant. Since the results of research into student retention were not conclusive, Roblyer recommends more research in this area.

Learning Time

Blaschke and Sweeney (1977) found that CAI has been helpful in decreasing learning time. Their study compared the learning times between groups of military recruits one of which used a computer assisted electronic training package and the other a similar programmed booklet commonly used in secondary education. The results from this study indicated that students learn the same electronics information in 10 percent less time using the CAI module. A study by Dence (1980) indicated that students who use CAI master content faster than students who use "traditional" instructional techniques. A study by Lunetta (1972) showed that students could master physics content in 88% less time using CAI, when compared to students who used "traditional" physics instruction.

Research into instructional time indicates that CAI is highly effective in certain situations. Science teachers need to identify those bodies of information that can be learned "quicker" using microcomputer than by traditional instructional methods. The time saved by learning low level information on the computer could be used to teach higher order thinking skills in the science classroom.

Attitudes

In addition to improving students' achievement, increasing retention, and decreasing instructional time CAI appears to improve student attitudes toward learning. The meta-analysis conducted by Kulik et al. (1983) indicated that in eight of ten studies reviewed, student attitudes toward subject matter was more positive after students had used CAI. A study by Bracey (1984) indicated that students react favorably to the use of the computer for instructional purposes. In fact, Bracey found that students who have worked on microcomputers in a subject area have a more positive attitude toward that subject than students who have not used the computer in that particular class. Other researchers (Foley, 1984; and Fiber, 1987) also have reported that students' attitude toward an academic subject improves after they have had computerized instruction in that class.

Computers have endless patience in tutorial situations. Science teachers need to capitalize on the microcomputers ability to consistently teach and reteach information. Simulation games, tutorial packages, and drill and practice software can be powerful instructional tools. Electronic learning frees the science teacher from the drudgery of endless drill and practice learning situations.

BEYOND DRILL AND PRACTICE

While drill and practice software is the most commonly used software in the science classroom, it isn't the only computer application available to the science instructor. Indeed, other types of software provide for higher levels of user inter-action on the transactional triangle (Fig. 1) more so than do CAI packages. Graphing packages, instrument interfacing kits, electronic simulation activities, statistical packages, data based simulations, and spreadsheet simulations provide for high levels of user interaction—interaction among the learner, microcomputer, and the "real" world.

Computerized Graphing Packages

Computerized graphing packages such as MECC Graph and Microsoft Works allow students to rapidly generate bar graphs, line graphs, and pie charts. The primary value of computerized graphics packages is that they change the level of questioning skills in the classroom from the lowest levels of Bloom's tax-onomy to higher levels of that taxonomy, e.g., analysis, synthesis (Sloyer & Smith, 1986). When graphing packages are introduced into the classroom, they allow students to spend time asking "What if?" questions rather than questions concerning the construction of the graph itself.

In a study into the effectiveness of graphing packages, Gesshel-Green (1987) found that students who use computerized graphing packages do not differ from "traditionally" instructed students on immediate recall scores. However, students treated with computer graphing packages do show significant positive gains when compared to these students on measures of long term retention, motivation, and cooperation. Results from a study into the use of computer-ized graphing packages to teach concept formation (Heid, 1988) indicated that students formulate concepts in significantly less time when using computer graphing packages.

Instrument Interfacing

Instrument to microcomputer interfacing is the domain of the science teacher.

Connecting thermistors, potentiometers, photoresistors, and electrically conductive styrofoam to microcomputers allows students and teachers to collect and analyze data in real time. Microcomputer based laboratory activities have been shown to improve students understanding of the operation of science laboratories (Nachmias & Linn, 1987). While conducting an experiment to compare traditional laboratory activities with computer based laboratory activities, these researchers discovered that eighth grade students in computer based activities showed greater understanding of scientific processes than did students in conventional science labs.

Krendl and Lieberman (1988) report that student motivation, involvement with the laboratory activity, and self-perceptions are improved as a result of computer based lab instruction. Instrument interfacing allows learners to focus on science processes rather than focusing on data collection and interpretation (Assetto & Dowden, 1988). Events that occur very rapidly, very slowly, which involve minute changes in temperature, light intensity, or which require extreme care in data recording can be easily observed with instruments interfaced to microcomputers (Barba, 1987). Laboratory activities that have traditionally been left from the curriculum because they required extreme care in observation can be conducted using microcomputer interfaces.

Simulations

Simulation activities, a form of CAI in which the learner assumes a role within a structured environment (Lockard, Many & Abrams, 1987, p. 398), can be conducted in the science classroom with single use packages; such as, *Oh, Deer!* and *Energy House,* or through the use of spreadsheets or simple computer programs. In reporting on the effectiveness of *CATLAB*, a genetics simulation activity, Krajcik, Simmons, and Lunetta (1988, p. 151) point out that computer simulations facilitate student problem solving skills by allowing students to: 1) generate their own questions, 2) control variables themselves, 3) gather, record, and interpret data, and 4) draw conclusions to support or reject hypotheses. Experiments which are dangerous or very time consuming can be performed quickly and easily by computer simulation.

Spreadsheets also allow students to engage in simulation activities. While investigating the use of spreadsheets as simulation tools, Dubitsky (1986) discovered that sixth grade students are able to transfer understandings and methods of solution from problem to problem. In assessing the value of spreadsheets in the science classroom, Pogge and Lunetta (1987) point out that spreadsheets allow students to spend extra time collecting and interpreting data, rather than spend time on routine computational work.

THE ROLE OF EDUCATIONAL TECHNOLOGY

In the immediate future, the most dramatic changes in instruction in the science classroom will be those facilitated by the application of educational technology to the science curriculum. Educational technology is "an approach that has been directed toward expanding the range of resources used for learning, emphasizing the individual learner and his unique needs, and using a systematic approach to the development of learning resources" (Definition and Terminology Committee of the Association for Educational Communications and Technology, 1972, p. 36). A systems design approach to instruction involves identifying and controlling the variables which lead to increased information processing on the part of the learner. The melding of audio-visual technologies and a systematic approach to courseware design holds potential for increased learning in the science classroom.

Systematic design of instruction, such as the systems approach model designed by Dick and Carey (1985), includes: 1) identification of instructional goals, 2) task analysis for each learning outcome, 3) identification of the entry behaviors, and 4) consideration of the learner characteristics of students in the science classroom. Dick and Carey's model of instructional design allows the program designer to communicate the performance objectives expected of students in the science classroom as a part of the software development process. Criterion referenced test items are used in systematically designed science software. Instructional strategies and instructional materials are clearly identified in systematically designed products. Formative and summative evaluation are valuable components in the process of software design when a systems approach is used. An instructional systems approach to the design of science software takes into account the resources, constraints, delivery system, teacher preparation and installation of software into the regular science classroom (Gagne, 1988, p. 31).

IMPROVING CAI SOFTWARE

The basic question in designing instruction is how to proceed in order to facilitate the acquisition of essential background information by the individual learner. There is a solid base of research on the variables that control learning readily available to software designers. CAI software of the future will exhibit a research based approach to learning. It will incorporate the principles of cognitive psychology and the characteristics of the learner.

Science software in the future will attend to the events of instruction (Gagne, 1988, p. 182) by: 1) gaining the attention of the learner, 2) informing the learner of the objectives of instruction, 3) stimulating recall of previous learning, 4) presenting the stimulus material, 5) providing learning guidance, 6) eliciting student performance, 7) providing meaningful feedback, 8) assessing student performance, and 9) enhancing retention and transfer. The software of the future will address the needs of the individual and will exhibit a firm pedagogical base in doing so.

Generative Cognitive Strategies in CAI

Microcomputer software and interactive video programs in the future will be designed to foster an information processing approach to science education. We currently have available the technological means to accommodate the true individualization of instruction; not via a mathemagenic model of programmed instruction, but through an information processing approach to learning in science. Wittrock (1978) in addressing the status of microcomputer software states that meaning for material presented by computer or interactive video instruction is generated by activating and altering existing knowledge structures within the learner. Learning is an active process, not a passive one. New learning comes as the individual consciously and intentionally relates new information to existing knowledge. Hence, learning materials need to facilitate this interaction through the inclusion of learning strategies.

Learning strategies generate not only learning about science, but information about how to learn science (Brown, Campione & Day, 1981). Generative activities, such as outlining, underlining, paraphrasing, summarizing, mnemonic devices, cognitive mapping, metaphors, categorizing, and notetaking are "mental operations or procedures that the student may use to acquire, retain, and retrieve knowledge and performance" (Rigney, 1978, p. 165). Such learning strategies may be taught explicitly or they may be embedded in the instructional

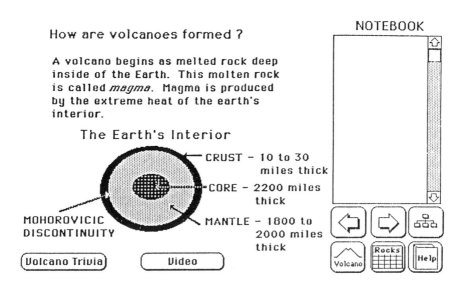

FIGURE 2. Electronic notebook embedded in a hypercard stack that allows learners to except text or take notes directly, and later print a hard copy.

materials themselves (Rigney, 1978). Embedding learning strategies directly into software can not only assist students with learning the material at hand, but also facilitate students development of metacognitive strategies for future learning.

We know from research (Carrier, 1983) that students who engage in active notetaking perform better than students who do not overtly organize information. Organizational strategies (Brunig, 1983), such as outlining and analyzing key ideas, and information integration strategies, such as paraphrasing and exemplifying, increase schemata formation in learners (Rumelhart & Ortony, 1977). Active study strategies (Dansereau, 1978) such as electronic notebooks can be embedded into science software and will lead to deeper information processing on the part of the learner. Electronic notebooks embedded in courseware, as in pictured in Figure 2, allows students to store, manipulate, organize, and retrieve information.

Visualization Strategies in CAI

Visualization strategies have been recognized for several decades (Day & Beach, 1950; Allen, 1960; Chu & Schramm, 1967; and Levie & Dickie, 1973) as being important variables that enhance student learning in all subjects, including science. Visualization increases learner interest, motivation, curiosity,

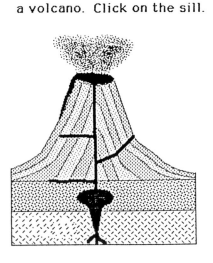
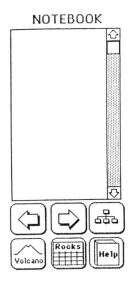

FIGURE 3. Visual questioning and line drawings embedded in courseware.

and concentration. It provides instructional feedback, facilitates information acquisition, spans linguistic barriers, increases the reliability of communication, and emphasizes and reinforces printed instruction (Dwyer, 1978).

Hockberg's (1962) research into the visualization continuum indicates that plain line drawings and shaded line drawings are the most meaningful for all classes of learners. When drawings are embedded into science software, as is pictured in Figure 3, they can focus the attention of the learner on critical attributes, and provide multiple channels (Dale, 1946, p. 37) of information to assist the learner in acquiring information.

Simple line drawings and shaded line drawings can be used for pictorial testing of students' knowledge and comprehension. In reporting on the advantages of pictorial testing, Gibson (1947) pointed out that pictorial tests are as reliable and as valid as verbal instruments. Pictorial testing has not been used in science classrooms in the past because it was cumbersome and time consuming to produce visual questions. Newer computer languages, such as Hypertalk and Linkway, make the possibility of visual testing as routine as traditional pencil-and-paper testing procedures. In the future, science software will routinely include questions in a visual format. Students will benefit from the change in questioning format because visual questions are: 1) more accurate and more easily understood, 2) reduce the emphasis placed on reading skills, 3) provide motivating situations, and 4) assess information not easily evaluated through a verbal format (Lefkowith, 1955, pp. 15-21).

The use of graphics in software also enhances student learning. Research (Dwyer, 1978; Koran, 1972; and Shapiro, 1975) has long recognized that visualization strategies embedded into textual material facilitates the learning of low verbal students.

Information Processing Strategies in CAI

In the past, computerized learning has followed the mathemagenic model established in the 1950s and 1960s. That model for computerized science instruction required all learners to follow the same basic linear program. Individualization of instruction, instruction based on the needs of the individual learner can be better met with a databased approach to learning. As the memory capacity of microcomputers increases, and as CD-ROM technologies develop, microcomputers will be able to hold the larger bodies of information in memory that are needed to individualize computer assisted instruction. Students who need additional information on a topic will be able to readily access glossaries, dictionaries, or information databases to assist them in the learning process. Students who do not need additional support or information will be able to complete learning activities in less time.

Naisbitt (1982) has pointed out that we live in an information processing age.

Yet, our pedagogical methods do not reflect the times in which we live. We do not teach students to actively access multiple sources of information as they engage in the learning process. Computerized science packages in the near future will allow the students to access on disk encyclopedias, dictionaries, pictorial dictionaries, and a multitude of reference sources with a single key stroke, or click of the mouse (see Fig. 4).

Feedback and Motivational Strategies in CAI

"Feedback or a knowledge of results facilitates meaningful learning cognitively, primarily through clarification and correction, rather than by reinforcing correct responses" (Ausubel, 1978, p. 310). Ausubel (1978, p. 310) states that feedback is "less important for meaningful than for rote learning because the internal logic of meaningfully learned material allows for more self-provided feedback than do inherently arbitrary association." In the design of computer software, feedback needs to be confirmatory, response correcting, or explanatory (Schimmel, 1986) depending on the nature of the learning activity. Fitting feedback to the learning activity, so that the feedback is meaningful will be a design feature of science software in the future. Gone are the days when courseware simply flashed the words "Right" or "Wrong, try again!" onto the screen. Future software will analyze student answers and will correct student responses so that practice is meaningful.

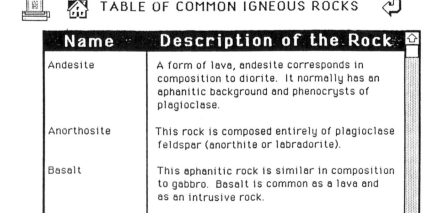

FIGURE 4. Databases embedded in software facilitate an information processing approach to courseware.

"Motivation is absolutely necessary for the sustained type of learning involved in mastering a given subject matter discipline, such as science" (Ausubel, 1978, p. 397). Attention to motivational variables in future science software design will make that courseware more effective as a pedagogical tool in the science classroom. The embedding of gaming techniques, the use of perceptual arousal, the use of relevant information, the use of learner control over the instructional program, and the use of positive feedback are motivational strategies which are currently being incorporated into science courseware (Dunne, 1984). Programming strategies, such as the use of instructional maps embedded in science software (see Fig. 5) can be used to provide students with a knowledge of their progress and results, and can function as motivational devices.

SUMMARY

We know from research that computerized learning can make a significant difference in student performance in the areas of achievement, retention of information, learning time, and student attitudes. Software in the future will feature a strong research base and rationale for pedagogical decisions. Transactional triangles, sets of interactions between learners, microcomputers and

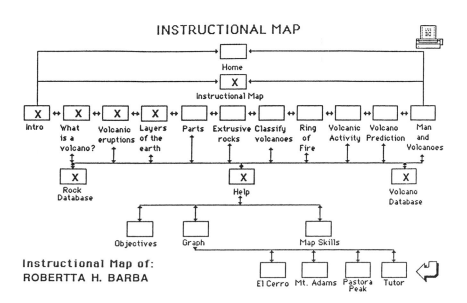

FIGURE 5. Instructional map from a hypercard stack.

the natural world will be established within the science courseware of the future. Technology will assist the teacher in providing individualized instruction that supports individual learners in the science classroom.

REFERENCES

Allen, W.H. 1960. Audio-visual communication. *In:* C.W. Harris (Ed.) *Encyclopedia of educational research.* MacMillan, New York, N.Y. pp. 115-137.

Anderson, C. 1962. *Technology in American Education: 1650-1900.* Report No. OE-34018. Office of Education, U.S. Department of Health, Education, and Welfare, Washington, D.C.

Assetto, A.R. and E. Dowden. 1988. Getting a grip on interfacing. *The Science Teacher.* 55(6):65-67.

Ausubel, D.P., J.D. Novak and H. Hanesian. 1978. *Educational Psychology: A Cognitive View.* Werbel and Peck, New York, N.Y.

Barba, R.H. 1987. In pursuit of the yeast beast. *The Science Teacher.* 54(7):30-32.

Blakely, R.J. 1979. *To serve the public interest: Educational broadcasting in the United States.* Syracuse University Press, Syracuse, N.Y.

Blaschke, C.L. and J. Sweeney. 1977. Implementing effective educational technology: Some reflections. *Educational Technology.* 17(1):13-18.

Bracey, G.W. 1984. *Issues and problems in devising a research agenda for special education and technology.* Paper presented at Special Education Technology Research and Development Symposium. Sponsored by U.S. Department of Education, Washington, D.C.

Bracey, G.W. 1982. What the research shows. *Electronic Learning.* Nov/Dec: 51-54.

Brunig, I.L. 1983. An information processing approach to a theory of instruction. *Educational Communications and Technology Journal.* 31:91-101.

Carnegie Commission of Educational Television. 1967. *Public television: A program for action.* Harper & Row, New York, N.Y.

Chu, G.C. and W. Schramm. 1967. *Learning from television: What the research says.* National Association of Educational Broadcasters, Washington, D.C.

Cohen, V.B. 1983. *A learner-based evaluation of microcomputer software.* Paper presented at the annual meeting of the American Educational Research Association, Montreal, Canada.

Dale, E. 1946. *Audio-visual methods in teaching.* Dryden, New York, N.Y.

Dansereau, D.F., K.W. Collins, B.A. McDonald, C.D. Holley, J. Garland, G. Diekhoff and S.H. Evans. 1979. Development and evaluation of a learning strategy training program. *Journal of Educational Psychology.* 71:64-73.

Day, W.F. and B.R. Beach. 1950. *A survey of the research literature comparing the visual and auditory presentation of information.* Air Force Technical Report 5921, Contract No. W-33-039-AC-21269. University of Virginia, Charlottesville, VA.

Definition and Terminology Committee of the Association for Educational Communications and Technology. 1972. The field of educational technology: A statement of definition. *Audiovisual Instruction.* 17(8):36-43.

Dence, M. 1980. Toward defining the role of CAI: A review. *Educational Technology.* 20(11):50-54.

Denenberg, S.A. 1988. Semantic network designs for courseware. *In:* D.H. Jonassen, (Ed.). *Instructional Designs for Microcomputer Courseware.* Lawrence Erlbaum, Hillsdale, N.J.

Dick, W. and L. Carey. 1985. *The systematic design of instruction.* Scott, Foresman, Glenview, IL.

Dunne, J.J. 1984. *Gaming approaches in educational software: An analysis of their use and effectiveness.* ERIC Document ED 253 207.

Dwyer, F.M. 1978. *Strategies for improving visual learning.* Learning Services, State College, PA.

Fiber, H.R. 1987. The influence of microcomputer-based problem-solving activities on the attitudes of general mathematics students toward microcomputers. Dissertation. *Dissertation Abstracts International.* The Pennsylvania State University, University Park, PA. 48/05A:1102.

Finn, J.D. 1972. The emerging technology of education. *In:* R.J. McBeath (Ed.). *Extending education through technology: Selected writings by James D. Finn.* Association for Educational Communications and Technology, Washington, D.C.

Filep, R. and W. Schramm. 1970. *A study of the impact of research on utilization of media for educational purposes sponsored by NDEA Title VII 1958-1968. Final Report: Overview.* Institute for Educational Development, El Segundo, CA.

Foley, M.U. 1984. Personal computers in high school general mathematics: effects on achievement, attitude, and attendance. Dissertation. *Disservation Abstracts International.* University of Maryland. 46/07A:1859.

Gagne, R.M. 1987. *Instructional Technology: Foundations.* Lawrence Erlbaum, Hillsdale, N.J.

Gesshel-Green, H.A. 1987. The effect of interactive microcomputer graphics on student achievement and retention in second year algebra in an academic high school. Dissertation. *Dissertation Abstracts International.* Temple University, Philadelphia, PA. 48/02A:326.

Gibson, J.J. 1947. *Motion picture testing and research.* Army Air Forces Psychology Program Research Report No. 7., U.S. Government Printing Office, Washington, D.C.

Gumpert, G. 1967. Closed-circuit television in training and education. *In:* A.E.

Koenig & R.D. Hill (Eds.). *The farther vision: Educational television today.* University of Wisconsin Press, Madison, WI.

Hannafin, M.J. 1985. Keeping interactive video in perspective. *In:* E. Miller (Ed.). *Educational Media and Technology Yearbook 1985.* Libraries Unlimited, Littleton, CO.

Heid, M.K. 1988. Resequencing skills and concepts in applied calculus using the computer as a tool. *Journal for Research in Mathematics Education,* 19(1):3-25.

Hezel, R.T. 1980. Public broadcasting: Can it teach? *Journal of Communication,* 30:173-178.

Hockberg, J. 1962. Psychophysics of pictorial perception. *AV Communications Review.* 10:22-54.

Jonassen, D.H. 1988. *Instructional Designs for Microcomputer Courseware.* Lawrence Erlbaum, Hillsdale, N.J.

Koran, M.L. 1972. Varying instructional methods to fit trainee characteristics. *AV Communications Review.* 20:135-146.

Kracjik, J.S., P.E. Simmons and V.N. Lunetta. 1986. Improving research on computers in science learning. *Journal of Research in Science Teaching.* 23(5):465-470.

Kracjik, J.S., P.E. Simmons and V.N. Lunetta. 1988. A research strategy for the dynamic study of students' concepts and problem solving strategies using science software. *Journal of Research in Science Teaching.* 25(2):147-155.

Krendl, K.A. and D.A. Lieberman. 1988. Computers and learning: A review of recent research. *Journal of Educational Computing Research.* 4(4):367-389.

Kulik, J.A., R.L. Bangert and G.W. Williams. 1980. Effects of computer based teaching on secondary school students. *Journal of Educational Psychology.* 75(1):19-26.

Lefkowith, E.F. 1955. *The validity of pictorial tests and their interaction with audio-visual teaching methods.* Technical Report, SDC-269-7-49. Special Devices Center, Office of Naval Research, Washington, N.Y.

Levie, W.H. and K.E. Dickie. 1973. The analysis and application of media. *In:* R.M.W. Travers, (Ed.). *Second handbook of research on teaching.* Rando McNally, Chicago, IL.

Lockard, J., P.D. Abrams and W.A. Many. 1987. *Microcomputers for Educators.* Little & Brown, Boston, MA.

Lunetta, V.N. 1972. The design and evaluation of a series of computer simulated experiments for use in high school physics. Dissertation. *Dissertation Abstracts International.* University of Connecticut. 33:2785A.

Merrill, M.D. 1988. Applying component display theory to the design of courseware. *In:* D.H. Jonassen, (Ed.). *Instructional Designs for Microcomputer Courseware. Lawrence Erlbaum, Hillsdale, N.J.*

Nachmias, R. and M.C. Linn. 1987. Evaluations of science laboratory data:

The role of computer-presented information. *Journal of Research in Science Teaching.* 24(5):491-506.

Naisbitt, J. 1982. *Megatrends: Ten new directions transforming our lives.* Warner Books, New York, N.Y.

Norman, D., S. Gentner and A.L. Stevens. 1976. Comments or learning schemata and memory representation. *In:* D. Klahr (Ed.). *Cognition and instruction.* Lawrence Erlbaum, Hillsdale, N.J.

Olsen, J.R. and V.B. Bass. 1982. The application of performance technology in the military 1960-1980. *Performance and Instruction.* 21(6):32-36.

Pogge, A.F. and V.N. Lunetta. 1987. Spreadsheets answer "What If...?" *The Science Teacher.* 54(8):46-49.

Rigney, J. 1978. Learning strategies: A theoretical perspective. *In:* H.F. O'Neil (Ed.). *Learning strategies.* Academic Press, New York, N.Y.

Roblyer, M.D. 1985. *Measuring the impact of computers in instruction: A non-technical review of research for educators.* Association for Educational Data Systems. Washington, D.C.

Rumelhart, D.E. and A. Ortony. 1977. The representation of knowledge in memory. *In:* R.C. Anderson, R.J. Spiro and W.E. Montague (Eds.). *Schooling and the acquisition of knowledge.* Lawrence Erlbaum, Hillsdale, N.J.

Sacttlcr, P. 1968. *A history of instructional technology.* McGraw-Hill, New York, N.Y.

Schimmel, B. 1986. *A meta-analysis of feedback to learners in computerized and programmed instruction.* Paper presented at the annual meeting of the American Educational Research Association. ERIC Document 233 708. Montreal, Canada.

Shapiro, K.R. 1975. An overview of problems encountered in aptitude treatment interaction (ATI) research for instruction. *AV Communications Review.* 23:227-241.

Sloyer, C. and L.H. Smith. 1986. Applied mathematics via student-centered computer graphics. *The Journal of Computers in Mathematics and Science Teaching.* Spring: 17-20.

Taylor, R.P. 1980. *The Computer in the School: Tutor, Tool, Tutee.* Teachers College Press, New York, N.Y.

Wittrock, M.C. 1978. The cognitive movement in instruction. *Educational Psychologist. 15:15-29.*

Science Education in the United States: Issues, Crises and Priorities. Edited by S. K. Majumdar, L. M. Rosenfeld, P. A. Rubba, E. W. Miller and R. F. Schmalz. ©1991, The Pennsylvania Academy of Science.

Chapter Fourteen

THE PROMISE OF COGNITIVE PSYCHOLOGY FOR SCHOOL SCIENCE

ROSALYN GATES

Doctoral Candidate
Curriculum and Instruction
The Pennsylvania State University
University Park, PA 16802

INTRODUCTION

In his recent book, *The Mind's New Science,* Howard Gardner, a psychologist affiliated with Harvard University, describes a fascinating and intriguing history of what has been called the "cognitive revolution"[1]. This revolution, dating back to an unofficial beginning in the mid-fifties with the birth of the modern computer era and the concept of information-processing, has within the last several decades produced a "new breed" of experimental researchers identified as cognitive scientists. Their background? Most commonly from the specific disciplines of psychology, philosophy, linguistics, computer science, and neuroscience. Their goal? To understand and explain human behavior in terms of the internal cognitive processes and structures of the human mind. Their significance to educators and researchers in school science? The offering of a new cognitive perspective from which to approach science teaching and learning in contrast to the traditional behavioral and developmental orientations.

The emergence of a cognitive perspective to science teaching and learning is actually part of a pervasive cognitive movement throughout the mainstream of education in general. In describing the breadth of this cognitive movement, F.J. DiVesta,[2] a respected educational psychologist and researcher, has written:

Although education remains dominated by behaviorist principles, a steadily increasing influence of cognitive science is apparent. The research literature is abundant with articles that are concerned with cognitive development, thinking and problem-solving, cognitive learning strategies and skills, and cognitively based instructional designs. The cognitive influence is finding its way into the teaching of specific subjects, particularly reading, but in mathematics, science, and social studies as well. (p. 203).

The focus of the present chapter is on the new cognitive approach to instruction as it is "finding its way into the teaching of . . . science." The potential of this new approach is the promise of cognitive psychology for school science. Since it is almost impossible to realize the potential in something that is not understood, the purpose of this chapter is to provide that basic understanding. This will be done by describing the cognitive approach in contrast to the more traditional behavioral approach and as an extension beyond the Piagetian developmental framework that has guided science education for many years. Also, a basic model of the human information-processing system will be described as the theoretical framework that underlies the cognitive approach. The concluding remarks will suggest some possible implications for change in school science resulting from approaching science instruction from a more cognitively-oriented perspective.

APPROACHES TO INSTRUCTION

There are two quite different approaches to an instructional program: behaviorist and cognitive. Each reflects a corresponding theoretical and research orientation from educational psychology. Since the 1960s, the orientation of school science instruction has been a curious blend of behaviorist principles and Piagetian developmental theory, which itself has been a stepping stone into the current cognitive orientation. This section describes and contrasts the behaviorist and cognitive psychological orientations and corresponding instructional approaches, and then indicates areas in which the cognitive approach to instruction is manifesting in science education and is extending beyond the traditional Piagetian emphasis.

The Behaviorist Approach

In the 1920s-1950s, behaviorism was the dominant orientation of educational psychology in America and therefore became the basis for many educational practices as well as the traditional approach to instruction. Guided by the learning theories developed by psychologists such as E.L. Thorndike and B.F. Skinner, the traditional or behaviorist (stimulus-response) approach to instruction assumes that teaching procedures (the stimulus) *directly* affect students' achievement (the response). Farnham Diggory[3] has pointedly expressed the underlying idea of the behavioral (S-R) approach to be "a stimulus goes in, a response comes out, and what happens in between is summarized by a hyphen" (p. 128).

An extreme behaviorist approach regards the learner as a passive recipient of information. Learning consists of acquiring appropriate responses to new stimuli. The role of the teacher is to present information for students to absorb and then reproduce at appropriate times. Correct responses are reinforced by an external stimulus or event. Teaching methods emphasize drill and practice, rote rehearsal, shaping errorless learning, and sheer transmission of information.[2,4]

The following scenario is a simple illustration of behavioral instruction. After a homework assignment to memorize a list of weather terms and definitions, the teacher asks a student, "What is a barometer?" (the stimulus). The student answers, "an instrument that measures atmospheric pressure" (the response). The teacher replies, "Correct" (the reinforcer). The teacher can conclude, according to the behavioral approach, that learning has taken place since the student was able to reproduce the appropriate response to the given stimulus. Note that the criteria for successful teaching is how often correct observable responses are given.

The behavioral approach is not necessarily wrong; in fact, it may be very appropriate in various situations. However, it is also very limiting. An alternative to behaviorist instruction is the cognitive approach which opens a much wider door for educators to include achieving such desirable instructional goals as conceptual understanding and creative problem-solving.[4]

The Cognitive Approach

Behaviorism dominated psychology in the 1920s-1950s because cognitive theorists, such as the Gestalt psychologists, lacked precise methodological tools and testable theories about the operations of the mind. However, in the 1950s-1960s, the development of the computer gave cognitive theorists the research tool they needed as well as powerful computational metaphors with which to theorize about the nature of the human mind. The effect of the computer and information-processing theory on psychological research was the cognitive revolution, in which behaviorism was "dethroned" and cognitive research was elevated to dominance. The change in the orientation of psychological research has led to the development of a new approach to instruction, based on cognitive research theory and findings in areas such as learning, memory, text comprehension, and problem-solving.[5]

Simply stated, the new cognitive approach, in contrast to behaviorism, maintains that teaching procedures have only an *indirect* influence on student achievement. The assumption is made that the *internal* variables of cognitive processing (how stimulus information is manipulated and transformed in the mind) and knowledge structures (what information is stored and how it is represented and organized in memory) mediate between the *external* variables of instruction (stimulus) and behavior (response). In other words, the cognitive approach replaces the "hyphen" of behaviorism (S-R) with mental information-processing and organized knowledge structures in memory.[3]

The cognitive approach to instruction views the learner as an active processor of information in an effort to construct a network of meaningful and useful knowledge in memory. Learning is characterized as changes in the learner's knowledge and occurs because the learner actively tries to understand the environment by relating new information to existing knowledge. The role of the teacher is to help students establish relevant and meaningful connections between new and prior knowledge so as to develop knowledge structures organized in memory to facilitate retrieval of needed information, promote knowledge transfer to novel situations, and support acquisition of useful new information.[4]

Instructional techniques, based on the cognitive approach, would be designed to influence how the material-to-be-learned will be processed (mentally manipulated) and consequently represented in memory by the learner. For instance, in the barometer scenario, the teacher may have asked students to illustrate in a diagram and explain in their own words how a barometer works. This assignment requires students to process the information both visually and verbally, resulting in two different types of representations of the same material in memory, which may aid later retrieval. Also, requiring students to explain the instrument in their own words forces them to integrate new information with prior knowledge, thus facilitating more meaningful learning. An analysis by the teacher of the student responses would enable the teacher to make useful inferences regarding the students' emerging knowledge constructions in memory, i.e., their learning about atmospheric pressure. Note that the criteria for successful teaching is students' conceptual understanding which can be measured by students' success in creative problem-solving.

In essence, the distinguishing difference between the behaviorist and cognitive perspectives is the attention given to the human mind. The behaviorist disregards it, considering the mind to be a "black box", unable to be scientifically studied because it cannot be observed directly. In contrast, the cognitivist regards the mind, particularly cognitive processes and knowledge structures, as *the* object of study. The challenge to cognitive researchers has been to devise methods of study that allow useful inferences to be made about internal processes and knowledge states of the learner.

Cognitive Beginnings in Science Education

As a result of the cognitive revolution in psychology, a number of science educators and researchers are beginning to change their views on the nature of the learner and the process of learning, the role of the teacher and the methods/goals of instruction. In particular, the new cognitive perspective has begun to emerge with regards to the nature of the learner.[6,7] G.M. Bodner, a chemistry educator from Purdue University, has summarized the new conception of the learner, with the help of a quote from psychologist Lauren Resnick:[8]

> Until recently, the accepted model for instruction was based on the hidden assumption that knowledge can be transferred intact from the mind of the teacher to the mind of the learner. Educators therefore focused on getting knowledge into the heads of their students, and educational researchers tried to find better ways of doing this. ...Most cognitive scientists now believe in a constructivist model of knowledge that attempts to answer the primary question of epistemology, How do we come to know what we know?" . . . This constructivist model can be summarized in a single statement: Knowledge is constructed in the mind of the learner. (p. 873).
>
> . . . Learners construct understanding. They do not simply mirror and reflect what they are told or what they read. Learners look for meaning and will try to find regularity and order in the events of the world even in the absence of full or complete information. (p. 874).

The view that learners construct their own understanding of the world "even in the absence of full or complete information" has helped to explain the presence of students' incomplete and often inconsistent "naive theories" or preconceptions about how the natural world works even before students' receiving any formal science instruction.[7] These naive theories have been found by research to differ from accepted scientific explanations, to be persistent and often resistant to change, and to frequently interfere with students' meaningful learning of science concepts during instruction.[9]

Extensive science education research identifying and describing the many naive theories held by students has led science educators and researchers to develop new instructional methods designed to expose, confront, and modify/change students' conceptual beliefs to agree more with those held by scientists. These instructional changes, which place a new emphasis on the content and organization of knowledge in students' memory, are significant in the fact that they indicate the beginnings of a more cognitively-oriented approach to science instruction.

One new instructional method which focuses on the changes in students' conceptual understanding in science is a type of instruction called "conceptual change teaching." James Minstrell, a high school physics teacher and cognitive researcher, has captured the essence of this type of instruction in the following metaphor"[10]

> Students come to the classroom with initial conceptions organized by their experiences . . . . [Their] initial ideas are like strands of yarn, some unconnected, some loosely interwoven. The act of instruction can be viewed as helping the student unravel individual strands of belief, label them, and then weave them into a fabric of more complete understanding. An important point is that later understanding can be constructed, to a considerable extent, from earlier beliefs. Sometimes new strands of belief are

introduced, but rarely is an earlier belief pulled out and replaced. Rather than denying the relevancy of a belief, teachers might do better by helping students differentiate their present ideas from and integrate them into conceptual beliefs more like those of scientists. (p. 130-131).

Thus, the new constructivist view of the learner and new instructional methods that focus on the important interaction of students' existing knowledge and content knowledge to be learned are two areas that indicate the influence of cognitive instructional psychology in school science.

In addition to these areas, another area in science education that is undergoing change due to the influence of theoretical changes in cognitive psychology is the emphasis on the Piagetian theoretical framework that has guided science education research and instruction since the 1960s. The next section highlights key changes in this area and indicates new directions in science instruction resulting from these changes.

The Cognitive Approach Beyond Piaget

Piaget's traditional theory of intellectual development has in the recent past enjoyed a major influence over the entire fields of developmental and educational psychology, and in particular, the field of science education. However, cognitive theory changes and methodological advances in cognitive psychology have led to revisions, re-interpretations, and in many cases the abandonment by developmental and educational psychologists of aspects of Piaget's theory. These changes in psychology have begun to be manifest in areas of science education through various departures from Piagetian theory to more current cognitive views.

One point of departure is the new cognitive conception of the learner. Although the Piagetian view of the learner as an active constructor of a representation of the world through interaction with the environment is consistent with the new cognitive view, the new view places greater importance on what the learner already knows, how that knowledge is organized in memory, and the appropriate integration of relevant prior knowledge with new information.[6,7]

Other points of departure for the new cognitive view from Piaget's theory of intellectual development are the notion of age-related stages of cognitive development and the role of general logical rules of reasoning. Piaget maintained that by early adolescence, all children have progressed through three and entered the fourth major stage of cognitive development. A child's reasoning is qualitatively different in each stage; that is, each stage is characterized by how the child is able to mentally represent the world and by the level of logical reasoning ability available to the child. Applied to school science, Piagetian theory has tended to result in practice that emphasizes the development of scientific reasoning skills apart from learning subject matter and in curriculum that imposes age-related constraints on children's conceptual learning in science.

In contrast, the new cognitive view places greater emphasis on the role of content-specific knowledge in the development of reasoning skills within that content area, and on the importance of relevant prior knowledge, not on stages of cognitive development, as a prerequisite for conceptual understanding.

James Minstrell's account of his own personal departure from a Piagetian view to a more current cognitive one is helpful to illustrate how the new cognitive approach has departed from the traditional Piagetian perspective in science instruction:[10]

Early in my experience as a high school science teacher, I became concerned about my ineffectiveness at teaching students to transfer their understanding. ...My students tested well...but I was amazed at the relatively little effect I had on my students' understanding of the ideas of physics. In the early 1970's, I, like others, thought the difficulty might be my students' lack of ability with formal operational reasoning. However, in my investigations of students' operational reasoning capacity, I found that the results of reasoning tests depended on students' conceptual knowledge. For example, consider a task involving two equal-sized balls of clay. After establishing that the two balls weigh the same, one of them is flattened into the shape of a pancake. Students are again asked to compare the weights. The task tests students' understanding of the operation of conservation, in this case conservation of weight. Many students fail the task not because they can't do conservation reasoning, but because their concept of weight or gravity involves air pressure, and a flattened ball of clay has more upper surface on which "air presses down." This represents a conceptual error rather than an operational reasoning error. Students are bringing to the situation content ideas that greatly affect their performance on questions that are supposed to test their operational reasoning. Even on problems that require the same abstract reasoning on similar quantitative data, the specific content of the situation affects the reasoning strategies students use. (pp. 129-130).

Summary

In essence, the cognitive revolution that has occurred in psychology is effecting change in science education research and practice. The revolution has replaced the previously dominant behaviorist orientation in psychology with a cognitive one in which mental activity is the focus of study.[5] This switch in orientation has and is affecting how science educators and researchers view all aspects of the instructional process.

Even further, the Piagetian theoretical framework that has, for many years, served developmental and educational psychology, is being re-interpreted, revised, extended, or replaced as a result of methodological and conceptual advances in cognitive research. In general, psychologists are moving away from traditional Piagetian theory and towards an information-processing (I-P) model for explaining and understanding cognitive development.[11] This theory change in psychology is filtering into many areas of science education research and instruction, bringing changes and opening new doors for researchers and educators.

The next section summarizes the I-P (information-processing) theoretical model of cognition and memory which underlies the new cognitive orientation in educational psychology and forms the framework for the cognitive approach to instruction. The "cognitive beginnings" in science education can be considered as pieces of the much larger picture of cognitive instruction being created as a result of educational research within the I-P framework. A basic understanding of this theoretical framework is necessary to gain a fuller appreciation of the potential in the cognitive approach to improve school science teaching and learning.

THE I-P FRAMEWORK FOR THE COGNITIVE APPROACH

Cognitive psychology has been defined as "the scientific analysis of human mental processes and memory structures in order to understand human behavior"[5] (p. 1). In order to analyze mental events, psychologists have constructed a general theoretical model that depicts the human being as an information-processing system partly analogous to a computer system. Both systems can take in information, operate on it, and generate some appropriate response. Hence, thinking can be described as the movement and processing of information through and among the components of the system. Bits of information can be represented and transformed in different ways, integrated with other information, "encoded" into memory, stored, and later retrieved when necessary.

The human-computer analogy, however, is not without its weaknesses. Human beings are clearly not computers. Computers have no emotions, no motivation, no "aliveness". Nevertheless, the analogy has provided cognitive researchers with a powerful conceptual framework in the form of an information-processing model of cognition and memory for characterizing and investigating thinking, learning, problem-solving, etc.

Memory Stores, Cognitive Processes, and Knowledge Structures

The basic information-processing model consists of several components and associated processes that describe the states and transformations of information between input (stimulus) and output (response). The main components are three memory stores: a short-term sensory memory that receives stimuli from the sense receptors, an active or conscious (short-term) memory that has limited capacity and is subject to overload, and a repository for long-term memory storage, like an organized storehouse holding vast amounts of information.

Basic processes include attention, rehearsal, encoding, search, and retrieval. Attention processes transfer selected information from sensory memory to conscious memory where information can be kept active by rehearsal processes. Encoding processes organize and transfer selected information from conscious memory to long-term storage. Search processes locate information in storage and retrieval processes transfer stored information back into conscious memory for additional processing and eventual output.

Besides the basic components and processes, another key aspect of the I-P system is the knowledge base that comprises long-term memory. In principle, this knowledge encompasses all of one's life experiences and is represented or stored in memory in some organized network of knowledge structures. These organized knowledge representations or knowledge structures have been variously referred to as schema (singular) or schemata (plural), semantic networks, production systems, scripts, frames, etc., depending mostly on the type of knowledge being represented; e.g., declarative (knowing that) or procedural (knowing how) knowledge.[2]

Importance of Knowledge Structures

The primary importance of knowledge structures lies in the fact that incoming information is processed in terms of the knowledge structures held by the learner. In other words, learners use their existing knowledge to understand, remember, interpret, and make inferences about new information and events. To illustrate, read the following paragraph and give attention to your cognitive efforts to understand and make sense of the procedure described.

THE PROCEDURE

The procedure is actually quite simple. First you arrange things into different groups. Of course one pile may be sufficient depending on how much there is to do. If you have to go somewhere else due to lack of facilities that is the next step, otherwise you are pretty well set. It is important not to overdo things. That is, it is better to do too few things at once than too many. In the short run this may not seem important but complications can easily arise. A mistake can be expensive as well. At first the whole procedure will seem complicated. Soon, however, it will become just another facet of life. It is difficult to foresee any end to the necessity for this task in the immediate future, but then one never can tell. After the procedure is completed, one arranges the materials into groups again. Then they can be put into their appropriate places. Eventually they will be used once more and the whole cycle will then have to be repeated. However, that is part of life.[12]

Did you experience some frustration in trying to comprehend the passage? Would a cue that permits retrieval of relevant information from your existing knowledge aid your understanding? Try it. Read the passage again, only this time change the first sentence to read: "The procedure *for washing clothes* is actually quite simple."

Now compare your initial frustration to that of a learner in science who either lacks appropriate and sufficient prior knowledge or lacks relevant cues as to what knowledge to access for interpreting and understanding a science text or classroom activity. If the material cannot be linked in a meaningful way to appropriate, existing knowledge, then no understanding is possible and the learner will usually default to non-meaningful rote memorization.

SUMMARY

In summary of this section, how does the I-P system with its memory components, cognitive processes, and stored knowledge structures fit into a cognitive approach to instruction? Briefly stated, the cognitive approach to instruction emphasizes the importance of variables internal to the learner and the I-P theoretical model provides a framework in which to identify, describe, and manipulate those variables to more effectively achieve desirable performance goals.

The importance of internal variables in obtaining instructional goals can be seen even more clearly in Mayer's[4] basic description of the cognitive approach to instruction: instructional manipulations (stimuli) + internal learner characteristics affect the learner's cognitive processing, which results in knowledge structure changes, which manifest in observable performance behaviors (responses).

A further elaboration of Mayer's description of the cognitive approach points out the intimate functioning of the I-P system within the approach. Instructional manipulations (e.g., teacher strategies, methods, materials, etc.) + learner characteristics (e.g., prior knowledge, learning strategies used, etc.) affect how the material-to-be-learned is cognitively processed, i.e., what information is attended to, how the selected information is reorganized (constructed) by the learner, and what relationships or links, if any, are established between new and existing knowledge. The cognitive processing results in the acquisition of new or reorganized knowledge structures in long-term memory. The resulting change in knowledge structures is the cognitive learning outcome of instruction, and that affects changes in observable performance behaviors (e.g., achievement tests) and makes additional learning possible.

CONCLUDING REMARKS

This chapter has been an attempt to present a basic understanding of the cognitive approach to instruction in order for readers unfamiliar with current cognitive research to realize the potential of this approach to affect the quality of school science teaching and learning. Several areas were mentioned as cognitive beginnings already in science education; however, there is much more to the potential of cognitive instruction than what these "beginnings" would indicate.

The potential of the cognitive approach is based on the research that supports it. Although it was not possible in this chapter to give details on the areas currently being researched in cognitive psychology that would apply directly to school science, a mention of some of the general research questions being asked would indicate the directions being pursued. Some of the questions are: How do we teach for science understanding? How do students' knowledge structures of specific science concepts change and develop over time? What affect does different types of instruction have on cognitive processing and the resulting representations of science concepts? How do students reason about new information, such as observation data? What are the differences in knowledge and knowledge use between successful and less-successful problem-solvers in science? How does the structure of science text affect how new information is processed by the reader? How can we teach students to learn how to learn science? How do we enable students to transfer and apply their science knowledge in creative problem-solving?

Considering these areas of current cognitive research and the resulting potential of cognitive instruction, what changes would this approach to instruction imply for school science? At least three areas would be affected considerably: science education research, teacher education at all levels, and curriculum development and materials.

In regard to science education research, there would be the necessity of increased research collaboration between science educators and cognitive psychologists, as well as other cognitive scientists. Each research group would bring needed methodological expertise and specialty knowledge to apply in a cooperative effort to maximize the potential of cognitive instruction.

The second area, teacher education, would need to be refocused at all levels to employ and teach a cognitive approach to instruction predominant over a behavioral one. This would include the incorporation of cognitive instructional psychology into science teaching methods coursework.

Curriculum development and materials is a third area where changes would be witnessed. These changes would affect science text materials and test construction, in addition to what content is presented, how much is presented, and how it is presented.

In conclusion, the potential of the cognitive approach to science instruction, based on current research theory and findings from cognitive psychology, holds tremendous promise for greatly improving, and even reconceptualizing, school science teaching and learning. However, the cognitive approach will remain only a potential with promise unless it is taken seriously by many science educators and converted to action.

REFERENCES

1. Gardner, H. 1985. *The Mind's New Science.* Basic Books, Inc. New York, NY.
2. DiVesta, F.J. 1987. The cognitive movement and education. *In:* J.A. Glover and R.R. Ronning (Eds.). *Historical Foundations of Educational Psychology.* Plenum Press, New York, NY. pp. 203-233.
3. Farnham-Diggory, S. 1977. The cognitive point of view. *In:* D.J. Traffinger, J.K. Davis, and R.E. Ripple (Eds.). *Handbook of Teaching Educational Psychology.* Academic Press, New York, NY.
4. Mayer, R.E. 1987. *Educational Psychology: A Cognitive Approach.* Little, Brown and Company, Boston, MA.
5. Mayer, R.E. 1981. *The Promise of Cognitive Psychology.* W.H. Freeman and Company, San Francisco, CA.
6. Linn, M.C. 1987. Establishing a research base for science education: Challenges, trends, and recommendations. *Journal of Research in Science Teaching.* 24(3):191-216.
7. Resnick, L.B. 1983. Mathematics and science learning: A new conception. *Science,* 220:477-478.
8. Bodner, G.M. 1986. Constructivism: A theory of knowledge. *Journal of Chemical Education.* 63(10):873-878.
9. Driver, R., E. Guesne and A. Tiberghien (Eds.). 1985. *Children's Ideas in Science.* Open University Press, Milton Keynes, England.
10. Minstrell, J.A. 1989. Teaching science for understanding. *In:* L.B. Resnick and L.E. Klopfer (Eds.). *Toward the Thinking Curriculum: Current Cognitive Research* 1989 Yearbook of the Association for Supervision and Curriculum Development. pp. 129-149.
11. Flavell, J. 1983. Preface to volume III. *In:* J. Flavell and E. Markman (Vol. Eds.); P. Mussen (Series Ed.). *Handbook of Child Psychology. Vol. 3, History, theory and methods.* Wiley, New York, NY.
12. Bransford, J.D. and M.K. Johnson. 1972. Contextual prerequisites for understanding: Some investigations of comprehension and recall. *Journal of Verbal Learning and Verbal Behavior.* 11:717-726.

cluded the factors revealed by Sia and Hines. By those goals and competencies, when STS is integrated into school science (or social studies) it should include:

1. STS Foundations activities that give learners an understanding of issue relevant natural science and social science concepts, as well as an understanding of the nature of science and technology, and their interactions in and with society;

2. STS Issue Awareness activities designed to make learners cognizant of how science, technology and society frequently interact to yield issues that are best resolved through active citizen participation;

3. STS Issue Investigation Skill instruction to enable learners to investigate science and technology-related societal issues from a number of different perspectives prior to developing and judging the value of a number of different possible resolutions paths;

4. STS Issue Action Skill development activities to give learners the tools to act individually or in groups toward the resolution of science and technology-related societal issues; and

5. Opportunities to apply STS investigation skills and action strategies in the resolution of a science and techology-related societal issue relevant to the learners and their community.

STS issue investigation and action instruction that follows from the goal structure (Rubba and Wiesenmayer, 1988) is, in essence, a project approach comprising four to six week units. A unit may begin with activities in which students examine the nature of science and technology, and characteristic interactions among science, technology and society. Next, critical science and technology-related societal issues may be identified, and analyzed to identify what makes them issues, for relevant science and social science concepts, and to identify the prominent value positions of the various sides of the issue. Case studies might be used to demonstrate that only through responsible citizenship action is there a possibility for science and technology-related societal issues to be resolved and that we can have an impact.

A community and student relevant science and technology-related societal issue may then be selected by the class or a number of issues may be selected by groups of students within a class, to serve as a theme or focus in the investigation and action activities. Students learn issue investigation skills as they apply them to the issue. That might include, the study of science and/or social science concepts, library research, securing data and information from governmental and private agencies, the collection of natural science data on sight, and use of questionnaires and opinionaires within the community to collect data. Information that is collected is analyzed by the students and used to propose alternative solutions to the science and technology-related societal issue. The pros and cons of each resolution strategy are weighed and a course(s) of action decided upon. Lastly, students carry through with the action plans they helped compose, as members of a group or individually, and evaluate their efforts.

An STS Unit on Trash

The STS unit was developed during a two-week summer workshop by five teachers from Roosevelt and Keith Junior High Schools in Altoona Pennsylvania—Tom Hite, Elaine Glashauser, Charles Guyer, George Mahon, and Kasey Prokop. A trash disposal/management issue that was under study by a county commission at the time was selected as the science and technology-related societal issue theme for the unit.

In the unit, students first examined the interactions among science, technology and society that lead to science and technology-related societal issues, in general, and those particularly associated with the trash issue in their community. Next, they completed a thorough investigation of the issue from scientific, technological, and social perspectives. This included: tracing the sources and paths of trash within the school and community, analysis of trash samples from home and school, hands-on investigations of relevant science concepts, reviewing newspaper and magazine articles on the trash issue that were a part of the unit, completing additional library research on their own, and the design and administering of a community opinion survey on solutions to the trash issue.

Findings from these investigations were used by the students to structure alternative plans that might be used to resolve the community's trash problem. The pros and cons of each plan were weighed in group discussions. Finally, the students took action on the issue in concert with their own personal decisions and evaluated the impact of those actions. The unit took a little less than five weeks to complete.

Three of the teachers who developed the unit taught it in their 10 sections of seventh grade life science. The other seven sections, taught by the other two teachers, received the regular life science curriculum. Statistically significant pre- to post-test gains were found within the 10 sections that completed the STS unit on the three dependent variables studied—STS content achievement, number of actions taken on science and technology-related societal issues, and life science achievement. Each dependent variable was assessed using a valid and reliable paper and pencil instrument. The seven class sections that completed the regular life science instruction showed statistically significant gains only in life science achievement. In addition, posttest comparisons across the STS unit and life science groups on STS content achievement and number of actions taken on science and technology-related societal issues were statistically significant.

The findings showed that issue investigation and action STS units can help middle/junior high school students develop an understanding of the interrelationships among science, technology and society, the ability to take informed action on science and technology-related societal issues, as well as foster the affective qualities needed to do so. They also indicated that issue investigation and action STS units can encourage meaningful science concept development.

instruction. *Education and Urban Society.* 22(1), 40-53.

Ramsey, J. and Hungerford, H. 1989. The effects of issue investigation and action training on environmental behavior in seventh grade students. *The Journal of Environmental Education.* 20(4), 29-34

Ramsey, J., Hungerford, H. and Tomera, A. 1981. The effects of environmental behavior of eight grade students. *The Journal of Environmental Education.* 13(1), 24-29.

Rubba, P. 1989a. The effects of an STS teacher education unit on the STS content achievement and participation in actions on STS issues by preservice science teachers. A paper presented at the 1989 Annual Meeting of the National Association for Research in Science Teaching, San Francisco, CA: March 30-April 1.

Rubba, P. 1989b. An investigation of the semantic meaning assigned to concepts affiliated with STS education and of STS instructional practices among a sample of exemplary science teachers. *Journal of Research in Science Teaching.* 26(8), 687-702.

Rubba, P., McGuyer, M. and Wahlund, T. (in press). The effects of infusing STS vignettes into the genetics unit of biology on learner outcomes in STS and genetics: a report on two investigations. *Journal of Research in Science Teaching.*

Rubba, P. and Wiesenmayer, R. 1988. Goals and competencies for precollege STS education: recommendations based upon recent literature in environmental education. *The Journal of Environmental Education.* 19(4), 38-44.

Sia, A., Hungerford, H. and Tomera, A. 1986. Selected predictors of responsible environmental behavior: an analysis. *Journal of Environmental Education.* 17(2), 31-40.

Simpson, P. 1990. The effects of an extended STS case study on citizenship behavior and associated variables in fifth and sixth grade sutdents. A paper presented at the 1990 Meeting of the National Association for Rearch in Science Teaching, Atlanta, GA: April 8-11.

Stapp, W. et al. 1969. The concept of environmental education, *Journal of Environmental Education.* 1(3), 31-36.

Wiesenmayer, R. and Rubba, P. 1990. The effects of STS issue investigation and action instruction and traditional life science instruction on seventh grade students' citizenship behavior. A paper presented at the 1990 meeting of the National Association for Reseach in Science Teaching, Atlanta, GA: April 8-11.

Zielinski, E. and Mechling, K. 1989. The effects of infusing STS vignettes into elementary science methods classes on learner outcomes in STS and STS attitudes. A paper presented at the National Association for Science, Technology and Society Fourth Annual Technological Literacy Conference, Washington, DC: February 18-19.

Science Education in the United States: Issues, Crises and Priorities. Edited by S. K. Majumdar, L. M. Rosenfeld, P. A. Rubba, E. W. Miller and R. F. Schmalz. ©1991, The Pennsylvania Academy of Science.

Chapter Sixteen

REFORMATION IN SCIENCE TEACHER EDUCATION

HANS O. ANDERSEN

Professor of Science Education
President of NSTA
Indiana University
Bloomington, IN 47405

INTRODUCTION

The National Science Teachers Association (NSTA) recently published a paper describing initiatives that are needed to bring about change in science education. Entitled *Science Teachers Speak Out,* it is the product of two years of work that involved many people. Because I will be basing this discussion of science teacher education paper on it and another paper developed by the NSTA Executive Director, Bill Aldridge (1989), which describes some necessary curriculum restructuring, I will spend some time describing the process used to develop *Science Teachers Speak Out* and recommendations for teacher preparation it suggests.

SOURCE #1

It began in the spring of 1988. Election reports were in and NSTA had a new president-elect elect. He was summoned by two of the Presidents to a retreat to discuss NSTA. One result of that retreat was a decision to prepare a paper, not to exceed two pages in length, to describe what science teachers, science educators, science supervisors, and scientists thought ought to be done in the 1990s to turn around the crisis in science education. Approximately 1000 people were interviewed to provide an information base. The resulting paper was then discussed by the executive committee and by the board of directors of NSTA at the summer board meeting and approved it for distribution. That paper was titled *NSTA's Science Education Initiatives for the 1990s* (Appendix).

As soon as the Initiatives paper was approved, it was given to a newly created Task Force that had been charged with the responsibility for preparing guidelines for science education for the year 2000 and beyond. It was their point of departure in synthesizing a longer document, *Science Teachers Speak Out,* which would carry our thinking into the 21st century. This was a small task force, but it incorporated considerably more input by reaching out to reference groups and requesting reactions to successive drafts of the evolving paper. Upon completion of the drafts, the paper was approved by the NSTA Board of Directors and named its "Lead Paper."

NSTA's Lead Paper is the organization's major policy/position statement. NSTA specifies two major goals for science education in it. The first is to provide science education curriculum and instruction that will render all adults scientifically literate. The second goal is to provide enough students to fill the science pipeline that leads into careers in science, mathematics, and engineering. To attain these goals, initiatives are needed in five areas: commitment to science education, curriculum development, instructional support, staff development, and research support.

The following list of things teachers ought to be able to do was drawn from the recommendations made in the paper *Science Teachers Speak Out.*

1. Involve *all* students and adults in the study of real life, personal, and societal science and technology problems.

2. Implement science curricula for *all* students, including special populations such as under-represented groups, talented, and "at risk" students.

3. Implement science programs which are responsive to students' personal needs in areas such as bioethics, reproductive biology, nutrition, drugs, human values, and self-esteem.

4. Implement science curricula organized around themes such as stability, energy, evolution, systems, and inquiry.

5. Involve *all* students in searching for, organizing, originating, and communicating scientific information.

6. Implement science curricula that develops awareness of science career opportunities and prepares students to pursue careers in science and engineering.

7. Implement curriculum which will prepare persons who are presently under-represented in science and engineering for careers in science and engineering.

8. Use appropriate technology effectively to deliver science instruction, e.g., satellite and cable T.V., video tapes, video discs, video microscopy.

9. Use evaluation and assessment tools that reflect the goals of science education.

10. Use evaluation materials which measure different levels and types of knowledge, thinking, and manipulative skills.

11. Manage, equip, and supply appropriate and safe science teaching facilities.

12. Continually update teachers' science and science teaching background.

13. Supervise laboratory assistants.

14. Actively pursue discussions of science teaching which enhance student learning with colleagues and students.

15. Teach students to think about how they learn. That is, provide students the "learning tools" which will empower the learner to achieve high levels of

meaningful learning. (Novak, 1990).

16. Lead the development of active student science organizations or clubs such as Future Science Teachers, JETS, (Add to this listing).

17. Encourage and direct students through science research planning and participation.

18. Effectively communicate explanations of both the nature of and need for quality science education to parents, laypersons, and administrators.

In addition to specifying what teachers completing teacher education programs ought to do, *Science Teachers Speak Out* also addresses the question of who should be involved in teacher education. At the present time, at most institutions preparing teachers, the responsibility for the preparation of science teachers lies in the hands of on-campus professional educators. Scientists and practicing teachers who are involved with preservice teachers are usually volunteers with other full time classroom or college/university responsibility. There are many who believe that the preparation of science teachers is too important to be left entirely a responsibility of the college/university level. They feel, as is stated in the paper, that programs should be designed and implemented "...cooperatively by education faculty, scientists, and practicing classroom teachers of science." Thus, the recommendation is that scientists and practicing teachers, as well as professional educators, have, as part of their salaried duties, responsibility for planning, implementing, and evaluating their roles as teacher educators.

SOURCE #2

In the second paper, *"Essential Changes in Secondary Science: Scope, Sequence, and Coordination,"* Aldridge (1989) discusses needed curricular restructuring and offers a model that compensates for the major flaws that exist in the way we presently deliver science instruction. I shall lead into the Aldridge paper by reviewing the background which stimulated its production.

In 1893, the NEA Committee of Ten examined the science offerings of 40 schools that they felt were representative of the schools in existence at the time. They discovered that there were over 40 different science courses being offered and that these courses ranged from a few weeks to, occasionally, a full year. As a result of these deliberations, the committee drafted a position statement that must be considered the most significant position ever taken by any agency considering science education. Why? The three major recommendations were adopted across the entire land in what was record time. Their recommendations were:

1. There should only be four science courses taught in the secondary school.
2. These courses should be full year courses.
3. Beginning in grade 9 and proceeding through grade 12, the courses should

be Physiography, Biology, Physics, and Chemistry.

That was in 1893 when the steam engine, telephone and telegraph were the newest technologies and people were experimenting with a horseless carriage. Since then, science and technology have grown by leaps and bounds, but we essentially have the same science curriculum. Oh yes, there have been a few changes. In about 1900, the Commission on College Physics convinced the Committee of Ten that Physics had grown up. It had become a theoretical study and therefore ought to be taught in 12th rather than 11th grade. Then, in 1920, another group decided that Physiography was too technical for 9th grade students and they invented a substitute which was to be less complex and contain fewer technical descriptions and fewer words. They invented General Science, which over the years has become science's greatest storehouse of words.

Now it is 1990, almost 100 years later. Science and technology have invaded every aspect of our lives, yet school science education remains largely unchanged. Now *all* science disciplines have grown up. While science and technology have each day in the past 93 years played a more significant role in our lives, we still teach the same courses in largely the same way. Change is needed and one significant change to the scope and sequence of the school science curriculum that is suggested in the Aldridge paper, is illustrated in Figure 1.

According to the Aldridge paper three fundamental changes are needed to the school science curriculum. These are:

First, earth/space Science, Biology, Chemistry, and Physics should be taught every year to all children in grades 7-12.

Second, instruction should begin with concrete/phenomenological science and systematically proceed through empirical science. From there instruction should move on to abstract theoretical science which would not become a prominent curriculum inclusion until approximately grades 11 and 12.

Third, eliminate all tracking of students until at least grade eleven. (It appears that tracking eliminates continued science study by minorities and women.)

Grade Level	7	8	9	10	11	12	
Subject	Hours Per Week						Total Time Spent
Biology	1	2	2	3	1	1	360
Chemistry	1	1	2	2	3	2	396
Physics	2	2	1	1	2	3	396
Earth/Space Science	3	2	2	1	1	1	360
Total Hours Per Week	7	7	7	7	7	7	
Emphasis	descriptive; phenomeno- logical		empirical; semi- quantitative		theoretical abstract		

FIGURE 1. Proposed Example of a Revised Science Curriculum for Grades 7 through 12 in the United States.

IMPLICATIONS FOR SCIENCE TEACHERS PREPARATION

Preparing teachers who can implement a model that calls for science courses that are to be extended over six years will offer a new challenge. However, it is patently obvious that all sciences, not only physics, have theoretical/abstract dimensions. All sciences also have concrete and empirical dimensions. Hence Aldridge's model offers a logical structure for delivering all science to all students.

Teaching the middle level student presents yet another problem. There is ample evidence suggesting that the successful middle level teacher is a unique individual who must daily deal with students who are dynamos of change, and who have the capacity to act age 3 and age 21 seemingly within moments. Therefore, other curriculum models are emerging that are consistent with the Aldridge paper. One such model is represented in Figure 2.

According to this model, one teacher teaches all four sciences in grades 7 and 8. Grade 9-12 science, e.g., biology, chemistry, earth science, or physics, would each be taught by discipline specialists. Figure 2 presents one variation

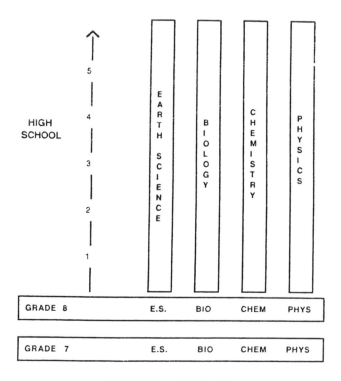

FIGURE 2. A SSC Model

of the theme, but it is by no means the only or best way to restructure the science curriculum. The best way will only be uncovered after considerable research and what may be the best way in School A could spell disaster for School B.

For example, what should be done in small schools with only two science teachers? There are many small schools where new teachers may essentially become half or even the entire science department and teach a variety of science courses. Preparing teachers for small school teaching assignments demands that the teachers receive a broader science education. A solution for a two teacher school would be to offer two four-year long science courses in the secondary school. One course could be a Biology/Earth/Space Course and the other course could be a Physical Science Course and teachers could be prepared with joint majors. This pattern, which is illustrated in Figure 3, might even be a more effective means of delivering science instruction because it would cause fewer scheduling problems and it could well contribute to breaking down the barriers among disciplines.

Another significant NSTA initiative, its Science Teacher Certification Program, was designed in response to findings which reported that a large number of our nation's science teachers are teaching science courses they never intended

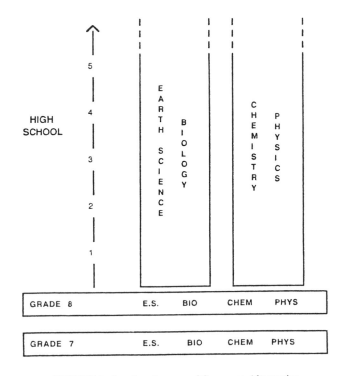

FIGURE 3. Another Scope and Sequence Alternative.

or prepared to teach. Furthermore, it was felt that offering such a certification would be a way to recognize that there are people who are certified to teach science and those who are qualified. Unfortunately, both are often paid the same. But perhaps if we could separate them, the separation might lead toward greater recognition of the gifted and talented teacher who has worked harder to become prepared to teach students science.

A growing number of science teachers are applying for NSTA certification because it attests to the fact that one is truly prepared, rather than simply certified. The initial NSTA Certifications intentionally defined the minimal level of competence and education needed. The success of this program has been so encouraging that NSTA has begun development of a Professional Level Certification to offer the most highly qualified professionals.

As we talk about certification, thoughts about what we are preparing teachers to teach must enter our minds. These thoughts always lead me to think about the tools we provide the beginning teacher. The number one tool continues to be the textbook and all who have read Tyson-Berstein (1988), or followed the reviews published in *Bookwatch,* realize that textbooks, for the most part, have become a series of one-liners without adequate explanations and without challenges to students to ever use higher level thinking skills. Alan McCormick begins his exciting science magic performance by bellowing *"BORING, BORING"* and proceeds to demonstrate just the opposite. I've always had the urge to bellow *"SCIENCE TEXT-SCIENCE TEXT"* and see how an audience reacts. The essential point is this: teacher preparation should help teachers use effective tools effectively. We must provide teachers the effective tools, good textbooks or the next generation of effective interactive electronic media, and help them develop the knowledge, skills and understanding to use them well.

The changing demographies also must be addressed. At the present time, 80% of the practicing scientists and engineers are white males. Between the year 1985 and 2000 only 15% of the people entering the workforce will be white males. A much larger number of white males will retire, leaving as many as 700,000 unfilled science, mathematics, and engineering jobs in the USA alone by the year 2010. To cope with this problem, we will have to prepare teachers to do something we have not been able to do. We must prepare them to make science friendly to minority and female students.

The real significance of this problem lies in the fact that scientists and science teachers come from the same pool of people. Today it is common to find a science teacher leaving the teaching profession for a career in business and industry where demand is high, salaries are higher and the work load often easier. The interface between science and science teaching careers is typically crossed at least once by science teachers during their undergraduate educations and professional life. As the pool of people entering science and engineering shrinks, who will be available to teach science to our teachers? Will science teachers have access to science courses that meet their needs? It is highly probable that the science teaching profession will feel the impact of the shrinking scientist population to a greater extent than other science professions. It will be necessary to develop instructional technologies and teachers capable of creating learning opportunities for much larger groups of students than at present.

Currently, education has the lowest level of capital investment of any major industry. Education provides only $1,000 in capital for each student whereas the average capital investment made by other industries is $50,000 per worker. (Perelman 1990). This data forces the conclusion that education has spent far too much time continuing the old labor intensive practices and not enough time searching out technological solutions that are less labor intensive. New teacher education programs will, of necessity, need to prepare people who are capable of using technology and managing learning opportunities for several hundred students.

"IF IT AINT BROKE, DON'T FIX IT."

No one seems to be arguing that it isn't broke but many seem to be saying that it cannot be fixed. Fixed it can be, but not without a lot of effort. Here are some of the most obvious changes which are necessary.

1. Responsibility for science teacher preparation must be shared equally by science educators, practicing teachers, and science professors. Shared equally means that each person has teacher education as a part of their defined assignment. Teachers never have enough time for their teaching and science professors never have enough time for research so adding teacher education to their already burdensome workload generally means something gets short-changed. Generally the something is teacher education. Here are some things that each must begin doing consistently.

Practicing science teachers must know the research knowledge base and demonstrate the use of this knowledge base in making decisions about which teaching strategies they are using with a given target audience on a given day. We must move beyond the point where we use a given teaching strategy because we did it that way last year or that is the way we learned it. That is, the teacher in training must see the practicing teacher draw from a knowledge base in making teaching decisions. The knowledge base will stay on campus until that time when the novice teacher sees experienced teachers go to it, use it, and thereby be an effective teacher.

College science professors must also draw upon the science teaching knowledge base in making teaching decisions. People teach as they are taught. Often the only teacher model a preservice teacher sees in science courses is a lecturer who assumes the students know nothing and that it is their job to fill students up with as much "good knowledge" as possible. College science professors must become good models of effective science instruction. A single science methods course, even a great one, cannot undo sixteen years of experience. Students, even very young students, learn that memorizing leads to success in school. Once the memorizing path to success is learned, learners are not interested in learning other ways to learn. Why should they? Who argues with success?

Science educators must also change. I think our role ought to be to assist both the college science professor and the practicing teacher to access appropriate science teaching knowledge bases as efficiently as possible. I argued earlier that practicing teachers and scientists both have full-time jobs, hence science educators should try to make the involvement of these persons in science teacher preparation as time effective as possible.

2. Science teaching must become activity oriented and learner centered. It has now been convincingly demonstrated that the two most important characteristics of learners is what they know and their cognitive self-esteem. It is also known that learning is an active process in which the learner must be engaged. In previous years much of the focus in teacher education has been placed on preparing teachers who master this or that model—teachers who look like good teachers. Now it is known that the teacher must first engage the learner, discover what the learner knows, and then provide the learner the next engaging activity. Success must be measured in terms of student, and not teacher, performance.

3. More technology must be employed. Perlman's (1990) statement that the average industry has a capital investment supporting its employees which is fifty times greater than the capital investment education provides its teachers is revealing. Have we, as he argues, avoided innovation and fought to preserve old, outmoded ways? Frankly, if Perlman wasn't correct, we would be trying to fix rather than completely restructure. Apparently we have begun to think about extensive applications of technology in education. For example, the computers are now in the schools. However, they are commonly only one to a science classroom and used for other than instructional purposes (word processing, grade books, stockroom inventory), when they should be used to collect data and other activities that facilitate higher order thinking in science instruction.

Another reason for employing more technology in the teaching of science is related to the pending shortage of science teachers. The time we spend arguing for smaller classes could probably be better spent thinking about how we can deliver more effective instruction to larger numbers of students while preserving the value of the small class environment. Studies of teachers continue to indicate that they spend most of their time as information providers, which is a very inefficient use of teacher time. Of course, if all one is doing is providing information, it is just as effective to provide information to 1000 or more as it is to provide information to 24.

4. Students who do not abandon the traditional teacher-based definition of learning for the learner-based definition should not be certified to teach. The

traditional cultural transmission approach (Rogers 1969), which focuses most attention on perfecting teacher performance, must be replaced by approaches that recognize that learning is not passive, that the student is not an empty vase, and the teacher's role is much more than lecturing. Rogers (1969) called for learner-based learning and the research supporting this paradigm switch is overwhelming (Cleminson 1990; Mandler 1990).

Learner-based learning places the student, not the teacher, at center stage. When instruction is learner-based it begins with a question which is, "What does the learner know?" Once that is ascertained, the teacher proceeds to help the student learn (Novak 1990) by providing the student the appropriate next experience. Learner-based instruction also carries with it the conviction that all students can and will learn and that all students, given appropriate experience, will modify and gradually perfect their conceptions. The teacher thus becomes an arranger of conditions that permit the student to become actively involved in learning.

5. It is well-documented that tests drive the curriculum. Students do study for tests, and they should. However, the tests should be worth studying for and few existing standardized tests are. Tests must measure the attainment of significant performances and of higher order thinking. This is important for several reasons. First, interpretations of recent research continue to suggest that young children are capable of conceptual thought very early in life. (Mandler 1990). It has been suggested that instruction and testing that involves higher order thinking skills ought to begin in kindergarten. (Resnick 1987).

One important note is the fact that performance testing is growing up. More and more people are convinced performance testing is not only possible, but it is absolutely necessary. Douglas Reynolds (1990) of the New York Department of Education is able to point to performance tests administered to fifth graders that are having a dramatic effect on what is being taught. Science, good science, is now frequently taught regularly throughout the elementary grades. Why? Because science performance is being measured and reported to the public. FACT! Tests drive the curriculum. Let's make sure it is driven in the right direction.

6. Science instruction must be made more attractive to females and minorities. One way to do this is to replace competitive teaching models with cooperative learning. It appears that white males, but only white males, thrive on academic competition. Women and minorities tend to prefer cooperation to competition. Cooperative learning which provides a friendlier learning environment is advantageous for women and minority students without interfering with the success of the white males (Bean 1985).

This is but the tip of the iceberg. Moving from teacher-centered to learner-centered instructional models cannot be taken lightly. Teachers will teach as they were taught. Beginning teachers will also tend to adopt the teaching strategies used by the experienced, successful, teachers they encounter in the real world. The reality is that we must restructure the schools. For example, future teachers must experience innovations like cooperative learning first hand. And, experiencing it once in a science teaching methods class will not be enough. They must experience cooperative learning in their elementary, middle level, secondary, and college classes, including their collegiate science classes.

7. Teachers must become researchers. They must knowingly collect data on what they are doing to enhance the opportunity their students have to learn. The education knowledge base needs to come off the shelf and become embedded within the teacher, facilitating the teacher's decision making capabilities. Reflective teaching based upon an understanding of the literature and first-hand study of the teaching-learning process is a key to effective science education.

SUMMARY

The author began this chapter with the certainty that teacher education could be fixed. The focus has been on what needs to be done to fix it!

ACKNOWLEDGMENT

The author, Hans O. Andersen, acknowledges the editing assistance of Gayle Reiten, an advanced graduate student at Indiana University, Bloomington, Indiana.

REFERENCES

Aldridge, B.G. 1989. Essential Changes in Secondary School Science: Scope, Sequence, and Coordination. *The NSTA Report.* January-February 1989.

Andersen, H.O. 1990. Reality is a Rut. *The Hoosier Science Teacher,* Vol. XV, No. 4. May 1990.

Bean, D.B. 1985. *Mathematics and Science: Critical Filters for the Future.* The Mid-Atlantic Center for Race Equity. The American University, *BOOKWATCH,* Published by the National Center for Science Education, Inc.

Cleminson, A. 1990. Establishing an Epistemological Base for Science Teaching in the Light of Contemporary Notions of the Nature of Science and of How Children Learn Science. *Journal of Research in Science Teaching,* Vol. 27, No. 5, pp. 429-445.

CORETECH. 1990. *Meeting the Needs of a Growing Economy: The COR-ETECH Agenda for the Scientific and Technical Workforce.* Council on

Research and Technology, 1735 New York Ave., N.W. Suite 500, Washington, D.C. 20006.

Holden, C. 1989. Wanted: 675,000 Future Scientists and Engineers. *Science*, Vol. 244, p. 1536-1537, 30 June 1989.

Mandler, J.M. 1990. A New Perspective on Cognitive Development in Infancy. *American Scientist*, Vol. 78. May/June 1990.

Matyas, M.L., K. Tobin, and B.J. Fraser. Eds. 1989. *Looking into Windows: Qualitative Research in Science Education.* American Association for the Advancement of Science.

McBay, S. 1990. *Education that Works: An Action Plan for Education of Minorities.* Project Director, Quality Education for Minorities Project, Cambridge, Mass.

Novak, J. 1990. Helping Students Learn How to Learn: A View from a Teacher Researcher. A paper presented as the opening address of the Third Congress on Research on Teaching Science and Mathematics. Santiago de Compostela, Spain. Sept. 20, 1989.

Resnick, L. 1987. *Education and Learning to Think.* National Academy Press.

Perlman, L.J. 1989. Learning Revolution. *CHALKBOARD.* Indiana University School of Education, Alumni Association. Fall/Winter 1989.

Perlman, L.J. 1990. *The "Acanemia" Deception.* Hudson Institute, Indianapolis, IN.

Pope, M. and J. Gilbert. 1983. Personal Experience and the Construction of Knowledge in Science. *Science Education.* 677(2):193-203.

Tyson-Bernstein, H. 1988. *America's Textbook Fiasco: A Conspiracy of Good Intentions.* Council for Basic Education.

Tyson, H. 1990. Reforming Science Education/Restructuring The Public Schools: Roles for the Scientific Community. A background paper prepared for a meeting sponsored by the New York Academy of Sciences and the Institute for Educational Leadership. March 1990.

Vetter, B.M. 1989. Manpower Data Need an Overhaul. *The AAAS Observer.* 7 July 1989.

APPENDIX

Science Education Initiatives for the 1990s

Prepared by the
National Science Teachers Association
1742 Connecticut Avenue, NW
Washington, DC 20009
(202) 328-5800
September 7, 1988

To resolve the problems inherent in science education today, four areas require immediate initiatives:
1. Teacher preparation and staff development
2. Curriculum development
3. Instructional support
4. Research and dissemination

I. Preparation and Staff Development Initiatives

- Development of research-based preservice teacher preparation programs for elementary, middle, and high school teachers that are designed cooperatively by science educators, scientists, and practicing classroom teachers of science.
- Implementation of staff development programs for teachers of science who have a need to reinforce or enhance their science knowledge and science teaching skills.
- Recruitment of a greater number of highly qualified and competent individuals into science teaching (especially minority populations) and retention of these people in the science teaching profession.

II. Curriculum Development Initiatives

- Development and implementation of more unified, in-depth, hands-on science curricula for preschool, elementary, middle/junior high, and high school students.
- Development and utilization of evaluation and assessment tools that measure student achievement of higher order thinking skills.
- Production of materials designed for instructional administrators and lay people (e.g., principals, superintendents, school board members, and parents) that would provide better understanding of science education needs of students.
- Implementation of curricula for preparing science laboratory technicians to assist teachers.
- Development of curricula that would instruct teachers on the appropriate uses of technology in the classroom.
- Development of curriculum models that integrate science with the learning of other elementary school subject matter areas.

III. Instructional Support Initiatives

- Provision of appropriate electronic technologies to science teachers at all grade levels.
- Provision of funds for the construction of adequate science teaching facilities (e.g., activity centers and laboratories).

- Development of regional science centers that would make the following available to local teachers:
  A. models of effective teaching practices
  B. science updates
  C. research opportunities
  D. media
  E. science equipment and supplies

IV. Research

- Establishment of long-term funding for regional science education research centers that would conduct and disseminate research on:

  A. designs of science-teaching facilities
  B. appropriate uses of technology
  C. science curriculum for all students
  D. instruction
  E. science teaching practices that are taking place outside the United States

Science Education in the United States: Issues, Crises and Priorities. Edited by S. K. Majumdar, L. M. Rosenfeld, P. A. Rubba, E. W. Miller and R. F. Schmalz. ©1991, The Pennsylvania Academy of Science.

Chapter Seventeen

POSTSECONDARY SCIENCE FOR NONSCIENTISTS: STS, THE NEW APPROACH TO THE LARGEST POPULATION

FRANZ FOLTZ[1] and RUSTUM ROY[2]

[1]Doctoral Candidate
STS Program
Rensselaer Polytechnic Institute
Troy, NY 12180
and
[2]Evan Pugh Professor of the Solid State, and
Professor of Geochemistry
STS Program
The Pennsylvania State University
University Park, PA 16802

THE FAILURE OF POSTSECONDARY SCIENCE

We must begin this chapter with a discussion of the gross failure of post-secondary science education. Think about the ghetto dweller in the Bronx, or if you prefer, the usual cocktail party crowd in New York or Washington. Or take George Bush and his Cabinet. Could these citizens follow the issues involved in global warming, BGH use, or the asbestos controversy at the level they need for effective action? Could these citizens vote intelligently on the supercollider or whether to permit RU486, or encourage estrogen replacement therapy to be introduced in the light of new findings? We know the vast majority of citizens including the latter groups noted above, most of whom graduated from college, cannot and do not feel comfortable or competent dealing with such issues. The question is, Who failed? The "students?" Their teachers, the science faculties? So-called "science" courses for nonscientists in college compose no more and no less than an "entitlement program" for a few departments (who have erogated to themselves a monopoly on the word "science"). Mandated science courses earn them large numbers of student credit hours. This makes it possible for research-oriented science faculty to teach the minimum number of courses and spend most of their time writing proposals and, as time allows, doing research. It also enables science departments to hire graduate students to help teach the larger enrollment required courses.

But that is only the first of our failures. Let us summarize what is wrong with the *system* of teaching science to nonscientist. In the typical major university:

1. It is a job nobody wants to do, cares about or is specially rewarded for. Hence it is done unimaginatively—large lectures typically without recitations and labs.
2. The subject matter taught is ludicrously narrow and irrelevant to the interests of most students. Why do we teach mainly "abstract science,"—the notorious PCB's of education (Physics, Chemistry and Biology) as though they were all of science? Why not the more applied sciences: earth, materials, medical. By what reasoning does one conclude that it is important to teach a student about pulsars, when they haven't the foggiest idea of how their lights, stoves, or cars work. Who concluded that technology was less important to most citizens than science, not only from a utilitarian viewpoint but from the view point of esthetics, from learning to reason and connect reason to reality.
3. The "science" taught is not connected in any way to the science that every *citizen* needs to understand: acid rain; rain forest destruction: job offshoring, etc., the science they need to participate in society.

We cannot make any progress on improving the situation until we recognize clearly who has to be blamed for this situation. We believe one can show that

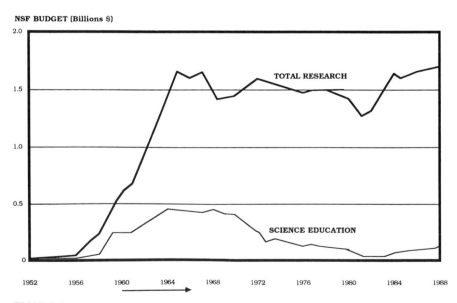

NSF BUDGET (Billions $)

FIGURE 1.

it principally has been the PCB science community itself. The first contribution has been the community's single mindedly focusing on getting more money for research at the expense of science education. Figure 1 shows how even while the NSF budget *expanded tenfold* the percentage the National Science Board gave to *science* education *dropped tenfold.*

Second, in the most successful sleight-of-tongue since Jacob and Esau, the science community successfully befuddled the public (including the congress) by using the term "science" when talking of the *benefits* of technology (jobs, comfort, security), but made sure that funding flowed to science *alone.* Thus, U.S. technology has not been the beneficiary of huge R&D budgets.

A major concern of STS includes the reform of present day education which has led to technological illiteracy for the vast majority of the population in the U.S. Regrettably, the recent emphasis in reform proposals has been on "science education" for the pre-professional cohort group. The goal is epitomized as "more and better scientists." Yet it is *this* wrongheaded approach that has brought our capacity to utilize this area of knowledge to a critical condition in our country!

Lastly, interest and attention has been lacking from the science community for the science education of nonscientists, even after the enormous ballyhoo about the parlous state of science education. Prof. Frank Westheimer of Harvard, in his acceptance speech for the 1988 Priestley medal of the American Chemical Society (its highest award), noted that when he approached his colleagues to take on the task of teaching "science" to Harvard non-scientist undergraduates, they "didn't really want the responsibility of teaching the unwashed". If Harvard undergraduates are "the unwashed," in what class are Roxbury ghetto dwellers or philosophy majors at Northern State College? The fact is that in the vast majority of post-secondary institutions this task of teaching the unwashed is relegated to the lowest on the academic totem pole. The fault lies clearly not in the stars but in our comtemporary scientist selves, obsessed with fundable and publishable research.

Repentance or Ruse?

Tides ebb and flow. In the last two years, the tide of the single-minded pursuit of "research" in the University *appears* to have turned. Teaching is said to be in; the single goal of more and better scientists as the nation's greatest need appears to have been overturned. Whether this is a genuine change of goals or merely a ruse to buy time till research regains favor in congress will have to await the event.

The new direction is towards "technological literacy for the masses", "science for all", "citizen empowerment in a technological culture". The following recent quotations mark this turning point.

We must now begin to make the case for a stronger and more sustained

Such specialization on the part of scientists clearly carries the potential to cause many societal problems. Since in reality all things are interconnected, a scientist could because the neglect of other concerns often causes major negative effects on a part of society outside his or her area of understanding. Science, in its practice qua science, inherently must focus on the details of the small picture, and ignore any broader social connections. Three books, C.P. Snow's *The Two cultures* (1959), Ellul's *The Technological Society* (1964), and Rustum Roy's *Experimenting with Truth* (1981), address this potential for disaster.

Snow shows that the world has become divided into two communities or cultures, a scientific culture and a traditional non-scientific one. Neither group understands the other well or even knows how to interact and communicate effectively with it. The scientific culture controls the world's technology. The two cultures create the potential for disaster because of their inability to work together. Somehow these cultures must be brought back together. The only solution Snow offers is rethinking the educational process.[7]

Ellul looks at how "technique," a broader definition for technology that includes methods of doing as well as devices, has gained control of society. "Technique" is now leading society to the ends desired by mass values. Society has no effective way of controlling it. The average individual can no longer understand technology and therefore has lost any ability to control it. Nor can society as a whole any longer stop technique's progress. "Technicians," those that perform the tasks of technique, do their jobs without concern for society. They have no moral standards by which to make decisions about technology. Ellul calls for an affirmation of religious values to prevent these inevitable technological disaster.[8]

Roy specifically defines science as quintessentially reductionist process, and hence, inherently *unable* by its charter to provide a philosophy or worldview. He contrasts it with "religion" which provides the big picture, but cannot and should not concern itself with the details left to science. Hence, science and religion are *hierarchically* different, and science must always be under the hegemony of some worldview.

In the late 1950s and 1960s many other writers dealt with the theme of technologically caused disasters. Writers such as Ivan Illich, E.F. Schumacher and William Barrett have dealt with the grave problems in today's technological world. They reject the "normal" disciplinary approach to understanding technology as unresponsive to the needs of humanity.[9]

Where STS Stands Today

Science, Technology and Society—STS—has become the flagship of the movement for *integrative general education.* By definition STS is the most integrative subject matter one can define in higher education. In many ways it is the reinvention of the University within the "Multiversity." It is the very antithesis of the "distribution" requirements *among* the disciplines. It is the structured unification of these disciplines into a new melded general education—which is what the *Uni*versity was all about in the first place. Figure 2 attempts to portray this graphically.

STS as an intellectual field started as a response to the questions raised by various critics of science and technology's effect on society. Ellul's seminal work *The Technological Society* marked the kick-off of this campaign. But in the quarter century since that book was published (in English), STS has become a focal point in interdisciplinary general education in U.S. postsecondary education. STS pedagogy provides a structure to look at the grave problems that have arisen due to the compartmentalization of knowledge. Societal problems always cut across academia's arbitrary divisions of fields of knowledge. Addressing them requires input from many areas. STS examines these problems using an interdisciplinary study approach.

From a pedagogical and epistemological viewpoint, STS represents a unique change in western academic development. Instead of prolonging the fissiparous tendency built into academic science, STS attempts to synthesize the widest possible range of disciplines. "Macro-STS" is a new Gestalt or paradigm breaking down the dominations of the university paradigm of specialization and departmentalization of knowledge.[10]

The emergence of the STS movement has created the potential for a major paradigm shift in education. A radically new way of looking at the relationships between science and technology is developing. This new look also puts science and technology into the historical context of worldwide human society.

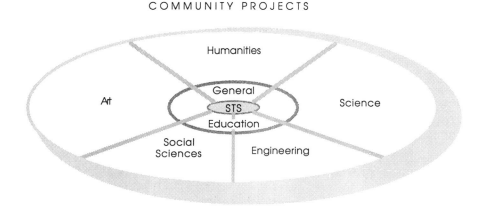

FIGURE 2. STS Forms the Interactive Core within General Education.

Knowledge is considered in integrative interdisciplinary terms instead of the present rigid, ever more specialized disciplines. Values, morals and concerns for the larger society are dealt with along-side of scientific equations.

STS responds to a study of the real problems facing the world by proposing an integration of knowledge. It is the systematic study of the interactions of science, technology and society outside of strict disciplinary bounds. It looks at science and technology as two independent but related social forces and examines how they impact each other, government, policy-making, business and everyday life in myriad ways. It uses biology as well as business, art as well as physics, engineering as well as philosophy. It is assumed that all disciplines have information of potential value to society.

STS constitutes a major advance in holistic thinking as it looks at societal implications as well as scientific laws. It calls for moral guidance or humanistic understanding of all work in science and technology. It reminds us that ignoring moral and humanistic standards is itself a choice with far-reaching implications. Mindful of this, STS seeks a thorough reform of scientific education as now practiced.[1]

The rapid increase in interest and support of STS in the United States and abroad shows that the STS "paradigm shift" has started and can probably not be stopped. STS has grown into formal programs on over a hundred college and university campuses, and into courses at two thousand. Most of these programs developed independent of any curriculum review process, RFP, special funding, etc. They were "grassroots" approaches to dealing with an important concern. Each has their own diverse objectives and reasons for why they have appeared.[2]

The unrest of the 1960s, particularly in the United States, helped to focus attention on the more humanistic elements of postsecondary education. Outrage with the Vietnam War manifested itself in an anti-technology theme. Professors and graduate students formed groups to study technology in its social context.[3] It was out of this renewed interest in science and technology that the STS concept emerged at the college level. By 1976 this new conception of the intrinsic interrelations between scientific and technological advancement, and the broader context of change in society as a whole, spread to the point where 178 STS-related programs existed in 108 separate institutions in the United States (Heitowit, 1977).[4] The 1980s have been the refinement years for university STS programs, through the start of liberal arts college program via the New Liberal Arts project, the introduction of STS to K-12 education, and the creation of the National Association for STS (NASTS). In a study of the origins of STS, Foltz[5] has shown the differences in motivation, structure, etc., which led to an increasingly coherent field. What is relevant in present context, is the fact the STS has become an alternative approach to teaching nonscientists the amount and kind of technology and science and science that most are interested in, can handle, and can put to use in their lives.

STS Origins in Postsecondary Education

Concerns over the role of science and technology in various aspects of society arose at the various universities through different media, and the exact concerns and aspects that were important to the programs differed among the universities. The actual concerns dictated the structure of the program and the composition of the faculty. These in turn helped define the problems and obstacles encountered by each program. STS programs with a strong origin in a single discipline or department tended to keep their affiliation with that discipline and structured themselves within that department. Those that had a strong interdisciplinary or multidisciplinary origin tended to be outside of the regular university power structure. Most of the interdisciplinary programs had and continue to have problems dealing with faculty with primary responsibilities to other departments, with the lack of resources and with their support infrastructure. The primary reason for the birth of STS was faculty interest. It can be assumed that the atmosphere of the late 1960s and early 1970s fostered faculty concerns parallel to the student unrest.

It took dedicated individuals to carve a place for STS in highly disciplinized universities. The arrival of new faculty helped in many cases. A person without previous ties in the university, was often the catalyst in STS's origin whether this person was a new professor in a department or a new dean open to innovative ideas. They came from outside the university's political structure and could perform with more freedom than those that were caught in it. Sometimes they were brought in just to create such a program. (*University of Washington, St. Louis*).

A second major factor to the development of an STS program was the existence of modest outside funding. Not all programs received outside funding, but a significant number owe a lot to it. The Sloan Foundation, with its focus on technology/society/policy established outside of the normal university structure, alone gave out $8 million dollars in 25 grants to start STS type programs.[16] The Sloan Foundation's, NASA, NEH and other funding help STS programs to weather the conflict between disciplines and interdisciplinary education. STS programs did not have to rely solely on university support, and so, had greater ability to shape an innovative interdisciplinary structure. STS owes a lot to these organizations.

More recently a few STS programs have had substantial university support as deans and/or planning committees saw STS as a means of improving education, particularly general education offerings. They easily recognized the link between STS and general education. STS provided a viable method of meeting institutional goals for general education (e.g., *St. Louis University*).

Another important initiator of STS programs has been the existence of earlier programs that dealt with *part* of what is now included in STS. Earlier programs or symposia leave footholds on which to build. In some cases the loss of programs, especially general education ones, have left a void that STS filled. Also, university evaluations of their programs in many cases, have pointed to a need for STS.

STS programs have been staffed mainly by Liberal Arts faculty. On the whole engineering and the natural sciences supplied a third of the faculties compared to about two-thirds from the social sciences, arts and humanities. This happens when the program emerges out of a social science look at technology, i.e., policy analysis (*George Washington University*) or values and technology (*Stanford University*). It can also happen because a technical institution's non-technical faculty wants to teach more than just introductory courses; STS can be seen as a way of being more useful in the institution (e.g., *Case Western Reserve*).

An STS program's structure reflects how it was conceived. A program inside a department usually originates within that department and takes a more disciplinary appearance. A separate interdisciplinary program usually emerges from differing backgrounds. Department or college status happens because of a greater perceived need for and acceptance of the program within the institution. This status usually affects the structure very little but does correlate to better funding and university support. The University structure then dictated the major obstacles to the program.

The main obstacle to the health growth of STS programs has been the primary "loyalty" of faculty to their home departments. The weak interdisciplinary structure that exists in virtually every American university does not allow for the hiring of faculty unique to the needs of the (STS) program.

Minuscule budgets leave all such cross cutting programs at the mercy of the very departments they could help, but which see them as competition. Most are forced to survive on outside funding. Opposition also comes from other faculty who see STS as a threat because of the outside funding it received, the faculty time taken up, and/or the attraction of students. Many universities use student attendance in determining funding levels. Any program that takes away a share of the students also takes away funding (e.g. *Virginia Polytechnic Institute*). In spite of *all* the national reports urging cooperation, integration, interdisciplinarity, etc., many faculty just have not thought through their rationale for dealing with the interdisciplinary concept.

The interdisciplinary nature of STS can cause problems. Many at the university level find it hard to think along disciplinary lines. The whole American university structure, including hiring, promotion, and rewards, is so completely and rigidly structured along disciplinary lines that STS threatens this fundamental university paradigm.

Other major obstacles come from administrative issues. Lack of funding typically allows for only part-time administrators of STS programs. Sometimes the university administrator has little understanding of the research needs of the faculty (*West Virginia University*). On the other hand, in some cases large amounts of money from outside sources can cause problems in administering the program, as the internal structure is not structured for its management (*St. Louis University*).

Yet even with all of these obstacles inherent to the university, the structures can be manipulated by astute faculty. The need for the various STS programs grew out of the desire to examine relevant social issues. Each group saw the value in examining how science and technology affects our society. Unfortunately, interdisciplinary programs have had difficulty fitting into the university structure of departments and disciplines.

The difference between the STS paradigm and the single-discipline professional paradigm is stark. The two have radically different goals and methods. STS wants to give citizens the power to act in today's society. To do this they need a new kind of integrated knowledge. The professional paradigm wants to keep special knowledge locked up with the specialists or professionals in the university. Yet, it appears in 1990 that the "paradigm -shift" is underway. STS, as Roy has pointed out, is a "megatrend".[17] It cannot be stamped out nor controlled because it has nucleated everywhere at once. Integration and interdisciplinarity are slowly being accepted in theory and practice.

Program Specific Analysis

STS exists at the college/university level in many forms. It can be just a single course, an interdisciplinary program, a center of learning, a department, or even a college. No matter what the actual structure, STS deals with issues concerning the interaction of Science, Technology and Society. STS has made a strong beginning with its foundation in many of the country's most prominent universities and colleges. A survey of these will provide the reader with valuable hints on where STS stands today in U.S. postsecondary education. It should be noted that not all of these programs identify themselves as STS programs. For example, *SUNY at Syracuse*'s program is environmental science not STS per se. All these programs deal with STS themes and most try to use a strong interdisciplinary approach. The Brookings Report[18] classified them all as STS-related.

The Programs run the gamut from Science and Engineering Policy, Environment programs, and History or Philosophy of Science and Technology, to pure Interdisciplinary STS. Most programs are unique, specially designed for their institution. A few grew out of early work at another school. Faculty moved or were consulted from already established programs. We describe below several such and include their goals to show better commonalities and differences.

Interest in STS first appeared at *The Pennsylvania State University* in the 1960s. Faculty members from diverse fields found that they shared a common belief in the need for an interdisciplinary approach to contemporary problem solving. One of the areas that strengthened the belief was the Science, Technology and Values aspect. Harold K. Schilling, Professor of Physics and then Dean Emeritus of the Graduate School, had a national reputation from two books authored from the point of view of a scientist concerned with the philosophy of Religion. Maxwell Goldberg, Professor of Humanities and Director of the Center for Continuing Liberal Education was a nationally known speaker and had organized symposia on the Impact of Technology on Human Values. Joseph Kockelmans, Professor of Philosophy, was concerned with the philosophy of science and its relation to religion. Rustum Roy, Professor of Geochemistry, had authored a book on ethics and was active in policy at the National Council of Churches, chairing its Science , Technology and the Church committee. He was also involved with science policy through the National Academy of Sciences, and the Pennsylvania Governor's Science Advisory Committee.[19]

In 1968, students and faculty met in a series of dialogues concerning the status of American society, and developed the idea of new courses. Professor Roy presented the question of creating such courses in a program to the Penn State Faculty Senate in the Fall of 1968. Senate members were initially enthusiastic and gave verbal encouragement. The initial concept was perceived as university wide general education at the undergraduate level and the broadening of technical curricula at the graduate level. To keep the courses' ownership from any one discipline and to note their universal value, the original course designation was *Univ.* 4XX.[20] The 4XX denotes junior/senior level courses that graduate students could also take for credit. The Univ. designation was later changed to STS.[21]

A faculty committee was established to discuss how to implement such a program. On December 15, 1969 the convener of this committee sent a proposal for a "Program on Science, Technology and Society" to University President Walker. This was forwarded to the Provost on January 12 and approved by the President on February 2, 1970.[22] By 1971 an undergraduate STS program was formally in place.[23] The Vice-President appointed Dean Theodore R. Vallance as the first chairperson. The program consisted of seven core courses, all at the 400 level (junior/senior).

Four factors led to the *Stanford University* conception of the *Values, Technology, Science and Society* (VTSS) program in the department of Humanities: 1) the sense that technology, a pivotal factor in social change, was not being reflected in departmentally organized undergraduate curriculum, 2) the lapse of general education requirements in the late 1960s in the context of Vietnam War protest, 3) the sense that there was a need for a new kind of liberal education adequate to the nature of the contemporary era, 4) the need for a forum to air responsibly issues related to the alleged charge of engineering collaboration in the war effort.

In 1971, 10 faculty from *Stanford University* founded the *Program in Values, Technology and Society* (VTS).[24] They used their Human Biology (teaching) program, an earlier interdisciplinary program, as a model. The leaders of the original VTS faculty group consisted of Eric Hutchinson, Chemistry, Stephen Kline, Mechanical Engineering, William Clebsch, Religious Studies, Philip Rhinelander, Philosophy, and Walter Vincenti, Aeronautics and Astronautics. The university's administration did not play a role in the formation of VTS. Though they supported VTS as it offered a new direction for a previously dismembered general education effort.

Edward Wenk, Jr., Professor, Engineering (Civil) and Public Affairs, originated the Program in Social Management of Technology (SMT) at the *University of Washington*. Wenk, a science policy expert from the federal government, came to the *University of Washington* in 1970 to offer a graduate seminar in science policy. The university president and the Dean of Engineering approved the proposal for a trans-campus department SMT Program. In June 1973 the Board of Regents gave approval.

SMT provided a focus for teaching, research, and public service in technology intensive public policy. It was created to train experts to deal with technology and public policy methods in federal, state and local agencies. Initial funding came from state funds and a grant from the Sloan Foundation. The Sloan grant made up a critical 40% of the operating budget. Still, institutional support of SMT is substantial. Faculty are hired directly by the program. The most important factors for the program's success include a strong program concept, faculty strength and public identity, including good visibility around the country.[25]

In the summer of 1968, Robert Morgan visited *Washington University* (St. Louis) and helped set plans for an *International Development Technology Program*. This program was developed with strong administration support with an emphasis on international development. In the fall of 1969 a grant was received from NASA to do interdisciplinary research on applications of communication satellites to educational development in the U.S.. Lack of international development research funding and the growing concern about the impact of technology in the U.S. prompted a reevaluation of the international thrust of the program. The program took on a domestic development emphasis.

In the fall of 1971, a masters degree program was initiated with a domestic thrust. This program was to serve students with either engineering and "hard" science backgrounds or social science backgrounds. In 1972 a bachelor's degree program was started. A grant was received from the Alfred P. Sloan Foundation in 1974 that permitted expansion of the faculty and assignment of people primarily to the program.[26]

Four key people to the program were Robert Walker, McDonnell Professor of Physics, James McKelvey, Dean of Engineering, who provided critical support, George Hazzard, Vice-President for Research, and Lattie Coor, Head of the Office of International Studies. There was no initial interest by the engineering departments in the program. Yet, the university administration played a key role in the program's formation.

The program had the following five objectives: 1) to provide a holistic education that showed the "big picture" STS views; 2) to provide a significant understanding of the technology-society interface; 3) to give marketable skills to students above the conventional ones; 4) to produce literate and articulate students in scientific and technological concepts; and 5) to provide a system understanding of STS on both the national and global level.[27]

The program was changed to the *Department of Technology and Human Affairs,* in the School of Engineering in December 1976, the program received significant institutional funding. Outside support was fundamental during the early years and remains substantial. The department has its own faculty that is awarded and promoted on the basis of teaching, research and service.[28]

The question of how best to improve the humanities/social science core in science and engineering education at *Rensselaer Polytechnic Institute* motivated the origin of the Center for the Study of the Human Dimensions of Science and Technology (HDC). Particularly, it was a 1970 administrative committee in the School of Humanities and Social Science (HSS) on how to improve undergraduate education that created HDC's interdisciplinary structure. In the Fall 1973 a planning committee made up of representatives from each school at *Rensselaer Polytechnic* and from each department in HSS was formed. Tom Phalen, Dean of HSS (History/Theology), Charles Sanford, Professor of Literature and Communication, and Robert Baum, then assistant Professor of Philosophy were instrumental in laying the groundwork for HDC.

HDC initially had the support of the majority of the faculty and the administration, who saw it as responding to the liberal education needs of science and engineering students and as being potentially beneficial to local business and industry. While HDC did not receive any direct institutional funding, departments donated faculty time and outside funding paid for the director. The program was not dependent on outside funding but had its scale of operation set by it. In July 1974, the A.W. Mellon Foundation gave $200,000 for the development of HDC. Due to the volunteer nature of the faculty, HDC's size fluctuates from year to year. The size of the faculty for the 1975-76 academic year was nine tenured and eight non-tenured members.[29]

The Department of Engineering and Public Policy at *Carnegie Mellon University* was an outgrowth of the joint efforts of the engineering school and School of Urban and Policy Affairs to educate engineers to work at the technology/society interface. The program emphasizes technology/public policy studies. The curriculum was originally designed as a sociotechnology curriculum which would "provide an integrated group of courses from the three significant areas: (1) the humanities . . . (2) understanding of and competence in . . . sociological, economic and policital structures and methods . . . and (3) . . . basic engineering."[30] It is a degree-granting unit that has a separate budget and has the right to hire its own faculty and to grant them tenure.[31]

SUNY College of Environmental Science and Forestry set up a separate interdisciplinary *Graduate Program in Environmental Science.* The structure kept the curriculum broad. The Brock Report (1975) noted that the SUNY College of Environmental Science and Forestry was designed:

> *To educate persons* for professional service in environmental science including development of knowledge and understanding of techniques, principles, and relationships concerning environmental systems and their components; and in the design, management, and operation of reconstructed or modified environmental system for human benefit.
>
> *To conduct research* in environmental science and its supporting disciplines and arts.
>
> *To collect, organize, and communicate* information relating to environmental science.
>
> *To perform* the above functions for and on behalf of the general public endeavor.[32]

The mission of the graduate program was to provide transdisciplinary education and train students to be effective environmental professionals through the use of:

(a) multidisciplinary approach-recognition of the necessity to approach environmental problems with input from several disciplines and professions;

(b) holistic perspective-awareness of and defense to the interdependence of elements within broadly defined ecosystems, including physical, biological, social, and economic systems; and

(c) topical grounding-competency to understand and apply the principles of a particular subject of environmental inquiry, in sufficient depth to interact with other disciplines and professional fields.[33]

Faculty interest sparked by "The Role of Liberal Arts in Engineering Education" (Walter Lynn, 1976) and long range planning by the Vice President for Academic Affairs led to the *Program in Science, Technology and Society* at *Michigan Technological University.* The department of Social Sciences housed the new program. This tended to reduce input from other fields and created a conflict between the four members of the program's faculty and other traditional social scientists. This conflict and disputes that arose over "ownership" of and rights of participation in the program became the main obstacles to the program.

Two documents, *Long-Range Planning: Michigan Technological University Committee C Report* (1977) and *Role Statement: Department of Social Sciences* (1978) were instrumental in the development of an STS program at *Michigan Technological University.* The needs of technological and scientific understanding in general education were stressed in the Long-Rang Planning Report. The report suggested that a program should be developed that stressed the impact of science and technology.[34] This call was taken up by the Department of Social Sciences which took on the responsibility for providing courses that dealt with STS issues in association with other departments.[35]

Interest in STS issues at Purdue University grew out of faculty concern with the role of science and technology in almost every aspect of our lives and interest in technological ethics explored through philosophical, sociological, historical and public policy perspectives around 1969. *Purdue University* set up an interdisciplinary program that tried to help students understand and deal with technology and science in this rapidly changing world. Their goal was to learn to use science and technology as instruments to achieve human ends, rather than as ends in themselves. Only through interdisciplinary methods could they accomplish this by giving a truly global view of society.[36] Key faculty members came from the schools of the Humanities, Social Sciences and Education. Set up as a separate program, Curriculum in Science and Culture had great flexibility in recruiting faculty for short periods, giving the program a broad scope. Problems came because faculty were only voluntary, with primary responsibilities to other departments which took priority over the program.

Western Connecticut State University has a capstone STS course for their various science programs that focuses on the interrelationships among a science field and STS, i.e., Biology, Technology and Society. Students engage in the process of analysis of problems and synthesis of solutions to STS issues from their background field of science. The following objectives are given for the course:

Students will participate in the communication process.

Students will understand the nature of science, its methods of inquiry, its strengths and its limitations.

They will understand the forces that shape institutions and societies and human behavior within them.

They will reflect on the human condition through an understanding and discussion of significant ideas and values.

They will understand and apply quantitative concepts.

Finally, they will develop attitudes and behavior that promote physical and mental health.[37]

Faculty and administration interest in exploring the societal context of science and engineering led to the STS Program at *Lehigh University*. In 1972 Lehigh University received a National Endowment for the Humanities (NEH) grant[38] that was used to start a program with the goal to "create educational experiences which bring humanistic perspectives to the application and evaluation of technology."[39] The original name of the program, the *Humanities Perspectives on Technology Program* reflected the emphasis on the humanities. The scope of the program quickly expanded into the social and natural sciences. In 1979 the title of the program was changed to *Science, Technology and Society* to reflect this broader domain.[40]

The Program's separate interdisciplinary structure was important as it did not place the program in competition with the departments. While the non-departmental status of the program allowed for course development, it made research difficult.

R.A. Rosenbaum, Professor of Mathematics and Acting Provost, and Earl

D. Hanson conceived the *Science in Society Program* at *Wesleyan University* out of a course for non-science majors they taught together, entitled "Ideas and Inquiry in Biology and Mathematics". The program was set up as a separate interdisciplinary department. While its structure was solid, stable staffing proved to be a major obstacle. Yet, some faculty members proposed establishing the College of Science in Society[41] because they believed "that sciences in general and the life sciences in particular have a central role in responding to questions *prior* to 'the power and greatness of man': the Who am I? Where did I come from? Where should I go? questions that have challenged man since before he ever knew pretention."[42]

STS at the Liberal Arts Colleges differ in many respects from the comprehensive universities and polytechnic institutes. The majority of their STS programs originated under funding from the Sloan Foundation New Liberal Arts program. It is useful though to take a quick look to see how these few example programs compare to the university STS programs described above.

The *Vassar* program emerged out of the 1970 brainchild of a few faculty members for a single course that would critique technology. A seed grant from the National Endowment for the Humanities (NEH) allowed two faculty to teach this course. An additional NEH funding allowed for two more courses to be developed. Success of these programs led to a Sloan Foundation Grant to start a program during the 1971-72 school year. The program had four upper-level courses that were team taught. By 1973-74 the program's name was changed to STS and its structure was designed for two divisions: Critical Thought and Urban Studies. A crises of low enrollment in 1981 led to major reform of the program. A NLA grant was used to change the program's alignment to a more "Hooray for Technology" outlook.[43]

Carleton College established the Carleton Technology Policy Project in 1981 through a Sloan Foundation Science, Technology and Public Policy (STPP) Grant. The program was designed to allow a small number of students to explore in detail a single issue. In the past, each year, six students were given a fellowship during which they usually had a summer internship dealing with that year's issue and then would spend the fall doing policy oriented research.[44]

The Program in Technology Studies at *Davidson College* was the result of a Sloan Foundation NLA Grant during 1981-82. The envisioned plan was to create a third year introductory course followed by a variety of upper-level in-depth courses.[45] Two levels of courses emerged: an Introduction to Technology and a group of Special Topics courses. By 1986 the program had developed four major components: Faculty Workshops, Summer Special Projects, Faculty Leaves, and Public Symposia.

Wellesley College developed its interdisciplinary *Technology Studies Program* with support of a NLA Grant in 1983. A curricular component was established that included infusing technology into existing courses and developing new courses. The program includes 18 courses, an internship program with the

Smithsonian Institution and college-wide symposia.[46]

SUMMARY

Three major attitude changes are imminent on the horizon for what has been called "science education" on the college campus. First, there will be more emphasis on (and reward for) general education that includes teaching some kind of "science" to *nonscientists*. Second, what is included under "science" for "non-scientists" will move away from the "PCB" (physics, chemistry and biology) towards including more applied science (earth, materials, medical) and technology. Third, the integrative principle of teaching science to technology and human concerns and values, which are expressed in the "STS" approach, will take over a substantial portion of the "science education" of most citizens. The postsecondary institution is enormous and has a great deal of inertia, but STS is a megatrend; its growth is likely to be exponential.

Despite the enormous growth and increasing importance of STS, the only comprehensive attempt to locate every STS program in the nation was done in the mid-1970s by Ezra Heitowit at Cornell University. The Heitowit Report (1977) listed over 100 STS type programs nationwide. It gives very useful information but is now over a dozen years outdated. In a Spiegel-Rosing and Derek de Solla Price also examined STS studies programs in 1977 in their book *Science, Technology and Society: A Cross-Disciplinary Perspective,* which looks at the social and policy dimensions of STS studies internationally.

Since this period only a sporadic handful of attempts have been made to keep track of the growing STS field. Among these are:
- A 1978 AAAS publication examining *Ethics and Values in Science and Technology,* which is a national directory of this subset of STS programs.
- A Paul Durbin book entitled *A Guide to the Culture of Science, Technology, and Medicine* examines the specific area in STS studies that deals with medical issues.
- A 1981 UNESCO *World Directory of Research Projects, Studies and Courses in Science and Technology Policy* dealing with a subset of STS studies.
- A Brookings Report, *The Status of STS Activities at U.S. Universities* by Rustum Roy and Joshua Lerner (*Bulletin of STS*, vol. 3, 1983), drew upon the Heitowit report to gather more in-depth information on those programs.
- Likewise a project headed by John Wilkes at Worcester Polytechnic (1987) overviewed a handful of previously-known well-established STS programs to provide models for institutions to use in developing new STS programs.
- *Graduate Education and Career Directories in Science and Engineering, Public Policy* (AAAS, Dec. 1980) surveyed the science policy subset of STS programs.

Examination of these reports and our own present overview survey lead to the following tentative conclusions:

•Current postsecondary science education has failed to capture the interest of most students and to import much knowledge or understanding to the majority who are forced to take these required courses. Improvements at the margin *cannot* help the system reach its true goal.

•Science, qua science, or as a liberal art as permitted in most colleges simply does not interest most students. The vast majority of citizens, from the President to the plumber can and do manage life very well without it.

•Technology which is almost universally ignored in postsecondary education is both more relevant and more interesting to most students.

•STS is not the total content of postsecondary science. But it has to be a major component of any attempt to teach something about technology (vastly more important) and science to every student.

•STS is by far the most effective approach to both getting the typical U.S. college student interested in T&S, and to retaining a life-long interest in such topics as citizen-activists. Nothing else has surfaced in 20 years.

•STS is especially significant in capturing the interests of the "science-averse" students—including women and minorities. In this respect it is the strongest ally for those interested in increasing the pool from which to draw S/T professionals.

FOOTNOTES

1. Hurd, Paul DeHart, "Perspectives for the Reform of Science Education", *Phi Delta Kappan*, Jan. 1986, p. 353.
2. Bledstein, Burton J., *The Culture of Professionalism*, W.W. Norton & Co. Inc., New York, 1976, p. 290.
3. Ibid., p. 298.
4. Ibid., p. 327.
5. Price, Derek J. De Solla, *Little Science, Big Science,* Columbia University Press, New York, 1963, pp. 63-91.
6. Roy, Rustum, "Interdisciplinary Science on Campus - The Elusive Dream," *Chemical and Engineering News,* Vol 5, August 29, 1977, pp. 28-40.
7. Snow, C.P. *The Two Cultures,* 1959.
8. Ellul, Jacques, *The Technological Society,* Vintage Books, New York, 1964.
9. Waks, Leonard J., *STS Education and Citizen Participation,* Science, Technology and Society Program, The Pennsylvania State University, April, 1987, p. 4.
10. Vanderburg, Willem H., "Macro-STS: The New Frontier?", presented at the Third TLC in Washington, D.C., Feb. 5-7, 1988.
11. Walker, Robert, untitled, unpublished, 1987.
12. "New Groups for Science, Technology and Society", *C&EN* March 1987, p.6.

13. Waks, p.5.
14. Heitowit, Ezra D., *Science, Technology and Society: A Survey and Analysis of Academic Activities in the U.S.,* Cornell University, Ithaca, N.Y., July 1977, p. 2.
15. Foltz, Franz, "Origin of an Academic Field: The Science, Technology and Society Paradigm Shift," unpublished, 1988.
16. *Planning Document,* Program in Engineering & Public Affairs, Carnegie-Mellon University, June, 1976. p. 4.
17. Roy, Rustum, "STS: The Megatrend in Education," The Proceedings of the 1984 International Congress on Technology and Technology Education, Pittsburgh, PA, October 8-10, 1984, p. 1.
18. Roy, Rustum and Lerner, Joshua, "The Status of STS Activities at U.S. Universities," *Bulletin of STS,* Vol. 3, 1983, Perganom Press, Ltd., pp. 417-432.
19. Bowers, Raymond, Kranzberg, Melvin A. and Tribus, Myron, *The Science, Technology and Society Undergraduate Program at Penn State:* A Background Paper prepared for the External Evaluation Committee, February 1978, pp. 1-4.
20. Ibid., pp. 3-4
21. Heitowit, p. 107.
22. Bowers, et. al., pp. 4-5
23. *Science, Technology and Society Program,* The Pennsylvania State University, pp. 1-2.
24. Adams, J. and Mann, V., "The Stanford Program in Values, Technology, Science and Society", *Conference on the State of Science, Technology and Society Programs in Western Europe, North America and Australia,* October 10, 1987, p. 1
25. Heitowitt, pp. 85-87.
26. Morgan, P. Robert and Hill, Christopher T., *The Technology and Human Affairs Program at Washington University (St. Louis),* Frontiers in Education Conference, 197X, p. 153.
27. "Technology and Human Affairs", *Washington University Magazine,* Fall 1977, p. 25.
28. Heitowitt, pp. 93-95.
29. Ibid., pp. 115-117.
30. *Planning Document,* CMU, pp. 1-3
31. Ibid., p. 5.
32. Smardon, Richard C., *Evolution of the Graduate Program in Environmental Science,* pp. 1-2.
33. Ibid., p.5.
34. *Long-Range Planning: Michigan Technological University Committee C Report,* 1977.
35. *Role Statement: Department of Social Sciences,* Michigan Technological

University, 12/19/18, pp. 1-5.

36. *Science and Culture at Purdue,* program brochure.

37. Objectives, Rationale and Course Outline, Draft, Western Connecticut State University, 1987.

38. Cutcliffe, Stephen H., *The Emergence of STS as an Academic Field,* March 20, 1986, p. 7.

39. Gallagher, Edward J., *Humanities Perspectives on Technology, Annual Report Year Five, 1976-77,* Lehigh University, Bethlehem, PA, 1977, p. iii.

40. Cutcliffe, Stephen., *STS Studies at Lehigh University,* revised version of paper read for the History of Technology's Annual Meeting - Milwaukee, WI, October 15, 1981, p.1.

41. Baker, Jeffery J.W., Barbara and Hanson, Earl D., *The College of Science in Society (CSiS): A Proposal,* Wesleyan University, Revised 12/1/73, pp. 4-8.

42. Ibid., p. 3.

43. Tavel, Morton, "The Origin and Evolution of Vassar College's Multidisciplinary Program in Science, Technology and Society (STS)" *Conference on the State of Science, Technology and Society Programs in Western Europe, North America and Australia.*

44. Casper, Barry M., "The Carleton Technology Policy Project", *Conference on the State of Science, Technology and Society Programs in Western Europe, North America and Australia.*

45. Brockway, John P., "Technology and the Liberal Arts", *Bulletin of Science, Technology and Society,* Vol. 6, 1986, pp. 240-245.

46. Ducas, Theodore W., Grant, James H., and Schuchat, Alan. "Medical Technology and Critical Decisions: An Interdisciplinary Course in Technological Literacy", *Bulletin of Science, Technology and Society,* Vol. 7, 1987, pp. 71-77.

Science Education in the United States: Issues, Crises and Priorities. Edited by S. K. Majumdar, L. M. Rosenfeld, P. A. Rubba, E. W. Miller and R. F. Schmalz. ©1991, The Pennsylvania Academy of Science.

Chapter Eighteen

SCIENCE AND TECHNOLOGY: A LIBERAL ARTS COLLEGE PERSPECTIVE

DAVID W. ELLIS[1,3] and MARION S. ELLIS[2]

[1]President and Director
Museum of Science
Science Park
Boston, MA 02114-1099
and
[2]Former Biology Teacher
Writer and Editor

INTRODUCTION

The viewpoint of this paper is that a person cannot claim to be well-and liberally-educated today unless that person has sufficient background in science to be an informed citizen, to ask intelligent questions related to science, and to make judgments on matters of science that affect that person and the populus as well. It is clear also that, if this country is to maintain a reasonable standard of living in the future, we must have both a well-informed electorate, literate in science, and a well- and broadly-educated workforce. In brief, education in science—more particularly, mathematics; theoretical, experimental, and descriptive science; and technology—must be a major component of the liberal arts curriculum for all students, regardless of major.

In addition, this paper will assert that the differences in aims between students preparing to be professional scientists and those preparing to be informed citizens but workers in other fields should be adequately reflected in the philosophy of the curriculum.

The latter category of students, those preparing to be informed citizens but workers in other fields, will be designated as "lay scientists". This term "lay scientist" is in contrast to the more common term of "non-scientist". The term

---

[3]Former President of Lafayette College.

"non-scientist" is consistent with a deficit model, one that describes these students by what they lack. The term "lay scientist" is consistent with a competency model, one that describes these students by what they can achieve. Such a reorientation within the distinction between the two categories of students, professional and lay, is crucial to the development of a curriculum philosophy that meets the needs of all students.

For clarity, it should be noted at this point that the concept of the lay scientist is different from that of the amateur scientist, a person who undertakes scientific work as a hobbyist, a non-professional. All future citizens need to be lay scientists; amateur science is a matter of inclination.

Finally, in this paper attention is given to the special case of students who will become the elementary and secondary teachers of not only future college students but also the general population. A profile of these students would show varying characteristics of all three categories mentioned above.

Background

The roots of a liberal arts education in the western tradition are found in the concept of the broadly educated person. Included in that breadth has always been an element that today would be called science. In ancient times philosophers were interested in the nature of the universe; in the Middle Ages, mathematics, astronomy and geometry were part of the curriculum.

The scientific revolution of the late 1600's and the 1700's led to the introduction of natural science, or science education as we know it. Suggestions began in the iconoclastic ferment of the 18th Century from people such as Franklin and Jefferson for the inclusion of "useful knowledge"; (or the "mechanic arts", as they were called in the 19th Century; or "technology" as it is called today). Since that time, whether to include in a liberal arts curriculum not only the "pure science" of more traditional roots but also technology (or for that matter other forms of "useful knowledge" as well) has been a point of some tension, particularly in many liberal arts colleges.

In the last decade, the place of math, science, and technology in our society, and thus in our educational systems, has been a topic of considerable urgency. The concern has at least two aspects: on the one hand, because our society has become more dependent on technology, work will require increasingly more people well-educated in math, science, and technology. On the other hand, studies comparing the achievement of U.S. children with that of children of other nations on standardized tests of math and science show that U.S. children rank near the bottom on most tests. With the U.S. falling behind in competitive world-wide markets today, many people fear that future generations of workers will not be properly prepared for the work place and world in which they will be living.

The situation is summarized in the report *Science for All Americans* published by the American Association for the Advancement of Science in 1989: "The terms and circumstances of human existence can be expected to change radically during the next human life span. Science, mathematics and technology will be at the center of that change—causing it, shaping it, responding to it. Therefore, they will be essential to the education of today's children for tomorrow's world."[2] If one accepts that proposition, then in the curriculum of liberal arts colleges, clearly, the study of math, science, and technology should have a prominent place.

These concerns over the last decade or two have led to many suggestions for the teaching of math, science, and technology at all levels. At the college level, two efforts are worthy of special note. In 1981 the Sloan Foundation initiated a program entitled, "The New Liberal Arts" which focused specifically on liberal arts colleges and embraced the view that the liberal arts experience should include, in addition to the traditional content, significant components on the applications of mathematics and science in society. Second, the AAAS published in 1990 another major report entitled, *The Liberal Art of Science: Agenda for Action,* which calls for treating science (broadly defined to include math, science, *and* technology) as a liberal art and for devoting more time to science within general education. A number of college courses currently in use are described in the report.[3]

Since the aims of the education for the science major and for the lay scientist are different, there should be educationally sound differences in their training experiences. Unfortunately, in many colleges and universities, the general education science courses for the future science laity have tended to be referred to, not totally unfairly, as "watered down" courses for science majors. Such courses may be easy to offer both for the faculty and for the institutions, but these courses fail to meet the full range of needs of the future lay scientist. At the same time, obviously there must also be on-going concern for nurturing those students who are our future professional scientists. An increase in effort on the one front should not imply a diminution of effort on the other.

The rest of this paper expands upon the concept that liberal arts colleges need to develop a philosophy for their science curricula that meets the differing needs of students, and briefly discusses some of the implications thereof.

Educating the Laity

If we were to consider a science curriculum for the student preparing to be a lay scientist as deserving a philosophy of its own, what would it include? First, every course must help the student to understand that science is a way of thinking, a disciplined way of approaching questions. It will not do to reduce this merely to a static description of what is commonly referred to as "the scientific method", often formulated as follows: state the problem, form the hypothesis, design and conduct the experiment, state the conclusion. To present this important material in some such simplistic or rigid manner, a practice which unfortunately can be found at all levels of introductory science education, not only ignores the significant variations of methodology required for descriptive and statistical sciences in contrast to experimental sciences, but also usurps a more meaningful exploration of the scientific mode of thought.

An example of a more far-reaching formulation—indeed, one that is useful in all of life, not just the laboratory—is that of Wendell Johnson.[4] Johnson described the scientific orientation as "a process versus a static notion of reality, adaptability versus rigidity, a language of clarity and validity, and finally, and very importantly, disciplined behavior in wording questions so that they are answerable."[5] He analyzed scientific thinking as employing three simple questions: "The first (for clarity) is, 'What do you mean?'; the second (for validity) is, 'How do you know?'; the third (for disciplined generalization) is, 'What then?' "[6] Within a more broadly conceived framework such as this, scientific methodologies become more understandable and usable by all students.

Second, there is a place for a meaningful laboratory experience in the education of a future lay scientist. The laboratory should be designed to expose the student to the skill of scientific thinking in action, but again not just as a series of steps to be followed. Accurate observation, devising controls when possible, logical deduction, valid inference, and clear communication of procedure and conclusions, all are important elements to be demonstrated and experienced. Current methodology and instrumentation and the role of mathematics and statistics in scientific work are also important to include.

In addition to instilling an understanding of the scientific way of thinking and giving experience with it, the course or courses for the preparing lay scientist should introduce a body of scientific knowledge, nomenclature and theory, in at least one traditional scientific discipline through which the student will come to see how investigators in the past have approached scientific problems and how they currently approach them. To include some history of the discipline is necessary if the student is to learn how a field evolves, and to include current work in the discipline helps the student to see the cutting edge, to experience the sense of exploring a frontier. This is the place to learn how scientists today approach discovery and validation.

Furthermore, this course of study should demonstrate how the discipline relates to other scientific and non-scientific disciplines. In this way the student begins to place the discipline in cultural perspective as well as historical. If examples are chosen well, they will illustrate some of the more exciting areas of scientific endeavor.

Finally, the student should be introduced to applications of the discpline under study. Since the future lay scientist will need to consider in later life issues of public policy, which frequently will involve the economic and environmental outcomes of the application of science, some introduction and consideration of application is essential. In addition, the courses should help students to appreciate that we do not live in a world of clear and simple issues, and that questions of ethics, international affairs, finance, and many other competing perspectives often must be considered as well as "pure" science and related technologies in determining societal priorities. The choices as to which applications to include in the curriculum will depend naturally on the discipline and the interests of the instructor; however, selections that relate to current issues will help to maintain a high level of student interest.

The development of a special philosophy for the education of lay scientists would constitute a major change in thinking about curriculum development. For example, it is likely that a minimum of five or six semester courses in math, science, and technology would be required, an increase of 50% to 100% for many institutions. According to Westheimer, however, this would still be less than was the norm in the mid 19th century.[7] Furthermore, such a change obviously would require shifts of priorities at all levels in order to find the time, the resources, and the commitment to design and implement the undergraduate academic experience specifically to meet the goals outlined above.

The faculty selected to teach this curriculum must develop special characteristics. They must have reasonable depth in the discline, *and* reasonable breadth across related disciplines, *and* knowledge of some applications of the discipline. They must be expert at communicating science as a way of thinking and at explaining how technology relates to our lives. This is asking for much more breadth than almost any graduate student or even most post-doctoral fellows would have today. In fact, since most new faculty today are very highly specialized, few of them would qualify.

Liberal arts colleges need a new approach to the task of obtaining faculty to teach courses for the science education of the future laity. One approach might be to develop summer programs or academic year sabbatical programs which would supplement the basic science education of experienced faculty. With the rapid advance of the scientific disciplines, and as some institutions increase the pressure for research, many experienced faculty may prefer the new challenges of specalized teaching as they move through the various stages of their careers. For some, however, the thought of expanding their spectrum of knowledge and interest in science into new areas, including the application of science, could be very attractive. For some purposes, team teaching with faculty from the humanities would be appropriate.

Such development obviously would require recognition and significant funding either from the institutions or from federal funding sources, such as the National Science Foundation. Institutionally, such changes will be expensive in personnel, time, continuing education, equipment and facilities. Although the sums involved are substantial, not to address more effectively the scientific general education of future citizens and workers would be short-sighted and unwise.

Educating Science Majors

The basic difference between the philosophy underlying the curriculum for the preparation of the lay scientist and that for the training of the professional scientist is linked to the basic difference between the goals of each. The would-be lay scientist is preparing to understand enough of science to be an informed citizen, and to understand enough about how scientists think and work to continue to evaluate what scientists communicate to them about their on-going work. The would-be professional scientist, on the other hand, is preparing to think and work as scientists think and work, within some specific content focus, *and* to communicate his or her work not only to professional peers but also to the lay public, the latter being a facet of professional development that we ignore at our peril.

It should be noted at this point that not all students who major in a science will become professional scientists; however, provided that the curriculum of the science major includes as much breadth and application as it should, these students should also be receiving adequate preparation to be informed laity.

Although the curriculum for science majors would include the same basic components as that for non-majors, the science major needs to become *skilled* in the scientific way of thinking and should have much more exposure in depth and breadth to the facts/theories/applications of a particular discipline than would the non-major. In all likelihood the applications would be more numerous and different and the laboratory experience would be very different; that is, much more extensive and designed with the assumption that the student operates as planner, implementer and evaluator.

At present, applications receive little attention, not because they are not important, but frequently because many faculty do not have first-hand experience with applications; thus the student is encouraged to focus almost exclusively on theory and its development. This prepares students well for graduate school but not well if the student should elect to go into industrial product development, industrial sales, some interdisciplinary areas and many applied areas of research such as environmental studies, or teaching. Many students have expressed the view that their education seemed disconnected from the exciting and challenging activities and issues that they read about in newspapers or magazines or hear or see on television. Students need to be helped to see how their work will eventually lead to something they believe will be "interesting".

Faculty to teach science majors in the type of curriculum suggested will need to be more broadly educated than are many new faculty leaving graduate schools today. Faculty will need to have a broader base of experience. Some approaches to meeting these needs might be the loaning of scientists by corporations to liberal arts colleges, opportunities for liberal arts college faculty to work in industrial or governmental laboratories during vacations and possibly on a part-time or consulting basis during the academic year, or team-teaching by college faculty with professionals from industry and government. Conferences and workshops may be another vehicle.

Science in the Elementary and Secondary Schools

There is a special case of students in a liberal arts college that requires special attention. This is the student who will become a teacher at the elementary or the secondary level. Obviously these college students have a special importance because they will provide the precollege preparation of future liberal arts students as well as the preparation for all the other post-secondary options. For that portion of the population that does not go on to further study after high school, these future teachers will provide the only science preparation some lay scientists receive. In view of the key function of these individuals, the lack of special attention given to their preparation on the part of liberal arts colleges and universities, and the lack of recognition of their importance by cultural and governmental institutions is stunning.

The actual situation today is a downward spiral. Many faculty observe that students come to college ill prepared to pursue careers in science. The reality of this is well documented in many studies.[1,2] In addition, many fewer students than just a decade ago have an interest in taking science courses, much less in majoring in science.

The aforementioned report, *Science for All Americans,* cites many different facets of the problem at the elementary and secondary levels. These include a lack of academic preparation in science for teachers, teaching loads too great for even well-prepared teachers to do a good job, text books and methods that actually impede science literacy, lack of the values that foster creativity and intellectual achievement. The report does not find, however, that there has been a dearth of effort; rather, the efforts are frequently misguided: "The present curricula in science and mathematics are overstuffed and undernourished ... Some topics are taught over and over again in needless detail; some that are of equal or greater importance to scientific literacy—often from the physical and social sciences and from technology—are absent from the curriculum or are reserved for only a few students."[8] Until the faculty of liberal arts colleges, as well as other institutions of higher education, recognize their obligation to do a better job of preparing the teachers of the very students they themselves complain about, it is unlikely that this downward spiral will be broken.

It is probable that in the elementary grades the younger the children the teacher teaches, the less likely the teacher is to have majored in science. For those teachers, their college education as lay scientists is particularly important and ideally would include an added component of a special commitment to the importance of early childhood science education, both for eventually preparing the lay public and for potentially inspiring future professional scientists.

Those primary teachers who do major in science as undergraduates, and all secondary science teachers who must do so are not likely to consider themselves to be professional scientists, although a few may be; however, many of them are amateur scientists, some of considerable note. Their example should be recognized; and opportunities should be fostered for them, and for others who might grow to follow their example, to practice science in some meaningful way.

There is the need to help science teachers at both the elementary and secondary levels through in-service work. Many colleges have already formed partnerships with local school districts through which teachers from different schools and college faculty have the opportunity to exchange ideas and undertake joint projects.[9] In addition, programs such as the NSF sponsored institutes for teachers that had such a positive impact in the 1960's should be reinstituted. The institutes would meet real needs of teachers to become current in their fields, to learn new techniques, to be reinvigorated, to help in building support networks, and to counteract the isolation that can exist for many primary and secondary teachers. Furthermore, if the colleges were to treat these in-service elementary and secondary teachers with respect as partners in a joint enterprise, the concurrent generation of college students would gain a respect for teaching in general and science teaching in particular without which, ultimately, the system cannot succeed.

Summary

The challenges are before us: the current situation calls urgently, even demands, significant change in the science curriculum of libral arts colleges and in the methods for teaching it. Solutions to these problems will require reorienting the general education component for future lay scientists away from a deficit model, toward a competency model; away from the current watered-down-major philosophy, toward a positive philosophy that reflects the vital importance of preparing a public of responsible lay scientists. The curriculum for science majors needs greater breadth and more real-life applications and better training in communicating to the lay public. Future elementary and secondary teachers must be prepared in new ways with a much greater emphasis on science as a way of thinking and experiencing, not merely as a body of facts and theories to be memorized. These changes will require of the institutions development of faculty reward criteria, marked increases in funds for faculty development, and the will on the part of all to focus on the issues and take appropriate action.

REFERENCES

1. Starr, Kenneth. Science Literacy: The Issue, pp 229-239. In *The Sourcebook*. American Assocation of Museums, Washington, D.C. 1990.

2. *Science for All Americans,* American Assocation for the Advancement of Science, Washington, D.C., 1989, Foreword.
3. *The Liberal Art of Science,* American Association for the Advancement of Science, Washington, D.C., 1990.
4. Johson, Wendell and Dorothy Moeller, *Living with Change,* Harper & Rowe, New York, 1972.
5. Ibid. p 54.
6. Ibid. p 37. Johnson's parentheses
7. Westheimer, Frank H., Education of the Next Generation of Nonscientists, C. & E. News. July 4, 1988, pp. 32-38.
8. Ob cit.(2) pp. 13 and 14.
9. Atkin, J. Myron and A. Atkin, Improving Science Education Through Local Alliances, Carnegie Corporation of New York, 1989, pp. 45-64.

Science Education in the United States: Issues, Crises and Priorities. Edited by S. K. Majumdar, L. M. Rosenfeld, P. A. Rubba, E. W. Miller and R. F. Schmalz. ©1991, The Pennsylvania Academy of Science.

Chapter Nineteen

GRADUATE EDUCATION IN SCIENCE AND ENGINEERING IN RESEARCH UNIVERSITIES

CHARLES L. HOSLER[1] and DEL SWEENEY[2]

[1]Senior Vice President for Research and
Dean of the Graduate School
Acting Executive Vice President and Provost
The Pennsylvania State University
University Park, PA 16802

[2]Special Assistant to the
Executive Vice President and Provost
The Pennsylvania State University
201 Old Main
University Park, PA 16802

INTRODUCTION

At no point in human history has there been a more universal consensus that science holds the key to the survival of society. That the environment is now at a delicate and critical juncture is becoming more widely appreciated. The world's population has increased to the point where we are adversely affecting the natural systems that sustain us. With little forethought or understanding of how the systems of which we are a part are interconnected, we have cleared the forests, tilled the land, extracted and burned fossil fuels, and introduced unnatural concentrations of chemicals into the environment. Our growing numbers have also exacerbated existing public health problems and produced new ones. The fundamental challenge is to improve the material standard of living for the billions of human beings already born and for the billions who will be born in the next century. Such a goal must be achieved without causing further deterioration of the environment and at a pace that will satisfy the expectations of developing nations.

To correct existing problems and to prevent further abuse of the environment, a more comprehensive understanding of natural systems is required. Developing this understanding will require dedication, hard work, and sacrifice on the part of the industrial nations of the world. If these nations are to provide the scientific and engineering base for such an effort, a much larger proportion of students must be willing to dedicate themselves to scientific studies than has traditionally been the case.

The stakes are high in this country as well. U.S. competitiveness in the global marketplace is being challenged and, with it, the standard of living to which Americans have become accustomed. The United States is no longer the acknowledged world leader in science and engineering. Although U.S. expenditures on research and development are, by far, the largest in the world, the rate of investment in non-defense-related research and development is only 1.8 percent of gross national product compared with 2.6 percent in Japan and West Germany![1]

Also of concern is the fact that the general scientific literacy of our population is low relative to that of our strongest competitors. In a recent survey, despite a strong belief that it is important to know about science, a majority of respondents could not correctly answer questions about scientific terms and concepts![1] As the world economy shifts to a "knowledge economy," our work force has lost its competitive edge and is matched or outstripped by Japan and members of the European Community.

A third issue, and the one with which this chapter is concerned, is the supply and quality of trained scientists in the Unites States. Recent studies project significant shortages of new Ph.D.s in science and engineering in the next twenty years. It is urgent, therefore, that graduate education in the sciences and in engineering be carefully evaluated to determine what new directions might be needed in the near future. Graduate education, however, cannot be considered in isolation; it is the last stage of a pipeline that flows upward from elementary school.

Human Resources

The National Science Board projects that by the year 2000 there will be a demand for an additional 600,000 science and engineering graduates. The growth in science and engineering employment is expected to be more than 33 percent, nearly four times the average overall growth in industrial employment of 9 percent![1] Employers will be adding positions related to technological innovation and will therefore need a highly trained work force. In examining the supply and demand for new Ph.D.s a recent estimate projects an annual shortfall of 9,600, or a cumulative shortfall of 153,000 Ph.D.s between 1995 and 2010.[2] Shortages in particular specialties have occurred during the past few years and are likely to continue, but whether there will be an overall shortage of scientists and engineers depends on the proportion of young people, particularly women and minorities, who can be attracted to technical fields.

On the supply side, college enrollments of "traditional" students—those 18 to 21 years old—are now declining, although not as rapidly as the total population of that age group, which has been decreasing since 1981. One reason that college-attendance rates have not fallen more rapidly is the increasing proportion of women attending college, 42 percent of all women aged 18 to 21 versus 38 percent of men of the same age.

The total number of baccalaureate degrees awarded annually in the United States more than doubled from 395,000 in 1960 to nearly 1 million in 1974, and has fluctuated around that level since then. The number of science and engineering baccalaureates increased at about the same rate, from 121,000 in 1960 to 305,000 in 1974, decreased until 1983, and then increased again between 1983 and 1986 to 324,000. In the past two years, however, the number has been declining again. The percentage of science and engineering baccalaureates has not changed over the past thirty years, remaining at about 30 to 33 percent of all baccalaureates.

The small increase in the total number of science and engineering baccalaureates in the 1980s is attributable to an increase in the number of degrees awarded to women. Whereas the number of science and engineering baccalaureates awarded to men decreased from more than 213,000 in 1974 to about 187,000 in 1988, the number of degrees awarded to women climbed from nearly 92,000 to 122,000. The share of degrees awarded to women increased from 30 percent in 1974 to 40 percent in 1988.[3]

If current and future needs for scientists and engineers are to be met, the *proportion* of students choosing science and engineering fields must markedly increase. Moreover, radical changes must be made in the gender and ethnic preferences for careers. Surveys of college freshmen have not been encouraging in regard to an overall change in the proportion of students planning careers in science and engineering. From 1982 to 1987 the percentage of college freshman intending to major in engineering decreased among men from 22 percent to 17 percent and among women from 4 percent to 3 percent. The proportion of freshmen planning majors in the physical sciences and mathematics remains very low—less than 3 percent of all freshmen. In contrast, 26 percent plan to study business.

The number of students who actually receive degrees in science and engineering is much lower than the number of freshmen planning to major in these fields. A survey of the high school class of 1982 found that whereas 70 percent of freshmen with planned majors in business completed the baccalaureate in business four years later, the retention rates for science and engineering fields ranged from a low of 38 percent in the physical sciences to a high of 58 percent in engineering (Figure 1). Freshmen intending to major in these fields are, on many indicators, "high achievers," yet they leave these fields in college at a rate twice that of students entering these fields from other majors.[4]

Students are increasingly aware that, in many fields, they need training beyond the baccalaureate level in order to be competitive. For research positions in industry and for faculty positions in colleges and universities, the doctorate is generally required. The percentage of surveyed freshmen planning to obtain a master's degree increased from 31 percent in 1982 to 39 percent in 1987. The number planning to work for a doctorate rose from 8 percent to 13 percent. Recognition of the importance of graduate study is even greater among freshmen interested in science majors. About 41 percent of the freshmen intending to study the physical sciences planned to obtain a doctorate as did about 25 percent of

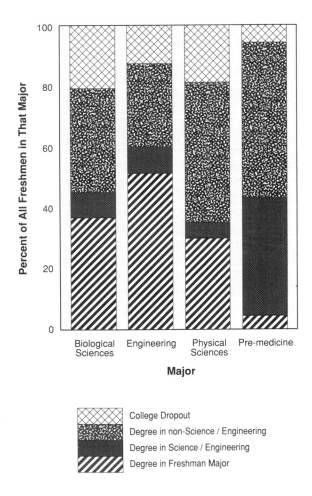

FIGURE 1. Natural science/engineering B.S. degree attainment, by freshman major, 1982 freshmen (Source: Reference 4)

those intending to major in the biological sciences. Currently, of those who receive a baccalaureate degree in a science or engineering field, just 3 out of 10 actually become full-time graduate students.[4]

The number of new doctorates is particular cause for concern. In the 1950s, spurred by the launch of Sputnik by the Soviet Union, U.S. government support for graduate students increased substantially. The number of doctorates awarded in science and engineering tripled, from about 6,000 in 1958 to 19,000 in the early 1970s. Since then, the number has fluctuated, reaching 20,000 in 1988.[3,5]

This apparent steady state masks three trends. First, the ratio of doctorates to baccalaureates is falling. In 1975 the ratio of doctorates produced compared with the number of baccalaureates produced ten years earlier was 1 to 10. The current ratio, based on the baccalaureate class of 1984, is estimated to have fallen to 1 to 20. Second, nearly half of the doctoral students who begin work in science and engineering do not complete their degrees, and few universities have retention programs aimed at graduate students.[4] Finally, the number of doctorates in science and engineering would have declined in absolute numbers except for the doctorates awarded to international students.

Between 1978 and 1988, the number of Ph.D.s awarded in science and engineering to U.S. citizens decreased from 13,086 to 12,847. In contrast, the number awarded to international students with temporary U.S. visas increased from 2,506 to 4,838. The proportion of doctorates in science and engineering awarded to international students increased during this period from 15 percent to 24 percent (Figure 2). In 1988 international students accounted for more than 40 percent of all doctorates in engineering and mathematics.

The number of women receiving Ph.D.s in science and engineering grew from 2,411 in 1975 to 4,114 in 1988, an increase of 71 percent. Of doctorates awarded in the sciences to U.S. citizens, the proportion received by women increased from 19 percent in 1975 to 36 percent in 1988 (Figure 3). Most of this change is accounted for by the greater number of Ph.D.s received by women in the life sciences and the social and behavioral sciences and by a leveling off of Ph.D. degrees obtained by men after 1970. The proportion of engineering degrees awarded to women increased from 2 percent to 10 percent between 1975 and 1988.[5]

The choices of fields made by men and women are significantly different. In 1988, 70 percent of the doctorates received by men were in science and engineering, but only 47 percent of the degrees received by women were in those fields. Eighteen percent of the men completed degrees in engineering as compared with 2.4 percent of the women. The physical sciences were selected by 12.7 percent of men receiving doctorates, but by only 4.7 percent of the women. On the other hand, only 9.3 percent of the men received Ph.D.s in psychology as compared with 14.2 percent of the women.[3]

Minority participation at the doctoral level does not appear to be improving significantly. The number of Ph.D. degrees in science and engineering awarded to Black students has fallen from a high of 288 in 1979 to 231 in 1988 (from 2.2 percent to 1.8 percent of all science and engineering Ph.D.s awarded to U.S. citizens). Doctorates in science and engineering fields awarded to Hispanics rose during the same period from 129 (0.9 percent) to 319 (2.5 percent). The number awarded to Asian-Americans increased from 192 (1.4 percent) to 441 (3.4 percent)! If international students with permanent visas are included, the

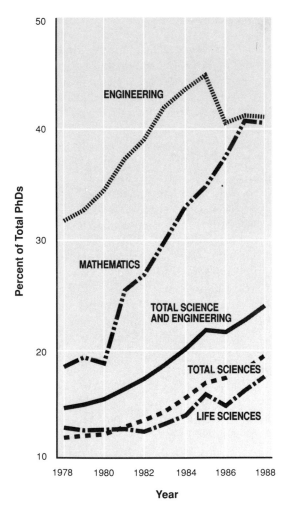

Note: Includes only international students
on temporary visas

FIGURE 2. Science/engineering Ph.D.s awarded to international students, 1978-1988 (Source: Reference 1)

proportion of Asians in 1988 rises to 6 percent in the natural sciences and 16 percent in engineering.[5] Figure 4 shows a selected racial/ethnic distribution of doctorates awarded to U.S. citizens.

Of particular concern is the fact that college enrollments of minority students have been decreasing since the mid-1970s as a percentage of high school graduates. Moreover, for a variety of reasons, minority students are graduating from college at lower rates than majority students. It is therefore not surprising to find that minority students are severely underrepresented in science and engineering majors.

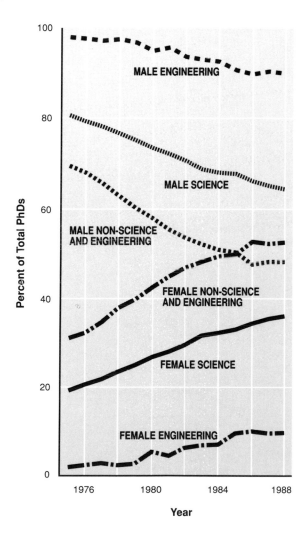

FIGURE 3. PhD.s awarded to U.S. citizens, by gender, 1975-1988 (Source: Reference 1).

Yet the proportion of minorities in the U.S. population is expected to increase sharply in the next generation. It is estimated that minority populations will constitute 33 percent of the school-age population by 2000 and 39 percent by 2020. Between 1985 and 2000, minorities will make up one-third of all net additions to the work force.[6] Increasing the number of minority graduate students—both men and women—is therefore a critical strategy in ensuring the future supply of scientists and engineers.

If the demand for Ph.D.s exceeds the supply, industry, government, and educational institutions will be competing for scientists, and levels of compensation are likely to rise. As a result more students may be attracted to these fields. There are, however, two limiting factors. First, because of the length of the pipeline (it frequently takes as long to train a scientist as it does to train a medical doc-

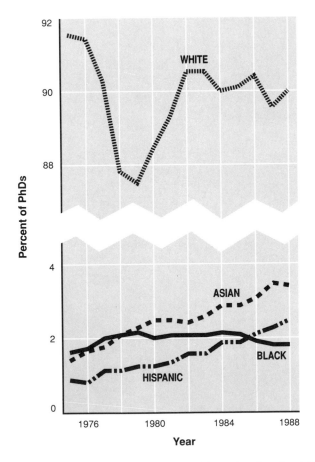

FIGURE 4. Science/engineering Ph.D.s awarded to U.S. citizens, by selected racial/ethnic group, 1975-1988 (Source: Reference 1).

tor), major shortages of scientists may develop before market forces fill the pipeline at the source. The second problem is the level of precollege preparation. Without an adequate grounding in mathematics, chemistry, physics, and biology in high school, students face severe difficulties if they decide to enter a scientific field during their early college years.

What needs to be done? Above all, we must stop the leaks in the pipeline. Students are frequently tracked even in elementary school, and a disproportionately large number of minority students are steered into classes where the pace of work is slower. The result is that minority students often have lower levels of mathematics achievement than majority students do by the time they reach junior high school. While girls have levels of achievement equal to those of boys in elementary and junior high school, their attitude toward science and mathematics is not as positive, and they have fewer science experiences. By the end of high school, girls have lower levels of achievement in mathematics and science than boys do.[7] At the college level, among declared majors in science and engineering, women leave those fields at a higher rate than men. The rate of loss is even greater for minority students.[2]

It has been suggested that race and gender-related differences in achievement and persistence in mathematics and science may be the result of "less access to the positive factors that work in favor of high achievement and continued participation generally."[7] An important initiative at the college level has been the development of special programs that encourage minority and women students to choose science and engineering majors and that provide advising support, both by faculty and by peers. Some successful, intensive programs in mathematics, designed to compensate for inadequate high school preparation, are being adopted outside their originating institutions. National societies such as the Society of Women Engineers and the National Society of Black Engineers have chapters at many universities and provide role models for aspiring engineers. Many universities, too, are introducing programs to sensitize administrators and faculty members to such concepts as the "chilly classroom climate" often encountered by women and minority students.

In addition, more rigorous programs in science and mathematics must be provided at precollege levels. In many high schools, with the exception of magnet schools in urban areas and affluent suburban schools, fewer mathematics and science courses are offered above the basic level than were offered thirty years ago. Students who are inadequately prepared in science and mathematics at the high school level often require a year or more to accumulate the courses needed to enter college-level science and engineering curricula. Thus, more advanced course content is delayed until the junior and senior years. It is the authors' opinion that today's average M.S. recipient has about the same degree of preparedness in a technical field as a B.S. recipient had in 1950. For this reason as well as the enormous growth in the body of knowledge in technical fields, the support of graduate study is more imperative than ever before.

Support of Graduate Students

A critical point in the pipeline is the decision to attend graduate school. Most students need some kind of financial support for graduate study. The sources of such support include direct federal funds (fellowships, traineeships, and research assistantships), institutional funds (fellowships and teaching assistantships), other sources including industry, and funds provided by students and their families. Institutional sources include a significant amount of indirect government funds. Figure 5 shows changes in the number of graduate students in science and engineering supported from various sources. In 1988, full-time graduate students in science and engineering reported the following sources of primary support: federal sources - 20 percent; institutional sources - 44 percent; self support - 27 percent[!]

In the category of federal sources, a significant change has occurred in the type of support over the past twenty years. In 1969 there were 60,000 federally funded fellowship for graduate study and 20,000 research assistantships (a ratio

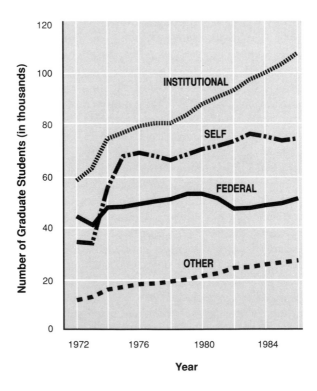

FIGURE 5. Major sources of support, science/engineering graduate students in Ph.D.-granting institutions, 1972-1986 (Source: Reference 4).

of 3 to 1; in 1989 there were only 14,000 federally funded fellowships, but research assistantships had increased to 35,000 (a ratio of 1 to 2). Thus, the total number of graduate students supported from federal sources fell from 80,000 to 49,000, a decrease of 39 percent, and the emphasis changed from unrestricted fellowships to financial support provided by government and industry-supported research.[2] Since 1980 the number of federal fellowships and traineeships has remained about the same, but federally supported research assistantships have increased by nearly 5 percent a year. Institutional sources also have become an important source of support for graduate students, increasing by about 4 percent a year.

The support of graduate students through sponsored grants and contracts raises some concerns. Graduate students working on sponsored projects perform research in areas of interest to the sponsoring agencies. On the one hand, this ensures that university research is related to the perceived needs of society; on the other hand, it may discourage work in novel areas which, although important, may not yet have generated significant financial support. The result is that one of the most desirable characteristics of basic research—spontaneity— is lost to some degree. Many graduate students still find ways of doing novel things while achieving the goals of directed research, but others are discouraged from working in areas in which they may have the most talent or curiosity. Students supported on teaching assistantships or university fellowships are not as constrained in this way, but they may still tend to work on teams that have the support of mission-oriented sponsors.

A second concern about the trend toward supporting graduate students through assistantships rather than through fellowship is that it tends to lengthen the time for completion of the doctorate. In 1970, the average total time-to-degree was 6.6 years; in 1987 it reached 8.6 years.[9]

One means of encouraging more diverse research goals on the part of graduate students and of shortening the time it takes to earn a doctorate is to make available a larger number of unrestricted fellowships. The Office of Technology Assessment has concluded: "Federal funding has a direct positive effect on Ph.D. production. Fellowships and traineeships in particular have been a straightforward way to increase Ph.D. production in science and engineering."[4] The National Science Foundation has been the largest source of fellowships in science and engineering. The primary source of traineeships, which are generally awarded to the institution rather than directly to the student, is the National Institutes of Health. Fellowship support appears to be especially effective in aiding women to complete graduate work. There is also some evidence that for minority students, who seem particularly deterred by the prospect of personal debt, the availability of fellowships makes graduate study more attractive.[4,9]

Unrestricted fellowships would work to the advantage of some faculty members as well. Faculty researchers who encourage independent research by their graduate students may be unable to attract students unless fellowships unrelated to a particular research objective are available.

Support of Academic Research

Since the 1950s, the nature of graduate study in the sciences has changed dramatically. The accumulation of knowledge has resulted in greater rigor in methodology and has necessitated greater specialization. Advances in electronics and computers have significantly improved the instruments available for the qualitative and quantitative analysis of natural phenomena. These instruments have made the tabulation, processing, visualization, and manipulation of data astonishingly easier and faster.

The cost of acquiring the measurement and computer systems essential to modern scientific research has considerably redefined the circumstances in which individuals and institutions contribute to knowledge and compete for research funds. Fifty years ago a graduate student geologist or biologist could operate alone. Access to a library and an optical microscope that cost the equivalent of two weeks' salary would render a scholar competitive. Today, training a molecular and cell biologist or a geochemist requires, in most cases, a large capital outlay for a laboratory and equipment, a substantial annual operating budget for technical assistance, supplies, travel, and electronic access to data-bases, and access to a supercomputer costing tens of millions of dollars. It is estimated that in 1988 the average annual expenditure (including operating expenses, equipment, and capital spending) per full-time academic investigator was $225,000.[5] It is also important for an investigator to be at an institution where there are enough colleagues to form a critical mass that fosters discussion, seminars, and general intellectual stimulation.

A successful graduate program in most, if not all, of the sciences, must therefore be located at a major research university that spends hundreds of millions of dollars annually on research and that has access to national centers and laboratories. A talented graduate student at a small institution, working under a single professor, can still read the literature, propose ideas or theories, and attempt to publish them, but without access to a considerable amount of data and large computers or to highly sophisticated experimental or observational tools, the opportunity to do original research is severely limited.

The growth in academic research and development since the 1950s has been enormous. In constant 1988 dollars, expenditures increased from $2 billion to about $13 billion. In 1988 federal support was about $8 billion, or 60 percent of the total. While the share represented by federal support declined from a high of 70 percent during the 1960s, the proportion of expenditures from institutional sources increased from about 5 percent in the late 1950s to nearly 20 percent in 1988 (Figure 6).[5]

Expenditures on research equipment in 1958 amounted to less than $200 million (in 1988 dollars). That level rose to $600 million in the mid-1960s, but was sharply reduced during the 1970s. In the 1980s, spending levels recovered, and the total for academic research equipment is estimated to have reached nearly

$900 million in 1988. In this category, too, the federal share has declined, from about 75 percent in the late 1950s to about 60 percent in 1988.[5]

It is in the area of facilities that the most serious cutbacks have occurred. Annual capital expenditures for science and engineering facilities (for research and instruction), in 1988 dollars, increased from $1.3 billion in 1958 to $3.5 billion in 1968. Expenditures then declined sharply over the next decade to $1 billion in 1979. In the 1980s, expenditures increased again, reaching $2 billion in 1988 (Figure 7). Thus, in real dollars, investment in science and engineering facilities is 43 percent less than it was twenty years ago. Federal support, which represented one-third of the total in the 1960s, has now declined to 11 percent.[5]

Current research facilities at institutions of higher education need substantial improvement. A 1988 NSF survey found:[10]

- Only 24 percent of science and engineering research space is suitable for the most scientifically sophisticated research in its field.
- 37 percent is suitable for most purposes, but not for the most sophisticated use.
- 23 percent requires limited renovation.
- 16 percent requires major repair or renovation.

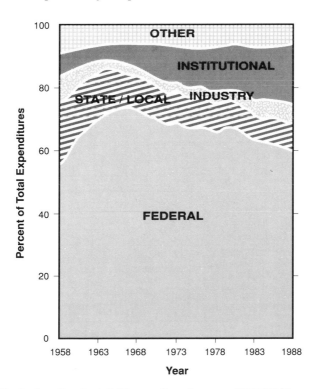

FIGURE 6. Distribution of academic R&D expenditures by source, 1958-1988 (Source: Reference 5).

Despite these great needs, universities are being forced to defer an estimated $2.50 of needed construction for every $1.00 of planned construction, and about $3.60 of needed renovation and repair for every $1.00 of expenditures. One reason for the deferral is the large increase in construction costs for technical facilities— from $192 per square foot in 1986 to $288 per square foot in 1988-1989, an increase of 20 percent per year.[10] This increase, which far outstripped inflation, is in large part the result of new technical and regulatory requirements governing research laboratories.

For the years 1986 and 1987 combined, the largest source of funding for the construction of new academic research facilities was state and local governments (38 percent overall, but 56 percent at public universities). Private sources contributed 24 percent (but 33 percent at private universities); debt financing accounted for another 15 percent; institutional funds, 14 percent; and the federal government, just 7 percent.[10]

If we accept the premise that it is imperative to foster original research on a larger scale, then as a society we must commit ourselves to a much larger capital

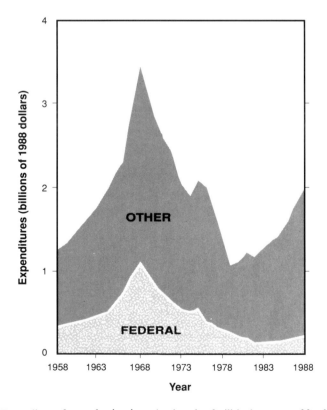

FIGURE 7. Expenditures for academic science/engineering facilities by source of funds, 1958-1988 (Source: Reference 5).

investment in research and education facilities than has traditionally been the case. This funding necessarily must come from a combination of sources, but it is critical that the federal government increase its share of support for the construction and renovation of academic research facilities.

Interdisciplinary Study and Research

Graduate science departments face a dilemma in determining how best to handle the relationship between the strict disciplinary approach to science and the problem-oriented research that must draw from many disciplines. Clearly, we must encourage fundamental research and study in all traditional disciplines, whether in physics, chemistry, biology, or mathematics. At the same time, some of the problems of most concern to society and of most interest to graduate students cut across many disciplines and require the melding of several traditional fields in order to converge on a solution.

One solution to this dilemma, which The Pennsylvania State University has tried with considerable success, is the creation of intercollege graduate degree programs. Committees of interested faculty from a variety of disciplines design programs centered on such areas as ecology, acoustics, operations research, and genetics. Students are mentored by a faculty member in a discipline-based department, but they take courses and have an advisory committee drawn from several disciplines.

Our experience has been that these interdisciplinary programs are of high quality and that well-trained students have been produced. However, there has not been great financial support for such programs. Competition for funds within large universities tends to favor traditional colleges and departments with large undergraduate enrollments. Advisory budget committees are usually made up of faculty from traditional disciplines or organizational units, which constitute the large majority of university interests. Since resources are always limited, new programs, often considered marginal, tend to be overlooked when resources are allocated. For this reason, universities may be slow to recognize and enter new fields of scholarly endeavor.

An example of a field that was considered a marginal interdisciplinary program forty years ago is materials science, which was shunned by physicists and chemists alike. Materials science, which embraces metals, ceramics, polymers, and composite materials, is now a prominent feature of many university programs. The field has attracted physicists and chemists and at some universities has become a legitimate department that can compete with departments of chemistry and physics. In the authors' opinion, the slow and difficult processs of legitimizing materials science impeded progress in the field. More radical changes in university structure and funding may be necessary to facilitate the creation and development of nontraditional areas of study. Unless the pace of change is accelerated, we risk the slow development and utilization of new knowledge in many scientific areas.

CONCLUSION

Graduate education in science and engineering at the leading research universities in the United States is indisputably of high quality. The rate at which international students apply for admission to our graduate schools attests to this fact. Nevertheless, we face significant problems in graduate education in the next two decades.

First, not enough students are undertaking graduate work in science and engineering, especially U.S. citizens. From past experience, we know that this situation can be remedied in large part by increasing financial aid for graduate students, especially in the form of fellowships. A combination of sources of support will be most effective—the federal government, state governments, industry, and universities themselves.

Second, too many students are being lost at much earlier points in the pipeline. By the junior high school level many minority students are being tracked into classes with lower expectations in science and mathematics; at the high school level disproportionate numbers of girls are dropping out of science and mathematics courses, and the attrition continues in the first college years. The only way to ensure that there are enough students in the pipeline is to intervene at much earlier school levels to provide strengthened science curricula, more out-of-classroom science experiences, and opportunities especially geared to girls and younger minority students.

Third, the problem of adequate research and instructional facilities is becoming acute. Increasingly sophisticated equipment and facilities are needed for academic research and graduate training. A sustained higher level of investment by federal and state governments is essential if research and training productivity is to be improved.

Finally, more support is needed for interdisciplinary programs at the graduate level. These programs are often on the cutting edge of scientific research and are of considerable interest to graduate students. If adequately supported, interdisciplinary programs would attract even larger numbers of graduate students and provide an intellectually stimulating environment for them and for their faculty advisers.

REFERENCES

1. National Science Board. 1989. *Science & Engineering Indicators—1989.* U.S. Government Printing Office, Washington, DC. (NSB 89-1)

2. Atkinson, Richard C. 1990. Supply and demand for scientists and engineers: a national crisis in the making. *Science* 248: 425-432. April 27.
3. National Science Foundation. 1990. *Science and Engineering Degrees: 1966-88.* Washington, DC. (NSF 90-312)
4. U.S. Congress, Office of Technology Assessment. 1988. *Educating Scientists and Engineers: Grade School to Grad School.* U.S. Government Printing Office, Washington, DC. (OTA-SET-377)
5. Government-University-Industry Research Roundtable. 1989. *Science and Technology in the Academic Enterprise: Status, Trends, and Issues.* National Academy Press, Washington, DC.
6. Commission on Minority Participation in Education and American Life. 1988. *One-Third of a Nation.* American Council on Education and Education Commission of the States, Washington, DC.
7. Oakes, Jeannie. 1990. *Lost Talent: The Underparticipation of Women, Minorities, and Disabled Persons in Science.* Rand Corporation, Santa Monica, CA (R-3774-NSF/RC)
8. National Research Council, Office of Scientific and Engineering Personnel. 1989. *Doctorate Recipients from United States Universities: Summary Report 1988.* National Academy of Sciences, Washington, DC.
9. U.S. Congress, Office of Technology Assessment. 1989. *Higher Education for Science and Engineering.* U.S. Government Printing Office, Washington, DC. (OTA-BP-SET-52)
10. *Scientific and Engineering Research Facilities at Universities and Colleges: 1988.* National Science Foundation. (NSF 88-320)

Science Education in the United States: Issues, Crises and Priorities. Edited by S. K. Majumdar, L. M. Rosenfeld, P. A. Rubba, E. W. Miller and R. F. Schmalz. ©1991, The Pennsylvania Academy of Science.

Chapter Twenty

THE FUTURE BEGINS TODAY: MATH, SCIENCE AND ENGINEERING EDUCATION IN THE 21st CENTURY

SENATOR
MARK O. HATFIELD

711 Hart Office Building
Washington, D.C. 20510

Mark O. Hatfield
United States Senator
Oregon

The wife of American physicist and Nobel laureate Robert Milliken was passing through the hall of their home one day when she overheard the family's maid on the telephone. "Yes, this is where Dr. Milliken lives," she said. "But he isn't the kind of doctor who does anyone any good."

Unfortunately, the Milliken family maid is like millions of other people in this country to whom algebra and chemistry and engineering are all part of the foreign language of some distant land. One recent study suggests that 92 percent of all Americans are "scientifically illiterate."

Scientific illiteracy poses an increasingly large threat to everything from American economic competitiveness abroad to the quality of life here at home. The urgent question: what can be done to reverse this dangerous trend as we look toward the 21st Century? In order to begin developing solutions to our national scientific illiteracy, however, it is first necessary to understand both the dimensions of the problem and its potential implications.

SUPPLY AND DEMAND

The fundamental problem involves supply and demand: the supply of qualified scientists and engineers in the United States is shrinking at the same time that demand for them is rising. The number of jobs for scientists and engineers has grown at an annual rate of 7 percent over the last decade, more than twice as fast as the number of jobs overall. Assuming current trends continue, the National Science Foundation predicts a shortfall of more than 600,000 scientists and engineers by the beginning of the 21st Century. The Office of Technology Assessment predicts a shortfall closer to 700,000.

Whatever the real number, the fact is that an increasingly small number of young people are choosing to pursue careers in science and engineering. The number of college freshmen choosing to major in engineering fell 21 percent between 1977 and 1987, while those choosing to major in computer sciences fell 75 percent in the same period. In 1988, only 1 percent of freshmen surveyed said they intended to major in mathematics, and only 1.5 percent said they intended to major in physics or chemistry. Two decades earlier, those numbers were 4 percent and 3 percent respectively.

The scores of American junior high and high school students on standardized tests given to students in industrialized countries suggest at least part of the reason for these dramatic declines. In 1985, American eighth graders ranked thirteenth among students from 17 countries in math skills. In a 1986 standardized test of college-bound seniors in 13 countries who had completed

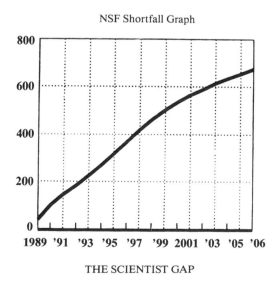

NSF Shortfall Graph

THE SCIENTIST GAP

Estimated cumulative shortfall between Bachelor's degrees in science and engineering awarded and U.S. requirements; in thousands. *Source: National Science Foundation*

two years of study, American students ranked ninth in physics, eleventh in chemistry and thirteenth—dead last—in biology. In 1987, eighth graders in the United States ranked well below the international average in math problem-solving abilities and among the lowest 25 percent in geometry. Today, the average Japanese senior in high school outscores 95 percent of his or her American counterparts on standardized math tests.

Although there are a great many talented and committed math and science teachers across the United States, one obvious cause of these dismal scores is the equally dismal state of math and science education in many of our secondary schools. A full third of all American high schools do not offer enough math course work to enable even the very best students to enter engineering school. 7,000 high schools in the United States today do not offer a single course in physics, 4,000 do not offer a course in chemistry and 2,000 do not offer a course in biology. And even in those schools which do offer science courses 75 percent of junior high school and 50 percent of high school science teachers today do not meet the minimum certification requirements of the National Science Teachers Association. According to the National Science Foundation, as many as 300,000 new math and science teachers will be needed by 1995 just to sustain education at current levels.

The statistics documenting this emerging math, science and engineering education crisis in the United States fill volumes. Indeed, as the Washington Post pointed out in a 1990 editorial entitled "Science and Science Appreciation," the number of reports on scientific illiteracy grew from 100 in October, 1988 to well over 300 by May, 1990.

Whatever the study, however, all the statistics point to one fact: our supply of scientists and engineers is shrinking by the minute. At the same time, our demand for scientists and engineers is *increasing* by the minute. Although some people think of science only as the frog they dissected in high school biology class (assuming they were lucky enough to attend a school which offered a biology class), math, science and engineering create the backbone of this nation. Everything from the nation's defense to health research to economic competitiveness depends on trained scientists and engineers.

A quick survey of several major research and development efforts underscores this point:

1. The Department of Defense, which receives the lion's share of federal research and development funding, already employs or has given contracts to an estimated 150,000 scientists and engineers just for the Strategic Defense Initiative.

2. Medical research coordinated through the National Institutes of Health receives the next largest share of federal research and development funding, and is used to develop treatment and cures for literally thousands of illnesses from Alzheimer's Disease to Sudden Infant Death Syndrome to AIDS.

3. American business and industry, which spends roughly the same amount

on research and development efforts as the federal government, has relied on technological innovation for as much as 80 percent of U.S. productivity growth since the beginning of this century.

Stiff competition for the dwindling pool of scientists and engineers has already begun, and will only grow in the years ahead. Although those organizations and institutions which have the most money to attract qualified scientists and engineers will come out ahead in the short run, nobody will win in the long run. An alternative simply must be found.

WINNING THE COMPETITION FOR ATTENTION

If that alternative is to be found, half the battle involves focusing political attention on this emerging crisis in math, science and engineering education. Just like the inevitable competition for scientists and engineers, the competition for attention between issues in Washington is stiff indeed. A thousand and one issues—virtually all of them critically important—compete for the limited time and attention of the White House and of Congress.

When Senator John Glenn and I introduced a resolution in June, 1989, calling on the Congress to make science, math and engineering education a top federal priority, our effort was met with little more than raised eyebrows. After President George Bush made the need for improved science and math education a top priority in his 1990 State of the Union Address, however, the attention focused on the entire issue increased dramatically. National news magazines and television networks began devoting attention to the issue, as did virtually every committee in Congress with jurisdiction over education or research and development programs. This heightened visibility paved the way for The Excellence in Mathematics, Science and Engineering Education Act of 1990, the most comprehensive math and science legislation ever introduced.

The critical importance of this heightened visibility cannot be underestimated. I certainly do not mean to suggest that the solutions to this emerging crisis lie exclusively with the federal government. But heightening the visibility of any issue in the White House and in Congress is a critical step toward developing solutions within both the private and the public sectors.

The challenge as we begin the last decade of the 20th Century is to seize the opportunity that visibility creates and develop lasting solutions to the emerging crisis—sooner rather than later.

SEARCHING FOR SOLUTIONS

Because virtually every profession and every level of government in every part of this country relies in some way on trained scientists and engineers, the solutions to this emerging crisis in math, science and engineering education must be both numerous and diverse. In spite of the increased attention focused on this issue in recent months and years, no completely comprehensive plan of action exists. A great many ideas and proposals have been put forward, however.

THE FEDERAL ROLE

Backing up the goals produced following the 1989 Nation's Governors for Education Summit and the declaration in his 1990 State of the Union address that "by the year 2000, U.S. students will be first in the world in science and mathematics achievement," President Bush used his FY 1991 budget request to propose a 26 percent increase in federal spending for math, science and engineering education. Underscoring the diverse applications of these disciplines, the Department of Education, the National Science Foundation, the Department of Energy, the National Aeronautics and Space Administration and the National Institutes of Health are given primary responsibility for the education mission.

By proposing an increase in funding in a time of federal budgetary constraints, the President made a very strong statement about the critical national importance of math, science and engineering education. The specific departments and programs he targeted, however, offer insight into the future role the federal government can and should play in enhancing education efforts.

Recently, the President's Science Advisor, Dr. Allan Bromley, convened a Federal Coordinating Council on Science, Engineering and Technology (FCCSET) to consider human resources issues. This Committee, which is scheduled to report in late 1990, is chaired by the Secretary of Energy and supported by the Secretary of Education and the Chairman of the National Science Foundation. This effort is clearly unique in that it strives not only to conduct an inventory of federal activity in mathematics and science, but it focuses on interagency cooperation and coordination in order to maximize federal resources.

There are also various legislative proposals pending in Congress, including S. 2114 - the Excellence in Mathematics, Science and Engineering Education Act of 1990, authored by Senator Edward Kennedy and myself. This legislation provides seed funding for a number of innovative ideas both within our formal educational structures and in our informal settings such as libraries and science museums. I have also authored legislation to enhance interagency cooperation in these areas and to provide fellowships for teachers to serve on committees in Congress which deal with education and science legislation.

CREATIVE IDEAS FROM THE PRIVATE SECTOR

Any shortage of trained scientists and engineers will affect the private sector in this country as severely as it will affect the government. From the automobile industry to companies like Hewlett-Packard and Upjohn, research and development are important elements of the work done by a wide range of businesses and industries. Without trained scientists and engineers, however, that work cannot be done and the competitiveness of the United States will suffer.

The task of improving the technological literacy of our population, therefore, cannot be assigned only to the government. In fact, while the current emphasis placed on math, science and engineering education programs by the federal government is vitally important, the realities of current federal budget constraints suggest that the government cannot do it alone. It cannot, and it should not.

Private sector leaders understand the importance of math, science and technology education as well as other disciplines, and have begun a wide range of innovative efforts to improve education at all levels. A genuine partnership is developing between all levels of government, individual school districts and teachers, and the private sector on this front. Nothing could be more exciting.

One way in which private industry can improve education is by supporting informal sector projects and programs. In Oregon, the Oregon Museum of Science and Industry brings math, science and engineering to life in front of children's faces, not just in its buildings in Portland but through outreach programs in libraries all over the state. Anyone who has been in OMSI or any other science museum understands how important these places are to the effort, but many of them could not operate without the practical help and financial support of the private sector.

Examples of other contributions being made today by the private sector include General Electric's involvement with the Manhattan Center for Science and Mathematics, Chevron's sponsorship of the Family Science Program at schools across the United States, the Regional Science School created by Upjohn in Kalamazoo, Michigan, and the Mobile Science Classroom sponsored by Southern California Edison. These initiatives, and many more like them all over the United States, are also designed to reach students early and to keep them interested in math, science and engineering.

A third area in which private industry has become increasingly involved is teacher training. Because teachers who understand how the skills and facts they are passing on to their students are actually used tend to be far more effective than teachers who do not understand the practical applications of these skills and facts, initiatives designed to teach the teachers are particularly important. Companies all over the country have developed programs which bring teachers into working laboratories during the summer to do their own "hands on" work.

IN THE CLASSROOM

All the programs and projects in the world will not make any difference without skilled and enthusiastic teachers. Conversely, however, just one dedicated teacher can make an enormous difference. As Andrew Kupfer wrote in a 1990 issue of Fortune magazine, "Want to help the dismal math and science skills of American kids? Just tuck away the textbooks, open up the labs, add an inspired teacher—and stand back." The fact that three current deans of medical schools in California all had the same elementary school teacher in Providence, Rhode Island underscores this idea.

As with all education initiatives, the drive to make our population technologically literate ultimately depends on teachers and on our strong support for them. That is why so many public and private programs focus on teacher training and support. And that is why increased attention is now focused on the creative initiatives teachers can take in the classroom to increase the interest of young people in technology of all kinds.

The 1990 Oregon Teacher of the Year, Stuart Perlmeter, a middle school science teacher from Springfield, offers an excellent example of what teachers can and should be doing in their classrooms. During a discussion of earthquakes, he had his class build structures with toothpicks and then test them to see where and why they were strained during ground movement. During a discussion of insects, he had his class make insect nests and actually eat chocolate covered ants. These "hands on" activities do not just teach young people—they *show* young people, and encourage them to become actively involved.

Of course there are some classroom activities that can be coordinated not only with the individual teacher but also with the government and with industry. The best example of this to date involves the tomato seeds flown in space by NASA's space shuttle Discovery and then distributed to classrooms and workshops throughout the United States for various experiments. With a little creativity and cooperation, initiatives like this one can be undertaken over and over again.

While the current concern over technological literacy is focused on the number of young people entering graduate school, attention must be first focused on grade school instead. In order for students to be interested in math, science and technology in graduate school, they must first be interested in these areas in grade school. According to the National Science Foundation, we must have had 750,000 high school sophomores with an interest in science and engineering in 1977 to produce just 9,700 PhD's in science and engineering by 1992. To get those 750,000 high school sophomores, no doubt at least double their number ought to be identified in grade school.

When it comes right down to it, the most important element of classroom education in math, science and technology is wonder. Teachers who can convey to their students—young and old—a sense of wonder and amazement about our natural world and about technology are invaluable. Exciting students and cultivating their natural curiosity are absolutely critical components of any effort to increase the finite skills of a new generation.

Again, the bottom line: teachers at all levels are the front line of our effort to create a technologically literate population. Their creativity, commitment and skills should be our first and foremost concern.

A NATIONAL CONSENSUS

When asked to define "science," an elementary school student in Missouri replied: "Science class is where we learn how tomorrow happens." Indeed it is. And our investment in math, science and engineering education is, very literally, an investment in the future.

Ultimately, the single most important element of a comprehensive plan of action to fight this nation's technological illiteracy crisis involves the creation of a national consensus that math, science and engineering education are important. A broad-based constituency must be developed which understands that math, science and engineering are disciplines which affect *all* our lives and that education in these areas at *all* levels is critically important today and in the future.

That constituency—which I believe has begun to emerge—will be the catalyst for real and lasting change.

Science Education in the United States: Issues, Crises and Priorities. Edited by S. K. Majumdar, L. M. Rosenfeld, P. A. Rubba, E. W. Miller and R. F. Schmalz. ©1991, The Pennsylvania Academy of Science.

Chapter Twenty-One

FEDERAL SCIENCE AGENCIES AS PARTNERS IN SCIENCE EDUCATION

W.R. GREENWOOD

Professional Geologist
1402 Earnshaw Court
Reston, VA 22090

INTRODUCTION

As many studies have made plain, we must radically improve science education in America to provide our citizens with knowledge to make informed decisions on crucial political and economic issues; to enrich their lives; and to prepare them to play vital roles as members of an increasingly technological society. What can federal science research agencies do to support this *perestroika* in American science education? What resources do they have, what responsibilities do they have, and how should they play their role? This discussion is based on my recent experiences in helping to expand educational support activities of the U.S. Geological Survey (USGS), but are my personal opinions and do not necessarily represent the views of that agency.

Federal Science Agency Responsibility for Education: As a major source of earth science information and the largest single employer of earth scientists, the USGS has an opportunity and a responsibility to support earth science education. The USGS and all science agencies have a responsibility to inform the public on research results, particularly those results that bear on environmental, economic, and political decisions. It is particularly important for science agencies to provide such information to young Americans. Furthermore, it is critical that federal science agencies help to ensure that all American children are aware of the exciting new discoveries in science. Scientists must convey their excitement to teachers and students in order to encourage an increased number of young women and men of diverse ethnic backgrounds to pursue careers in science.

Creative cooperation: Federal science agencies can provide science expertise, opportunities for teachers and students to participate in research projects, facilities, and information to augment state and local education projects that are funded by Federal education grant agencies. The participation of Federal science agencies can add immensely to the value of such projects, without adding to their grant costs. Teachers and students who have funding support from grants can be placed as science interns in science agency projects to gain invaluable personal experience. These teachers and students also add creative energy to the research projects. The opportunities for expanding cooperation among state and Federal science and educational grant organizations appear to be unlimited. Through such integrated activities, Federal science agencies can strengthen and support the existing educational organizations and associations that are responsible for American education.

Teachers the key: Teachers are the leverage point and the multiplier for Federal science agencies that seek to have maximum impact in enhancing education. Our goal, therefore, should be to enhance the scientific experience and knowledge, skills, public recognition, and self confidence of American teachers.

Build science into the whole school experience: Finally, earth science information is too important to future citizens to restrict its presentation to earth science classes. First of all, the earth on which we live provides children with many of their earliest experiences in asking the question "Why?"—the basis of science. Therefore, a general and scientifically-integrated understanding of earth processes can provide a natural preparation for subsequent detailed study of chemistry, biology, and physics. Furthermore, earth science information can be made available to teachers of physics, mathematics, chemistry and biology in forms that they can use to enhance students' practical understanding of these disciplines. Informative and provocative materials on earth science can be provided to English, history, and social studies teachers, who can encourage students to debate the types of environmental, resource, economic, and health issues that they will have to evaluate as voters, and to provide background understanding on events in history—from the viewpoint of the Old Testament to the Battle of Vicksburg. Art teachers can be provided materials for use in enhancing students' awareness of the beauty, power, and yet fragility of this earth. Earth science is too important and critical to understanding to be considered only in earth science classes; but then, so is all the rest of science.

RECOGNITION OF TEACHER CONTRIBUTIONS

Teachers are the ministers in the cathedral of education. They prepare the table for the feast of knowledge. There is no more demanding or crucial calling in America today than that of teachers. We scientists know how critical individual teachers have been to our growth and career paths. It is essential that we scientists and our science organizations indicate our gratitude and respect by publicly recognizing the contributions of teachers.

We scientists need to publicly recognize the accomplishments of science teachers who receive national recognition. The less visible accomplishments of each science teacher in every school also need recognition. We must help raise public appreciation for science teaching if we wish to promote public scientific literacy and expanded participation in science.

Federal science agencies can help enhance public appreciation of science teachers on a national level by augmenting support and recognition given to outstanding teachers identified by teacher organizations. The science agencies can also participate in local teacher recognition programs in school districts near their facilities.

Science agencies can also provide recognition to teachers by establishing education advisory boards, consisting of master teachers and representatives of education associations, to review and help guide educational programs in the agencies and in the activities of the Smithsonian system. Such guidance is essential to assure that Federal efforts will augment and not duplicate or interfere with teacher initiatives.

Teacher participation in all phases of science agency educational activities is essential for success. Assuring effective teacher participation takes constant attention and much teacher feedback. Teachers need to be given effective power to request, design, participate, and evaluate all activities with their research colleagues from science agencies. In short, an education campaign will serve teachers only if teachers have real power to influence all of its aspects.

RESEARCH EXPERIENCE FOR TEACHERS

The recently completed Final Report of the Task Force on Women, Minorities, and the Handicapped in Science and Technology (1989) calls for the opening of Federal laboratory facilities "to provide hands-on experiences to students and teachers, especially those traditionally underrepresented in science and engineering." Systematic incorporation of practical experience in science is especially important during pre-service and in-service education of science teachers. Teachers who have participated in Federal research projects are enthusiastic proponents of such opportunities. Authority springs from experience and knowledge. America's science teachers should be included as important participants in Federal research programs wherever possible.

Several Federal science agencies provide programs to provide teachers with research information and experience. The challenge is to expand the number of such opportunities to permit all science teachers to gain practical experience and the resulting confidence in the subject matter of their teaching.

One important solution can arise from recognition of the mutual benefit that teacher participation in Federal research can provide. Teachers get otherwise-unobtainable practical experience in research; research teams benefit from the curiosity and creative energy of the teachers.

Federal research laboratories and activities are operating on restricted budgets and generally are short of junior scientific staff. Therefore, these laboratories and activities have the capacity to productively include a large number of teachers in existing research projects.

Recently my research agency has begun working with Federal educational grant agencies to develop mechanisms whereby our research projects, facilities, information and personnel are available for participation in teacher enhancement projects organized by state and local educational organizations. We are responding to state and local requests to help design projects that include the expertise and knowledge of our scientists, opportunities for teachers to participate in ongoing research projects, and access to existing scientific information. Teacher stipends, expenses and other educational project costs are supported by the grant. The Federal science agency salary and other internal costs are not included in the proposals. We believe that such cooperation between Federal science agencies and Federal educational grant agencies is essential to increase the reach and effectiveness of Federal educational enhancement programs.

In addition, my research agency has expanded an existing volunteer program to focus on providing research opportunities to pre-college teachers. A book of opportunities has been developed to advertise volunteer research intern positions, conditions of work, logistical support available (including field and travel expenses where available), background required, name, address, and phone number of supervising scientist for each opportunity. These books have been distributed and advertised at teacher association meetings and at science association meetings that focus on teachers, such as national meetings of the American Association for the Advancement of Science (AAAS) and the National Science Teachers Association (NSTA). Teacher interest has been very high and teacher participation as volunteers is growing rapidly, especially among teachers who live near our research centers. Opportunities to obtain university credit for the research experiences are being negotiated for interested teachers.

TEACHER RESOURCE CENTERS

Federal research agencies can assist teachers by establishing *teacher resource centers* at each of their facilities. The National Aeronautic and Space Administration (NASA) has established an exemplary model in their network of teacher resource centers, located at each of their facilities. The NASA centers provide teachers with access to films, slides, videos, publications, graphic materials, teaching materials, information, use of slide and photocopying equipment, and the assistance of teacher support personnel.

Each Federal science agency would do well to establish its own network of teacher resource centers, following the NASA model. The next step would be to develop coordination and cooperation among the Federal science agencies to allow teachers to obtain information and access to all Federal teacher resources from whichever resource center is most convenient to them.

EDUCATIONAL MATERIALS

Communication — A Federal Science Agency Mission: Federal science agencies have traditionally addressed a particular audience that had a vital interest in their research results. That audience has ranged from industry and academia to environmental organizations. Each agency also has recognized that informing the public of research results is essential and it has established public affairs offices for that purpose. However, recent surveys of scientific literacy appear to indicate that little progress has been made in providing the American people with sufficient scientific information for them to make informed decisions on crucial public policy involving science. In a study of scientific literacy, Miller (1987) reports that only about 5% of American adults have the ability to recognize key scientific terms and concepts, the ability to follow basic levels of scientific reasoning, and a comprehension of scientific components of public policy issues. Miller's survey, done in 1985, shows no improvement from 1957. More public relations activities by Federal science agencies will not improve such scientific illiteracy. Federal agencies must find new and more effective means to educate the public on science, including new results of Federal scientific research, that can affect the lives and well being of the American people.

I conclude that there is a need for each Federal science agency to identify public education as a mission responsibility and for each agency to review and strengthen its educational outreach programs.

Sound Bites and Scientific Information: Providing scientific information for precollege education is a particularly important part of this Federal responsibility to educate the public on scientific results. This responsibility presents a difficult challenge to most scientists who are more versed in scholarly discourse than in effective packaging of information in "sound bites" and attractive graphics. The public, and especially young people, is accustomed to highly sophisticated information presentations by the advertising and entertainment industries. Unsophisticated presentations do not capture and hold the attention of today's public. A substantial body of research knowledge has been developed to understand and improve the success of such information presentations. Unfortunately, few Federal science agencies have direct access to that body of knowledge to guide development of their communications to students and adults.

I suggest that Federal science agencies should consider cooperation with state and local educational organizations as an effective and economic means by which to accomplish their responsibilities of communication of research results. The Federal education grant agencies have substantial programs to fund development of science education materials. My agency is exploring possibilities of providing expertise and information to support a multimedia science education project organized by an educational institution and to be proposed to a grant agency for funding. Such cooperation would combine the knowledge and latest published information of Federal scientists with the expertise of top educators and communication experts. The bottom line is that we need more effective information, not just more information, for science education. Cooperation among Federal science agencies, Federal grant agencies, and state and local educational organizations appears to be the best available means to accomplish that goal.

Computerized Information: Recent advances in computer technology present an important new opportunity to provide American schools with access to the latest scientific information and analytical tools that Federal science agencies are developing to understand natural and manmade processes that effect our living world, for example, global change. This technology can be used to provide information to identify and mitigate earthquakes and other natural hazards, and to follow the course of planetary exploration. Many science agencies are now publishing the results of their latest research on CD-ROM disks which can contain as much as 700 million bytes (or characters) of data. These disks commonly include public domain computer programs to view and analyze the published data. The USGS, NASA, and the National Oceanographic and Atmospheric Administration (NOAA) are examples of Federal research organizations that are expanding CD-ROM publication of scientific information.

An interagency project is underway to assemble selections of available Federal science agency digital information on an experimental science education CD-ROM disk. Teachers are helping select data for the disk and will work with agency scientists to prepare teacher application manuals for use in classrooms and science projects. English, history, social studies, mathematics, chemistry, biology, physics, and earth science teachers are all expected to participate in developing cross-course applications of the experimental disk. For example, data showing the frequency of sun spot intensity since 1610 could be compared in an earth science class to historical global warming and cooling, and in a history class to consider effects of such solar cycles on historical events, as during the

"Little Iceage" in Europe. Representatives of the computer industry have indicated interest in contributing equipment to participating schools for this project. University credit is being negotiated for the participating teachers.

FEDERAL SCIENTIST PARTICIPATION IN ENHANCING THE PUBLIC SCIENCE AGENDA FOR EDUCATION

Federal science agencies should encourage their scientists to participate in dialogues with scientific societies and teacher associations to identify the central scientific issues and related concepts, the understanding of which is required for all citizens to make informed decisions on the major policy questions of today. Such questions inevitably require understanding basic scientific arguments regarding healthy and sustainable stewardship of the planet and its resources.

Scientists should provide their expertise to identify such issues and concepts to help teachers to develop a prioritized agenda for science teaching. The Geological Society of America, Penrose Conference on "Geological Decisions for the 21st Century," held at Steamboat Springs, Colorado, in 1987, developed such a proposed agenda of important geoscience issues (Horten 1988). The American Association for the Advancement of Science, "Project 2061—Science for All Americans" is currently engaged in developing a focused and integrated agenda for science education (Anonymous 1989). Such an agenda must be dynamic and capable of adjustment as unforeseen understandings, opportunities, and challenges present themselves. For this reason, scientists must stay engaged in dialogue with educators on these issues.

SCIENTIST PARTICIPATION IN PARTNERSHIPS WITH SCHOOLS

Federal Employees: Federal science agencies should encourage and support the direct participation of their scientists and other employees as education partners in local schools, libraries and museums. Agencies can support the development of networks of partnership activities that are being developed by scientific societies such as the Geological Society of America and state coalitions being established by the Triangle Coalition. Agency scientists should be provided with materials and guidelines to assist in educational partnering and should be recognized by the agencies for their educational support activities.

Retired Federal Employees: Federal agencies can support retired employees to serve as educational partners in schools, museums, and libraries, as well as to prepare educational materials. Retired scientists represent a great untapped resource to support science education.

RETIRED FEDERAL SCIENTISTS AS TEACHERS

A large number of scientists in Federal science agencies are now or will soon be eligible for retirement. These scientists should be encouraged to consider becoming precollege science teachers under the alternate certification procedures that are being instituted in several States. Federal education grant agencies should consider initiating special programs to assist retired scientists to obtain the educational courses that they may need to enter a second professional career as a teacher. The Nation cannot afford to ignore the rich potential supply of science teachers that these retired scientists represent. In the short term, this group has the greatest potential to rapidly bring scientific expertise into American classrooms.

SUMMARY

I conclude that Federal science agencies can and should play a major role to enhance science education in America. I argue that to effectively and economically play this role these agencies must cooperate both with State and local education organizations and with Federal education grant agencies. Finally, I argue that scientific information should be provided to students as much as possible in the context of how they will be expected to use it to make decisions as informed citizens. Science is a quest for knowledge. Knowledge, as we know, provides us with the power to influence our future. Federal science agencies need to help all young Americans to gain the power of scientific knowledge so that they can fulfill their political, economic, and intellectual potential in the increasingly technological world of the future.

BIBLIOGRAPHY

Anonymous. 1989. Science for all Americans. A Project 2061 report on literacy goals in science, mathematics, and technology. American Association for the Advancement of Science. 217 p.

Horten, H.A. 1988. Report of the July 1987 Penrose Conference held at Steamboat Springs, Colorado, in GSA News and Information, May 1988. Geological Society of America, Boulder, CO. pp. 109-112.

Miller, J.D. 1987. Scientific literacy in the United States. *In:* Communicating Science to the Public. David Evered and Maeve O'Conner (Eds.). Wiley, London. pp. 19-37.

Task Force on Women, Minorities, and the Handicaped in Science and Technology. 1989. Changing America. The new face of science and engineering. Office of Science and Technology Policy. 46 p.

Science Education in the United States: Issues, Crises and Priorities. Edited by S. K. Majumdar, L. M. Rosenfeld, P. A. Rubba, E. W. Miller and R. F. Schmalz. ©1991, The Pennsylvania Academy of Science.

Chapter Twenty-Two

ROLES OF THE NATIONAL ASSOCIATION OF ACADEMIES OF SCIENCE IN SCIENCE EDUCATION AND LITERACY

GEORGE C. SHOFFSTALL

Past President NAAS,
PAS Executive Secretary
502 Misty Drive, Suite 1
Lancaster, PA 17603

George C. Shoffstall
Former President: NAAS and PAS

INTRODUCTION

Across these United States of America calls for action are resounding at every level of society to reverse the decline in science, mathematics, and technology education, and, specifically, the disappointing trends in scientific literacy among the populace.

Ironically, one of the perplexing problems facing this decline of American education in the sciences is a lack of clearly defined roles of government— local, state, or federal—and society, in general, demanding a reversal of these alarming trends.

A paramount role of the National Association of Academies of Science and its affiliate academies will be a drive to increase their efforts to ensure the youth of America are prepared for the jobs and societal responsibilities in an advanced technological age. A commitment to improve education in the sciences with special emphasis on programs to enhance scientific literacy, will significantly strengthen and enlarge the scientific R&D manpower pool and the technology potential of our nation.

NATIONAL ASSOCIATION OF ACADEMIES OF SCIENCE

The National Association of Academies of Science (NAAS) celebrated in 1990, a 64-year-old alliance of state, regional, and municipal academies of science affiliated with the AAAS. The affiliate academies are membership organizations with programs directed toward academic and industrial scientists, students, teachers, agencies of state government, and the general public. NAAS exists to promote common aims of the individual academies and the AAAS. The NAAS Articles of Incorporation in the District of Columbia state that the NAAS "shall promote scientific interest, inquiry, and education by meetings, projects, papers, reports, discussions, and professional contacts, thereby fostering public welfare and education." The NAAS arranges academy-related symposia, workshops, meetings, and junior academy activities at the AAAS Annual Meeting. Staff from the AAAS Directorate on Education and Human Services serve as liaison to the NAAS. The aims of the NAAS and the academy affiliates parallel the constitutional objectives of the AAAS -- to further the work of scientists, facilitate their cooperation, and increase the public understanding of science.

HISTORICAL RELATIONSHIPS WITH AAAS

With few exceptions, the state, regional, and municipal academies of science have come into existence since the founding of the AAAS in 1848. Maryland Academy of science pre-dates the AAAS by a half century and a third of them were founded in the latter half of the 1800s. The Pennsylvania Academy of Science was organized in 1924 and its Junior Academy of Science in 1934.

In 1926, it was suggested that a federation of affiliate academies with some coordination of their activities would be desirable. At the 1927 AAAS meeting in Nashville, an Academy Conference was authorized by AAAS Council to serve as a standing committee on relations among the affiliated academies and administratively between them and the AAAS. The Academy Conference met for the first time at the New York meeting in 1928, and has presented a program at every annual meeting of the AAAS since then.

In 1969, the name of the Academy Conference was changed to the Association of Academies of Science and in 1979, it was formally changed to its current status—the National Association of Academies of Science (NAAS).

The NAAS has a strong interest in the work of junior and collegiate academies of science and the encouragement of young people interested in science. At each annual meeting of the AAAS, the NAAS sponsors an American Junior Academy of Science where representatives of various state junior academies of science present scientific papers, orally and in poster session format. The Pennsylvania Junior Academy of Science (PJAS) is represented at these meetings annually.

NAAS HISTORICAL PERSPECTIVES

The National Association of Academies of Science (NAAS) has a unique capability to proffer, through its many affiliate academies of science, expansive variety in science training experiences. Distinct visibility is one of the best public relation tactics the NAAS has to market rational scientific values, attitudes, and skills to the American public. This can be accomplished via the municipal, regional, and state academy/junior academy network.

Historically, state academies and their junior academies abound with creative ideas, experience reports, science workshops, symposia, annual programs, press releases, and scientific publications heralding the virtues of basic education in science, mathematics, and technology. Alas, a void exists in communication—the results of the aforementioned statistically elude most Americans. Regardless, it is incumbent upon academies to do a better job of marketing this scientific expertise to the legions of citizens in these United States. Moreover, society is remiss in not capitalizing on the achievements and leadership potentials of the affiliate academies.

FUNCTIONAL ILLITERACY CRISIS

According to the U.S. Department of Education, approximately 27 million adults in the United States currently are functionally illiterate and 45 million adults are marginally illiterate. Moreover, each year 2.5 million Americans join the pool of adults who are unable to process the written materials necessary for safe, successful, and independent roles essential to function in our society.

Obviously, an excoriating and pervading crisis exists. The staggering number of disadvantaged adults negatively impacts every community and business in this nation. Within the community, especially a rural community, it is difficult to identify and serve adults who are educationally disadvantaged. The ghettos of major American cities compound the issues. The long term affects, in a rapidly evolving high tech international arena, will be regressive to the United States.

NATIONAL SYMPOSIUM ON THE CRISES

In response to one aspect of these crises, a "SYMPOSIUM: The Crises in

Science and Mathematics Education—Perspectives from the National, State, and Local Levels" was presented by the NAAS in May 1983 in conjunction with the 149th Annual Meeting of the American Association for the Advancement of Science in Detroit, Michigan. Particular attention was given to the inadequacies of science education at the grassroots level. In a more positive vein, initiatives were presented as to how academies of science can strengthen and/or modify the policies and practices responsible for the present status of science education and scientific literacy.

Changes in national science policies, the advent of New Federalism, conflict situations—especially creationism versus evolution biological science, apparent weaknesses in precollege mathematics and science training, illiteracy, gaps in science communication processes, and either lack of or obsolescence in university and secondary school laboratory equipment have been focal points for discussions and heated debates in the 1980s. All have been a contributing factor, to some degree, in the overall crisis in science, mathematics, and technology education confronting America.

THE RANKINGS

During this past quarter century, secondary schools across the nation have entertained a potpourri of science and mathematics curriculum changes. However, studies presage a schism in science, mathematics, and technology abilities when American students' achievement scores, in general, are compared with the higher score rankings of their international peers. Significantly, America's top high school students rank far below those of nearly all foreign countries according to an assessment conducted by the International Association for the Evaluation of Educational Achievement. Moreover, the study woefully found little improvement since 1970 in spite of increased attention being paid to the quality of education in science.

The recent survey sponsored by the Carnegie Foundation for the Advancement of Teaching Survey, "The Condition of the Professiorate: Attitudes and Trends, 1989", reveals, "Public education despite recent years of reform, is still producing inadequately prepared students." Three quarters of the 5,450 college campus faculty polled in this report consider their students seriously unprepared in basic skills. A lamentable situation for our nation.

SOCIETY AND SCIENTIFIC LITERACY

Assuredly, scientific literacy is a matter of national concern, yet the concept, lacking a consistent definition, is wanton. Undoubtedly, the greatest challenge beckoning in this last decade of the 20th century is the establishment of realistic goals for developing a scientifically literate society and the enhancement of the science pool for research and development.

With escalating drop-out rates in secondary school populations, a more national approach to assuage this phenomenon is prerequisite. Society is obliged to curb this travesty by encouraging these drop-outs to return to school in order to improve their basic educational skills and broaden their literacy levels. As a consequence, this may impact upon the attitudes these drop-outs have about school, in particular, science. It is often said, "If schools fail our children, it is only a matter of time before society will fail all of us."

Because of the universal and compulsory factors that force the nation's youth to remain in school until a specific age, education—particularly that segment of education referred to as schooling—has received national criticism from all levels of society during much of the 1980s. The nation's schools for our youth are criticized for many different reasons: their compulsory nature, their failure to guarantee learning, their laxity in assigning homework, and the short school year. Since almost all Americans have personally experienced schooling to some level, most feel they have a legitimate basis upon which to make judgments and to assume positions.

Like institutions of science and technology, education is a social institution that does not stand apart from the rest of society—all are integral parts of it. Anathema to these institutions are the very people who make the rules— governing board members and legislators—spending most of their working lives outside of the fields of discipline or have little expertise in either education or the sciences, mathematics, or technology.

Notably, as a process, education, per se, clearly is not as important as learning. Education is merely the tool that enables the person to learn. Whether a person learns through self-education or through group education, the focus should be on the person being educated.

What then are the proscribed antecedents contributing to the lack of scientific literacy and basic educational skills being ascribed to our nation's youth— Course content? Rote memorization of abstract course concepts without practical and/or meaningful application? Inefficacy of methodology? Lack of teacher training? Lack of ideal core curricula? Student ennui? or Bureaucracy?

In response, there are no valid reasons—economic, intellectual, or social— why the United States cannot resolve the aforementioned situations and solve its national literacy problem. Moreover, it is incumbent upon the nation as a society to postulate more innovative programs and techniques in the recruitment of the physically handicapped, minorities, and women to seek careers and economic job satisfaction in science, mathematics, and technology.

Obviously, you cannot achieve cultural, scientific, mathematical, or technological diversity unless there is a commitment. The multitude of professional organizations and their respective memberships must become more involved and committed to a national effort of providing resources, formulating new and attainable goals to ensure education in the sciences is accountable.

ORGANIZATIONS AND SOCIETY

Professional organizations are attempting to develop national models for science, mathematics, and technology reform in education. However, the inevitable concerns of society-at-large are endemic to the mere existence and function of the infrastructure in formal organization. Too often, from society's standpoint, organizations are working at cross-purposes to each other or fail to meet vital social and educational needs. Indeed, one of the most important organizational problems is to be more responsive to the needs of those very individuals that the organization has been established to serve—without considerable overriding bureaucracy and inefficiency.

Unfortunately, the term bureaucracy is often used as a synonym for all formal organizations—including those of the sciences. Paradoxically, the term may be used as an epithet for any rigid, rule-bound, inefficient, and labyrinthine organizational form, i.e., features of bureaucratic life such as "red tape" and "officialese"—the banal language of bureaucrats.

Commendably, science, mathematics, and technology organizations appear to have "cut the red tape" and are directing their resources in an effort to evolve the goals, the strategies, and the mechanisms to achieve quality education in the sciences. The AAAS, NAAS, and their affiliate societies are initiating vigorous action programs—science olympiads, workshops, symposia, visiting scientists, R&D experiences, etc.—in an attempt to stem the crisis in science education in the private and public sectors.

At this writing, a most praiseworthy, speculative program is being developed by the American Association for the Advancement of Science. Namely, PROJECT 2061—Science For All Americans, a long term, multiphase undertaking designed to help reform science, mathematics, and technology education in the United States of America.

SCIENTIFIC LITERACY

Strategies regarding the improvement of scientific literacy are of national concern. The need for scientific literacy for most Americans has been demonstrated if we are to remain competitive in the world market without being overtly jingoistic. Deficiencies in the mere appreciation of the sciences due to a lack of literate understanding is inadmissible with today's technological revolution. America can and must do better in the development of its young people in instilling positive attitudes toward learning, science, mathematics, and technology while enhancing their quality of life.

What should be deemed an appropriate level of scientific literacy for an individual to responsibly participate in civic affairs, read and understand medical, scientific, and technological publications, and to lead a rewarding professional life?

In October 1989 in Arlington, VA, the AAAS Forum for School Science addressed the topic, "Scientific Literacy: It's not a Trivial Pursuit". The Forum compared views on scientific literacy and the implications the different perceptions hold for the curriculum in grades K-12. A review and an analysis revealed many protocols for those nationally involved in education in the sciences to contemplate and implement.

THE COMMITMENT

In 1965, President Lyndon B. Johnson said, "It is education that places reason over force". On September 27, 1989, President George H.W. Bush, addressing the nation's Governors at the University of Virginia, said, "Today, we do not meet in a spirit of immediate crisis. The nation is sound", declared the President, "but the decline of our educational system is one problem that threatens to endanger the very leadership position in the next century".

The crisis is real, commitments are absolutely critical from all levels of society with no guarantee of immediate success. More global long range perspectives are paramount to an ultimate solution of this national problem.

Unfortunately, the resolution of the crises in science, mathematics, and technology education, and the improvement in scientific literacy will be very costly.

Educational reform will take time, national commitment, pooling of resources (financial and human), and leadership—a specific role that the NAAS and its affiliate academies can embrace.

CONCLUSION

The National Association of Academies of Science is cognizant of its role and shares in the responsibility of solving the many challenges that remain to be ameliorated in the current national crises in science education and scientific literacy.

As an approach, the NAAS and its affiliate academies must embark in terms of synthesizing knowledge gleaned from the enervating crises confronting the United States. This information will serve as a paradigm in the integration of local, state, and federal policies networking long range solutions and reforms in science, mathematics, and technology education.

The ultimate future of scientific literacy will depend, in part, on how other segments of society commit resources to resolve these profound and regressive declines in our educational system. The demand for scientifically literate citizens will impact on the leadership of the next century. The youth of America must be educationally prepared for the challenges of the 21st century—the NAAS and its affiliates are committed to this end.

REFERENCES

American Association for the Advancement of Science—Coalition for Education in the Sciences. 1989. Science, Mathematics, and Technology Education in the Federal Budget — FY 1990. Sc. Ed. News. 7:6.

American Association for the Advancement of Science—Constitution. 1973. *In:* Nineteen 89/90 handbook: AAAS Officers, Organizations, Activities. AAAS Publication. 89-13, pp. 143-159.

American Association for the Advancement of Science—Forum for High School Science. 1989. Scientific Literacy: It's Not A Trivial Pursuit. Arlington, VA.

American Association for the Advancement of Science—PROJECT 2061: Science for All Americans. 1989. A Project 2061 Report on Literacy Goals in Science, Mathematics, and Technology. AAAS Publication. 89-01S, pp. 217.

International Association for the Evaluation of Educational Achievement. 1988. Science Achievement in Seventeen Countries: A Preliminary Report. Pergamon Press, Oxford.

National Association of Academies of Science—Constitution. 1974. *In:* The Proceedings, Directory and Handbook of the NAAS. 1989-90. pp. 107-116.

National Governors' Association. 1987. Making America Work. The Association, Washington, D.C.

National Governors' Association. 1989. Address by: President George H.W. Bush—Decline of American Educational System. Charlottesville, VA.

Rutherford, F.J. 1983. The Dangerous Decline in U.S. Science Education. The Chronicle of Higher Education. April 13, 1983, p. 64.

Shoffstall, G.C. et al. 1983. SYMPOSIUM: The Crises in Science and Mathematics Education: Perspectives from the National, State, and Local Levels. *In:* The Proceedings, Directory and Handbook of the NAAS. S.K. Majumdar, Ed. 1983-84, pp. 11-48, National Association of Academies of Science.

Science Education in the United States: Issues, Crises and Priorities. Edited by S. K. Majumdar, L. M. Rosenfeld, P. A. Rubba, E. W. Miller and R. F. Schmalz. ©1991, The Pennsylvania Academy of Science.

Chapter Twenty-Three

PENNSYLVANIA'S ROLE IN SCHOOL SCIENCE EDUCATION, AND SCIENCE TEACHER PREPARATION AND CERTIFICATION

G. KIP BOLLINGER[1] and FREDERICA F. HAAS[2]

[1]Science Adviser
[2]Director, Bureau of Teacher Preparation and Certification
Pennsylvania Department of Education
333 Market Street
Harrisburg, PA 17126

INTRODUCTION

President Bush at the Governor's Conference on Education in February 1990 issued the challenge that by the year 2000 U.S. students will be first in the world in math and science achievement. To meet this challenge requires intensive support of K-12 science programs and extensive training of science teachers at all levels. Pennsylvania's role in science education encompasses a rich tradition of statewide science initiatives that address the needs of an extremely diverse state. Regulatory functions from Chapter 5 of the State Board Regulations and science teacher preparation by the colleges and universities supply directives for the education and certification of professional staff and the direction of planned courses, which form the organized framework of curricular offerings, within the state.

HISTORY OF SCIENCE PROGRAMS IN PENNSYLVANIA

In 1968 Dr. Irvin T. Edgar, who was then a Science Education Adviser of the Pennsylvania Department of Education, invited a group of science educators to meet and discuss the proposed direction for science teaching in the elementary schools of Pennsylvania. These educators were Francis Alder, Elementary

Science Coordinator, Keystone Oaks School District; Dr. Dorothy Alfke, Professor, Elementary Science Department, the Pennsylvania State University; Dr. Roy W. Allison, Elementary Science Consultant, Marple Newtown School District; Dr. William Chamberlain, Professor, Science Education, Clarion State College; Dr. Thomas V. Come, Professor, Elementary Education Department, Edinboro State College; and John J. McDermott, Science Education Adviser, Pennsylvania Department of Education. This committee developed materials and activities to be adapted by classroom teachers. The lessons were designed for use in elementary school classrooms and emphasized the use of concrete manipulative activities using everyday materials.

In 1970 an interactive television component was added and Dr. Duane R. Smith, Professor, Elementary Education, The Pennsylvania State University, Capitol Campus; Dr. Paul W. Welliver, Professor of Education, Pennsylvania State University, and Dr. Robert L. Shrigley, Professor of Education at Pennsylvania State University joined the committee. The classroom teachers, using supplied lesson plans, would lead discussions by the class, and, at a signal, the television would continue with a reinforcement of the concepts just discussed. The television programs were not complete lessons in themselves but were introductions to a lesson plan which the classroom teacher was to use as a follow-up. Ten primary and ten intermediate lessons were produced. These lessons were used statewide by public television stations for a number of years. "Science for the Seventies," as the project was entitled, (SFTS) was included in a book by Albert Piltz of the U.S. Office of Education and Robert Sund of the University of Northern Colorado entitled *CREATIVE TEACHING OF SCIENCE IN THE ELEMENTARY SCHOOL.*[1]

Historically, the Department of Education has provided outstanding leadership in science curriculum development materials. The SFTS project was further developed under Dr. Irvin Edgar's leadership into "Investigating Science in Elementary Education" (ISEE) and then under John J. McDermott's leadership into *"Science Unlimited."* This curriculum project provided forty outstanding hands-on activities in inquiry science and a rich variety of videotape supplementary materials. This series, still much in demand, emphasizes the current priority of "hands-on, minds-on" science. Over 1,000 teacher training workshops were held. Also, undergraduate and graduate courses were developed and offered. *Science Unlimited* became the most commonly used science program in Pennsylvania. It has been adopted by Delaware and is used in Florida, Texas, California, and Ohio.

One of John McDermott's early projects was *Nuclear Science* and the collegiate preparation of teachers of nuclear science. The Pennsylvania Nuclear Science Program was a direct offshoot of the Atomic Energy Commission/ National Science Foundation Summer Institutes in Radiation Biology. Participants at these institutes were trained in basic concepts in nuclear and radiochemistry, nuclear physics, radiation biology, radiation health, physics, and radioisotope applications. Participants took a complete equipment package back to their classrooms.

The program was piloted in Pennsylvania, then spread across the United States and to other nations. The course was introduced into the United Kingdom, Denmark, the Netherlands, Germany, East Germany, Poland, Hungary, Japan, and Korea. Today *Nuclear Science* is going strong, especially in Japan and Korea.

As the number of nuclear science teachers grew, the teachers formed the Pennsylvania Nuclear Science Teachers Association, affiliated with the Pennsylvania Science Teachers Association. As the number of teachers from other states increased, the name was changed to The American Nuclear Science Teachers Association, and became affiliated with the National Science Teachers Association and the American Nuclear Society.

The nuclear science project trained about 700 teachers from 40 states and 12 foreign countries. At least 10,000 students have received the course over the years. Over 200 program graduates have continued in related professional fields, such as nuclear engineering, including the first woman to receive her degree in nuclear engineering from Penn State University.

RECENT INITIATIVES IN PENNSYLVANIA'S SCIENCE EDUCATION

The Pennsylvania Science Teachers Education Program (PA-STEP) provides unique and extensive in-service programs to upgrade teachers of science and math. Since beginning in 1983 the program has had over 10,000 participants. The program, which was developed by the Pennsylvania Higher Education Assistance Agency (PHEAA), is directed by Dr. Kenneth Mechling at Clarion University. Thirteen centers offer "Programs for Improving Elementary Science" (PIES). Other course offerings include "Creative Integration of Science in Elementary Education," "Microcomputer Science Laboratory," "Science, Technology and Society," "Microcomputer Specialty Courses," and courses to enhance leadership in elementary science and the supervisors effectiveness. These courses are offered regionally and emphasize useful skills and strategies. They have been recognized nationally by the National Science Foundation and the National Science Teachers Association. Waiting lists for each course substantiate the high value of these programs to science teachers in the state.

Another science project in the state is the Commonwealth Elementary Science Teaching Alliance (CESTA). CESTA is administered by the Franklin Institute and directed by Dr. Wayne Ransom. With funding from the United States Department of Education and the Pennsylvania Department of Education (PDE), CESTA operates one day symposia and two week leadership training seminars for teachers across the state. This project operates with very close ties to the science education office at PDE. Additionally, leadership teams will be trained in delivering hands-on elementary science in-service programs. Through this project the well trained elementary science teachers will be able to network regionally. CESTA also supports varied follow-up activities to support the participants.

The first regional workshop to serve rural schools was held on March 30 and 31, 1990, at Chestnut Ridge High School with joint sponsorship from Pennsylvania Department of Education, the Pennsylvania Science Supervisors' Association, and the Pennsylvania Science Teachers' Association. This workshop, called Science on Saturday (SOS), drew 120 teachers from many rural districts of Central Pennsylvania.

Science Olympiad competitions began in Pennsylvania in 1985, directed by the Office of Science Education. Since then, Richard Smith and Susan Zamzow serve as co-directors of this statewide activity. In 1990 there were three regional competitions and a state competition. The national competition was held at Clarion University May of 1990. Nearly 3,000 students were involved. This competition showcases problem solving by teams of students in grades 6-9 and 9-12. The events follow classroom lab tests, popular board games, TV shows and Olympic games. Some of the training costs for this are subsidized by the Dwight Eisenhower funds from PDE.

Title II of the Education for Economic Security Act (ESEA) enacted in 1984 funds activities to strengthen the skills of teachers of mathematics, science, foreign languages and computer learning, and that increases the access of all students to that instruction. Funds for math and science are available through the Dwight D. Eisenhower Mathematics and Science Education Act as entitlement grants to local school districts, competitive grants, and state programs.

The entitlement money supports a variety of local science education needs. Some districts pool their funds to offer programs within the intermediate unit. Many schools use their allocation to enable teachers to attend professional conferences. The competitive grants provide resources for projects having statewide regional and local applicability such as: summer intensive science and math for migrant students, staff development projects for rural science teachers, staff development of elementary science teachers, and the design of specific curricula for local use.

Pennsylvania provides Governor's Schools of Excellence in Science at Carnegie Mellon University and in Agricultural Science at Penn State University. These five-week summer intensive learning experiences attract about 150 outstanding high school students each year. The science program provides experiences in molecular biology, modern chemistry and physics, discrete mathematics and computer science. Electives include biochemistry, nuclear chemistry, mathematics and philosophy of science. Students select one laboratory course in biology, chemistry, physics, or computer science and participate in a team research project.

The Governor's School for Agricultural Science provides experiences in remote sensing, land use planning, insect ecology, silviculture, wildlife management, nutrition, food biochemistry, annual physiology, veterinary science, animal breeding, and resource management. Additional schools in the arts, business, international studies and teaching are offered.

PENNSYLVANIA'S ENROLLMENT IN SCIENCE COURSES

State demographic statistics from the 1980 census show that the state's population is stable with a slight decline. The number of students in elementary school bottommed out around 1986 and has been slowly increasing since then. The secondary school total enrollment will continue to decrease until 1992 when small increases of about 2% per year are expected. Within this context, what does the science enrollment picture look like? Table 1 presents a summary of the numbers of students enrolled in secondary science courses between 1985 and 1989.

While trends are difficult to show with only four years of data, there are several noteworthy conclusions to be made:
1) Biology I, general science (junior high), physical science, and earth and space science are the most heavily populated science courses.

TABLE 1

Comparison of Students in Secondary Science Courses Over a Four-year Period (1985-1989)*

Science	Year			
	85-86	86-87	87-88	88-89
Bio I	134,226	123,379	119,251	111,049
Bio II	17,907	19,363	19,438	16,831
Chem I	65,799	64,516	67,808	63,518
Chem II	8,386	8,503	9,595	7,199
E & Sp. I	86,000	76,956	83,336	81,366
E & Sp. II	3,495	4,549	4,215	6,093
Gen. Sc. (Sr)	30,909	31,085	38,817	36,421
Gen. Sc. (Jr)	120,378	112,682	104,140	98,238
Life Sc.	77,197	69,346	79,363	72,438
Phy. Sc.	92,988	86,431	85,450	87,314
Physics I	34,185	33,383	40,629	34,184
Physics II	2,723	2,890	3,130	2,145
Env. Sc.	17,068	19,200	21,988	31,905
Total Enrolled in Science	691,256	652,283	677,160	648,701
Total Students in State Public Schools	1,683,221	1,674,161	1,668,542	1,658,335
Secondary	841,802	817,199	793,758	762,122
Elementary	841,419	956,962	874,784	896,213

*From Bureau of Informational Systems, PDE

2) Advanced level sciences are holding steady in terms of enrollment.
3) In spite of decreasing school enrollments at the secondary level, environmental science courses are showing large increases in their enrollments.

The driving force for environmental science in Pennsylvania is the Master Plan of 1984. This plan established networks for environmental education with the PDE Office of Environmental Education at the center. From this structure and through the efforts of the Director, Dr. Dean Steinhart, many interagency activities have become established. A special committee, composed of public and private agencies, exists to exchange ideas, outline new programs and seek cooperation in publishing new materials for the field.

Major environmental science activities include: an envirothon competition for high school students in the areas of wild life, forestry, aquatics, soils and a current issue, such as acid deposition; 46 county summer conservation schools sponsored by the Federated Sportsmen; six weeks of summer conservation leadership schools sponsored by Pennsylvania State University; in-service teacher training and pre-service (college seniors) training is provided for schools, colleges and universities. *Project Wild, Project Learning Tree,* OBIS, and *Aquatic Wild* are all available as requested by institutions; inner city nature programs are offered in Philadelphia to 6,000 to 10,000 students with conservation programs based on *Project Learning Tree;* and summer curriculum conferences are offered that stress environmental education and school site development.

Current state regulations in Chapter 5 mandate that elementary science be a planned course. Science is to be taught to every student every year in the elementary grades. It may be integrated with other appropriately planned courses. (A planned course is a written plan that includes learning objectives, expected achievement levels, procedures for evaluation and subject content for an instructional time of 120 clock hours.)

In secondary science Pennsylvania requires five planned courses, that may include laboratory science, three of which must be 120 clock hours each. Laboratory science courses must be offered in biology, chemistry, and physics.

SCIENCE CERTIFICATION IN PENNSYLVANIA:
OPPORTUNITIES AND RESULTS

The Commonwealth of Pennsylvania currently issues five instructional certificates in science for secondary teachers, grades 7-12: biology, chemistry, general science, earth and space science, and physics. State Board curriculum regulations require science to be taught by all elementary trained teachers. The focus of this discussion, however, will be on secondary certification only.

The Instructional I is the initial teaching Certificate. Over the last five years, Pennsylvania has averaged 315 Instructional I certificate issuances in biology yearly, 120 in chemistry, 199 in general science, 62 in earth and space and 50 in physics. Seventy-six (76) of the 86 state-approved teacher preparing colleges and universities in Pennsylvania support at least one of the five science certification programs. In fact, twenty-nine percent (29%) of the state-approved teacher training institutions currently offer programs leading to certification in all five science areas. An additional thirty percent (30%) provide certification opportunities in four of the five areas. The majority of the state university system institutions (the former state teachers colleges) and the state's largest state-supported institutions offer studies in all five areas leading to certification. This means greater accessibility and opportunity for students to pursue science certification.

All of the 76 approved science certification institutions in Pennsylvania offer an initial teacher certification program in biology. Almost seventy percent (70%) of them support certification studies in chemistry while 42% offer a general science program. Thirty-seven percent (37%) of institutions have programs in physics. A distant fifth is earth and space science. Of 76 science certification institutions in Pennsylvania, only about 11% provide an opportunity for students to pursue an instructional certificate in earth and space science.

TEACHER CERTIFICATION REQUIREMENTS: EDUCATION AND TESTING

There are two fundamental requirements for teacher certification in Pennsylvania. First, the candidate must successfully complete a state-approved teacher education program and meet the State Board approved standards for certification. If the student has met all educational program requirements of a Pennsylvania institution, the certification officer at the institution will sign the application and forward it to the Department of Education's Bureau of Teacher Preparation and Certification (BTPC) for processing. For those trained out-of-state or in-staters with multiple preparation programs, the Department's Division of Candidate Evaluation Services (CES) conducts transcript analyses to determine if the preparation meets the State Board's approved teacher education program standards.

The second certification requirement is to pass four tests designed not to predict teaching ability, but to measure knowledge of the subject, and grasp of basic skills, general knowledge and professional knowledge. Candidates for science teaching certification as of the Fall of 1990 are required to take the NTE (National Teacher Exam) Core Battery (Communications, General Knowledge and Professional Knowledge) and a subject-specific test depending on the major: NTE Biology; NTE Chemistry; NTE Biology and General Science; NTE Chemistry, Physics and General Science; NTE Physics; Earth and Space Education for Pennsylvania.

The BTPC determines if candidates meet both education and test requirements before issuing them Instructional I teaching certificates.

TEACHER TRAINING IN PENNSYLVANIA: PROGRAM APPROVAL

Teacher candidates in Pennsylvania must successfully complete a training program that has been approved by the state's Bureau of Teacher Preparation and Certification or is comparable to the state's preparation standards. Pennsylvania employs an accrediting system called "program approval." The State Board, a representative body of citizen leaders appointed by the Governor, authorizes the Department of Education to maintain certain programs standards across its teacher training institutions:

To be authorized to conduct programs that lead to certificates for professional positions, institutions and any of their off-campus centers engaged in the preparation of teachers shall meet all of the following requirements:

1) Be approved as a baccalaureate or graduate degree granting institution by the Department (of Education).

2) Be evaluated and approved as a teacher-preparing institution to offer specific programs leading to certification in accordance with the procedures established by the Department (of Education).[2]

Through its regulatory code, the Pennsylvania State Board of Education directs the Department of Education to approve teacher-preparing programs in the Commonwealth. The Program Approval Review is used to assess the extent to which each teacher education program is meeting Board approved program-specific standards as stipulated in Chapter 49 of the code. Reviews are conducted in each of the state's teacher-preparing institutions once every five years. Institutions may receive a five-year approval, a probationary period in which to modify practices and achieve approval (usually two years), or the program may be terminated.

Program approval standards are developed at least once every ten years by the Department of Education with the assistance of educators from across the Commonwealth representing both higher and basic education. Committee members are chosen by the Department of Education from lists of nominees submitted by professional organizations and associations, individuals from the Department of Education's Bureau of Curriculum and Instruction, school districts, intermediate units and institutions of higher education. While the Department of Education has no established criteria for selecting participants from among the many nominees submitted, those selected to serve are individuals engaged in leadership activities in their respective subject areas.

The standard-setting committees influence the course of professional preparation education in the state because the Department of Education traditionally has not played more than a facilitating role in the deliberations. The Department of Education assumes that as leading educators committee members are the most valid sources for identifying the standards necessary to drive teacher training programs in their fields. The State Board of Education does have the final approval of the standards established by these groups, however. New standards were adopted by the Board in 1985 and became effective in 1987.

PROGRAM APPROVAL REVIEWS: ASSESSING COMPLIANCE

Once the subject-area standards are in place, the Department's program approval review process is engaged. Using the program standards, teams of Pennsylvania educators from both basic and higher education visit the institutions and conduct reviews. Similar to other accrediting processes, the program approval team, led by a state designated chairperson, uses an institution's self-study report to organize and conduct the review.

The Division of Teacher Education of the Bureau of Teacher Preparation and Certification is responsible for organizing and conducting approval reviews. Teacher Education staff schedule the reviews, identify and train team members and chairs, (again from a cadre of names gathered through formal and informal networks of educational interests), accompany the teams on the reviews, study the team reports and ultimately determine from it whether the teacher preparation programs are to be approved, placed on probation, or terminated. No school in the Commonwealth has been dropped from the list of approved institutions as a result of a program approval review. Some have been placed on probation, however, until modifications are made that satisfy state officials that compliance with state standards has been achieved.

The emphasis of program approval is on input measures—the standards which drive the training experiences. (The state does not use terms such as courses or credits.) The program approval procedure is a verification procedure and no attempt is made by the state to measure either qualitatively or quantitatively the nature of the programs beyond assuring minimum standard compliance.

Conducting program approval reviews of science teacher training programs like all other subject areas involves team members examining course syllabi, interviewing faculty and students, observing classes in field experiences, and generally seeking by interview, documentation and observation ways to confirm that the State Board program standards are, in fact, being met.

The Department of Education must be confident that the program standards adopted by the State Board are reflected in all initial teacher certification candidates' preparation programs whether they are in-state or out-of-state applicants. Thus, all candidates for science certification in Pennsylvania must have at minimum those experiences or studies shaped by the program standards of their respective subject area. (See Appendix for the Biology and General Science

Standards). For those graduating from a Pennsylvania-approved program that confirmation is automatic since the institution has had its preparation program reviewed and approved by the Department of Education's program approval process. For out-of-staters, the process entails transcript analysis driven by the standards.

TRENDS IN SCIENCE CERTIFICATION IN PENNSYLVANIA: A BRIEF GLANCE

What are the trends in science teacher preparation in Pennsylvania as indicated by instructional issuances over the last five years? Before examining the figures, it is necessary to understand the nature of the data. Basically, they involve two general certificate types: a standard teaching type and a substandard teaching type. Included in the former category are teaching certificates called Instructional I's and II's. The Instructional II is called Pennsylvania's "permanent" certificate. All active teaching personnel must exchange their Level I certificate for a Level II after six years of service time, three of which must be documented as successful, twenty-four additional credits above the Bachelor's degree and a one-year teacher mentoring experience. A third standard teaching certificate is the Intern, a three year non-renewable license for non-educational baccalaureate degree holders. It is sometimes called the alternate route certificate.

Much attention in recent years has been paid to the alternate route to teacher certification. New Jersey's school-based alternate route program, for example, has received national attention. Pennsylvania has, for many years, supported an institutional-based alternate program called Teaching Intern. The intern program enables non-educational baccalaureate degree holders to enter the teaching profession while completing required educational studies. Alternate route programs are viewed by many to be an excellent way to tap a rich resource of well-qualified candidates for the teaching profession. Over the last five years many science teachers have entered the profession this way. In biology, for example, the Department of Education issued 107 Intern Certificates, 81 in chemistry, 19 in earth and space, 66 in general science, and 47 in physics. The Intern Certificate is valid for three years at which time the applicant must apply for an initial Instructional I certification.

From 1984 to 1989, Pennsylvania issued more standard teaching certificates (Instructional I, Instructional II, Interns) in biology (2012) and general science (1418) than the other three areas combined. During that same period, standard instructional biology certificates (I, II, Intern) averaged 402 per year while general science amounted to an average of 284. By comparison, the Department averaged 159 standard instructional certificates in chemistry, 80 in earth and space and 67 in physics. The number of total Instructional I, Instructional II and Intern Certificates for each of the areas fluctuated considerably during this time and suggests rather stagnant activity or slight declines in the five subject areas since 1987. Table 2 demonstrates this point.

Table 3 below illustrates the combined totals of actual new (Instructional I or Intern only) certificates during that same period.

The second type of certificate involved in the data is the emergency or substandard teaching certificate, so-called because it is issued to individuals who do not hold Instructional I or II's for the subject(s) they are teaching. Pennsylvania issues several kinds of emergency certificates but for this discussion no delineation between and among these substandard certificates is made. Table 4 shows the number of emergency certificates issued from 1984-85 to 1988-89 contrasted with the number of all standard teaching certificates during that same time frame.

Emergency certificates are issued for a variety of reasons. Supply of certified teachers and school location are two principle ones, although frequently many interpret their use as an indication of a lack of certified personnel only. It could be, however, that while there are sufficient numbers of properly certified teachers, they are not interested in or situated near a particular school or locality. A comparison of the instructional and emergency issuance data suggests that both general science and physics are particularly critical areas in terms of providing schools with sufficient numbers of certificated people. It is significant that there are large numbers of emergency issuances in general science even though the state issues a reasonable number of instructional certificates in this area. The same situation with emergency certificate occurs with physics, although it is less surprising since Pennsylvania, like many states, has experienced a downturn in standard instructional physics issuances over the last decade. Both location and demand do appear to play a role in the general science situation while supply and demand probably create the shortfall in physics. The other three areas are also affected by shortages, although the figures appear to indicate a less dramatic supply problem. Obviously, any time emergency certificates are used in place of qualified instructional certificate holders, questions concerning the quality of the instruction must be raised.

CERTIFICATION INTEGRITY: AN ISSUE WITH A FUTURE

Finally, the Bureau of Teacher Preparation and Certification through its Division of Staffing is responsible to preserve the integrity of all instructional certificates by cooperating with the Pennsylvania Auditor General's Office as it conducts certification audits of staffing practices in the public local education entities of the Commonwealth. Misassignments and certification violations are identified and reported to the Department by the Auditor General's staff. While there are a variety of locally-designed science programs and labels, the state must ensure that all secondary courses are taught by the appropriately certified persons holding one of the five official science certificates. If sustained by the Division of Staffing, audit citations can result in fiscal penalties levied against the offending educational entity. The certification audit is the primary means by which the state monitors and preserves certificate integrity. In addition, the differentiated science certificates used in Pennsylvania are becoming more problematic to school officials as the number of science teachers dwindles and certification issuances decline. This is very true in both urban and rural school districts where the bulk of the state's science emergencies are issued. It is becoming more obvious, however, that with the increasing changes in the roles and missions of schools, that subject-specific certification could dramatically change and evolve into some more generic type of certification, simply to allow educational entities to meet ever-changing needs of students and the demands of the market.

TABLE 2

Number of Standard Instructional Certificates Issued by Year

	Biology	Chemistry	Earth, Space	Gen. Sci.	Physics
84-85	308	111	54	241	42
85-86	385	146	78	290	56
86-87	531	211	89	366	71
87-88	381	164	93	273	89
88-89	407	166	87	248	79

TABLE 3

Instructional I or Intern Certificates Issued from 1984-85 to 1988-89

Biology	Chemistry	Earth, Science	Gen. Sci.	Physics
1585	602	308	996	249

TABLE 4

Standard and Emergency Instructional Certificates Issued 1984085 to 1988-89

	Biology		Chemistry		Earth/Space		Gen. Sci.		Physics	
	Stand.	Emerg.	Stand.	Emerg.	Stand.	Emerg.	Stand.	Emerg.	Stand.	Emerg.
84-85	308	120	111	65	54	11	241	193	42	29
85-86	385	141	146	68	78	26	290	296	56	35
86-87	531	85	211	60	89	17	366	256	71	31
87-88	381	80	164	61	93	12	273	241	89	34
88-89	407	76	166	57	87	17	248	247	79	44

CONCLUSION

Pennsylvania's teaching certification system is very traditional. In the sciences, it reflects conventional subject areas. The fact that the state's supply of standard certificated science teachers is inadequate to the demand is also indicative of the situation most states are currently experiencing as young people are attracted to other professions and few with the skilled scientific backgrounds are entering the teaching profession. As science education moves into the 21st century, programs such as STS and curriculum reforms can be expected to thrive. In this context it is very important to study the extent to which states like Pennsylvania should and can sustain the *status quo*. It may be that certification and teacher preparation as it now exist has a limited life and that substantive reform measures are only a matter of time.

REFERENCES

1. Piltz, A. and R. Sund. 1968. *Creative Teaching of Science in the Elementary School*. Allyn and Bacon, Inc., Boston, MA. pp. 192-195.
2. 22 Pennsylvania Code Chapter 49. 14(1):2.

PENNSYLVANIA STANDARDS FOR PROGRAM APPROVAL AND TEACHER CERTIFICATION IN BIOLOGY

Standard I

The program shall require studies of and experiences with living materials in laboratory and field experiences using investigation, inquiry, and experimental methods.

Standard II

The program shall require studies that provide analyses of the characteristics of organisms such as cellular biology, homeostasis, systematics, behavior, reproduction-embryology, genetics, evolution and ecology.

Standard III

The program shall require the studies of the interrelationships of organisms with the biotic and abiotic factors in their environment.

Standard IV

The program shall require studies of and experiences in general chemistry, organic chemistry, biochemistry, physics, earth sciences, and mathematics as they relate to biology.

Standard V

The program shall require studies of and experiences in designing, developing, conducting, and evaluating laboratory activities, using techniques, equipment and facilities that meet current technological standards for such laboratories. These studies should include computer applications to science teaching, emphasizing computers as a tool for (a) computation, (b) interfacing with lab experiences and equipment; (c) processing information, (d) testing and creating models, and (e) describing processes, procedures and algorithms.

Standard VI

The program shall include studies of the interaction of biology and technology and the ethical and human implications of developments such as genetic screening, cloning, organ transplants, etc.

Standard VII

The program shall require studies of and experiences in using contemporary biology curricula and the innovations in instructional practices.

Standard VIII

The program shall require professional studies distributed over the areas defined in General Standard XIV. The student teaching experience should require the candidate to demonstrate competency in these areas.

PENNSYLVANIA STANDARDS FOR PROGRAM APPROVAL AND TEACHER CERTIFICATION IN GENERAL SCIENCE

Standard I

The program shall require studies of the basic concepts and principles of biology, chemistry, physics, and the earth-space science.

Standard II

The program shall require studies of and experience in general and specific technological contexts in which scientific facts, concepts, principles, and processes are applied.

Standard III

The program shall require studies and experiences in the selection of recent research findings and processes for use in instructional activities.

Standard IV

The program shall require studies of and experience in the integration of instruction across scientific disciplines and the transfer of concepts and processes from one science to other sciences, technology, and non-science fields.

Standard V

The program shall require studies of and experiences in the use of the scientific processes such as observing, classifying, communicating, measuring, hypothesizing, inferring, designing investigations and experiments, collecting and analyzing data, drawing conclusions, and making generalizations.

Standard VI

The program shall require studies of science as a vehicle for developing capacities for inquiring, problem-solving, decision-making, critical and creative thinking.

Standard VII

The program shall require studies of the application of mathematical skills necessary for scientific problem-solving and decision-making including the basic operations with whole numbers, fractions and decimals, coordinate graphing, and simple, one-unknown equations.

Standard VIII

The program shall require studies of and experiences in the application of theory and research on the psychological development, behavior, and learning of children including exceptional characteristics and/or special needs that must be met in the science classroom, projects and laboratory activities.

Standard IX

The program shall require studies of and experiences in designing, conducting, and reporting of research in the science fields and in the teaching and learning of science.

Standard X

The program shall require studies of and experiences in use of the computer as a science teaching tool to facilitate student learning.

Standard XI

The program shall require studies of and experiences in identifying the nature and scope of science and technology related careers open to students of varying aptitudes and interests.

Standard XII

The program shall require studies of and experiences in designing learning events which address the applications and consequences of science and technology to the human race, and the environment.

Standard XIII

The program shall require studies of and experiences in planning and implementation of meaningful student-centered educational experiences outside the regular classroom, with emphasis on utilization of the immediate area—its local earth-space systems, ecosystems and examples of science-technology at work.

Standard XIV

The program shall require study of laws governing laboratory experiences and the instructor's moral responsibility for each student's safety.

Standard XV

The program shall require professional studies distributed over the areas defined in General Standard XIV. The student teaching experience should require the candidate to demonstrate competency in these areas.

Science Education in the United States: Issues, Crises and Priorities. Edited by S. K. Majumdar, L. M. Rosenfeld, P. A. Rubba, E. W. Miller and R. F. Schmalz. ©1991, The Pennsylvania Academy of Science.

Chapter Twenty-Four

TOWARDS THE GOAL OF AMERICAN SUPREMACY IN SCIENCE: PERSPECTIVE OF AN URBAN AND A SUBURBAN SUPERINTENDENT

LOUIS J. HEBERT[1] and PAUL D. HOUSTON[2]

[1]Superintendent of Schools
Abington School District
970 Highland Avenue
Abington, PA 19001
and
[2]Superintendent of Schools
Tucson Unified School District
1010 East Tenth Street
Tucson, AZ 85717

INTRODUCTION: Louis J. Hebert

When I asked Paul Houston to join me in addressing the President's goal to make the United States first in the world in math and science by the year 2,000, he accepted without hesitation. I wondered if he knew something I didn't. Would the views of an urban superintendent be different from those of us in the suburbs? It would be interesting to find out.

Rather than define separate topics for consideration each of us addressed the goal as presented. Crushing schedules prohibited an exchage of ideas during the writing so neither knew what the other had written until the drafts were complete. The result is to some extent a confirmation of expected differences between a city school district of 57,000 students and a suburb of 5,500 students. Perhaps less expected is the level of agreement on how the problems we face should be resolved. Both Hebert and Houston believe that if we are to be number one in science there must be:

- Solid academic preparation in the arts and sciences demanded of all prospective teachers.

- A major effort to teach existing elementary teachers enough science to present a meaningful science program.
- Financial support for ongoing elementary and secondary science inservice which goes far beyond traditional course taking.
- K-12 science programs which incorporate science and technology into activity based sequences.
- The resources necessary to implement those programs, i.e., equipment, supplies and enough laboratory space to ensure that each student is a participant, not merely an observer.
- An attack on illiteracy which will allow students to meet these new academic challenges and diminish increasingly serious divisions along racial, linguistic and economic lines.
- A national effort to remind aging Americans, union leaders, teachers, administrators and Board members that the quality of life that education makes possible is too precious to be bargained away.
- Recognition that the Presidential mandate cannot be met without tapping both the nation's noosphere (thinking layer) and its treasury.

MAKING AMERICA NUMBER ONE IN SCIENCE—
AN URBAN PERSPECTIVE
Paul D. Houston

The President has spoken. The United States will take a backseat to no one in preparing children in science and mathematics. We will lead the world by the turn of the century. As the great Pharaohs used to say, "so let it be written, so let it be done."

I would suggest that this one was easier written than it will be to carry out. We should be particularly concerned if we try to make this happen by pursuing our old methods: put money into training more math and science teachers, provide summer institutes for their enlightenment, buy some microscopes and hope for the best. This is also a very daunting challenge when one considers the numbers of students currently attending urban schools which are inadequately staffed and stocked to face the needs these students bring to the classroom.

Urban education is difficult under the best of circumstances. Changing demographics, poverty, ethnic politics, and economic decay are allied to inhibit progress. Urban districts are large, complex, and difficult to lead. Changes are swallowed up by the sheer number of buildings and personnel, and are thwarted by contracts, lack of funding, and the distractions of conflict and confusion. To improve the quality of science education in this context will require a focus of attention, increased resources, and a change in attitude by school board members, administrators, staff and community.

URBAN EDUCATION: THE GOOD, THE BAD, AND THE UGLY

The Good

While urban schools are often viewed as vast educational wastelands, there are certain advantages that urban districts enjoy that would allow real progress towards the President's goal.

First, size can be an asset. There are economies of scale that accrue to large organizations that are not available to small ones. For example, in Tucson we have a full-time staff member assigned to the University of Arizona's Flandrau Planetarium so that thousands of our children receive a hands-on educationally sound and relevant experience on fieldtrips to the planetarium with follow-up activities available to their teachers. We also have an environmental camp, Camp Cooper, which provides all day and overnight experiences for over 8000 students annually where environmental education, and the anthropology of the Southwest are taught. Smaller districts do not have the scale to justify these kinds of expenses.

Large districts tend to be more visible and be seen as being more representative of the community than smaller, suburban districts. Therefore, many contacts, offers of help and donated resources are available. There is interest and concern for large urban districts that lead to opportunities for staff development through foundations, from universities and the like that are unique.

As districts, urban schools often have large quantities of materials and equipment. Access is often a problem, but there is the possibility for use that would not exist in a smaller setting. Likewise, since there are many science teachers in the district, there is a pool of talent available with many different skills and knowledge. This talent can be shared and it provides a richness in human resources that can be stunning when brought to focus.

Historically, large districts have been able to provide specialized middle management positions which can provide leadership to certain areas such as science. Having one person whose only job is to worry about the teaching of science provides great opportunities for staff and students. Unfortunately, recently there has been pressure to "cut the bureaucracy" in large school systems and this kind of specialized help is often the first to go. In smaller districts and in larger ones which have failed to hold on to this talent, staff development and curricular improvements are left up to site administrators and teachers. While good programs can arise at this level, the options often tend to be limited. According to Ted Sizer, former Dean of the School of Education at Harvard and currently a national authority on school reform, it is difficult for teachers to teach and to think about teaching at the same time. It is like the old story of how difficult it is to think about draining the swamp when you're up to your armpits in alligators.

The Bad

Despite the advantages that urban districts enjoy, there are a great many problems confronting quality education. These problems loom large in the way of progress towards the President's goal.

First, facilities are a major issue. While urban schools may have a few modern facilities, most of them are antiquated and in shambles. It has been estimated that billions of dollars need to be spent on urban faciltiies. Age has much to do with it. Urban schools developed early. Many of them date back to the late eighteenth century. Aside from materials, these classrooms look a lot like they did when grandpa was a pup. Further, a lack of funding for maintenance has accelerated the decay. Scarce resources have been put into people rather than buildings because the people have tended to have unions and political clout. My district is a good example. Compared to many urban districts in the Midwest and East, Tucson is a relatively new district. Most of its growth came in the 50's and 60's. Even so, these building are now 30-40 years old and are showing their age. A recent audit indicated that it would take over $500 million dollars to bring the facilities up to current standards. Fortunately, the Tucson community recently approved a $400 million bond program so that great improvements will occur in the next few years including an entire revamping of our science labs. This is the exception rather than the rule for urban districts. Most continue to patch together facilities that are too old and too worn out. Obviously, under these conditions, modern science labs are rare. In fact, many schools have no labs. It is difficult to teach a hands-on experiential program under these conditions.

Urban districts are at the center of controversy and conflict. This results in competing demands being placed upon the district. This leads to a confusion in priorities. With so many things falling around the heads of administrators and teachers, science is often relegated to a place where it is taught when and if time and materials are available. This tenuous position in the program is hardly conducive to quality. If science is to be a national priority, it must start at the local level. Right now, it is well behind a number of other issues. In Tucson, science has only been taught one semester out of the two year middle school sequence. We have just revised our program to a three year plan with three full years of science being required. This kind of policy direction is critical if improvement is to be made.

In any school district, urban or suburban, the quality of education is directly dependent upon the quality of the instruction. Teachers make it or break it. Often at the elementary level many of those who are expected to teach science have little background in the subject. I'll speak more about this later. Even those who are well trained sometimes allow their knowledge to subvert the process. Like anyone who is expert in a discipline, the teacher can sometimes let the love of the discipline get in the way. There is a tendency to want to teach the child everything the teacher knows. Obviously, that is not possible. I would also assert that it is harmful to the teaching of science. Science more than most disciplines is a process and its power and potential comes from the process of learning and discovery. Teachers who would give their students the knowledge they have rob their students of the real opportunity to learn which comes from the process of discovery.

Resources and their allocation are a great source of problem in urban districts. Obviously, with so many students coming to school with very basic health and social needs, money that is used in many districts for learning is channeled into providing basic student support. For example, my district has about ½ of its students below the poverty level. For many of them family support and financial security are totally foreign. They come to school hungry, poorly clothed and in need of counseling and social and health services. All this comes from budgets which are often less than budgets in wealthier suburban districts. My current district has over 57,000 students. About 35% come from homes where English is not the first language. Our mobility rate is over 40% annually. We spend around $3500 per pupil. My former district in New Jersey had students from upper income homes with many advantages and yet that district now spends nearly $9000 per pupil. This gap between have and have not seems to be expanding.

Even the resources that are available are often not used as well or as wisely as they need to be. Principals decide to spend the money on some other parts of the budget rather than replacing test tubes or updating the microscopes. Again, the only real solution to this would be policy direction. Otherwise, every school is going to have a different situation with science taking the place it is given by the building administrator and department heads.

The ethnic diversity of urban districts mitigates against student interest in science, since access is a major problem. Students tend to become what they believe they can become. The dearth of women and minorities in science tends to send a message to urban youth that this is not a profession to think about or to aspire to join. Further, the diverse learning styles students bring with them can only be capitalized upon if schools are prepared to cater to them. As long as science is taught as a lecture and occasional lab course, many students with other learning modes will not see it as a place to be.

The Ugly

The underside of urban schools is not in their diversity. Treated properly, this can be a strength. It is not in their complexity. Complexity makes it difficult to get things done, but not impossible. It does not come from the grinding poverty or the lack of resources although these present constant problems and challenges. The ugly side of urban schools comes from the fact that over the years conflicts have become institutionalized and controversy has become a way of life. This has politicized the situation so that cynicism has become the coin of the realm.

It is difficult to charge up staff if they don't trust you or see it as an administrative plot. It is hard to get principals to take leadership when they have been undercut for years. It is hard to chart a sense of direction when different factions are lying in wait to subvert the process.

The politicization of urban schools through union conflicts, ethnic differences, and the grassroots nature of school boards has fragmented leadership, dispersed power and led to a gridlock of leadership that makes it difficult to chart or change the sense of direction to confront problems such as inadequate science education. This has led to a sense of inertia that, coupled with the sheer size and numbers represented by urban school districts, has made it nearly impossible to make major alterations and corrections.

This has rendered urban districts unable to affect the critical issue—functional literacy. Far too many students in urban districts come to school ill-equipped to make it in the academic setting. They spend too many years being only partially served and leave without the basic skills necessary to function in our increasingly complex society. Students who have trouble reading, writing, and computing tend to have problems in the science classroom and tend to make poor candidates for making a career of science. Before science can move to the forefront in urban settings, these issues must also be addressed.

In an effort to help make the President's bold goal a reality I would offer a modest proposal. When it comes to improving science education for K-12, I would suggest we turn things upside down and inside out.

Upside Down

Our normal approach to dealing with this kind of problem is to work on the end of the pipeline. Not enough kids going into science and engineering in college? Simple, improve the teaching of high school math and science. I would suggest that the weakness identified is an effect, not a cause. Although I think we have many wonderful science teachers in high school, I am certain that more effort in increasing their numbers and competence would be a good idea. However, a much better idea would be to make significant efforts to strengthen the teaching of science at the early grades.

I would start with Kindergarten and work my way up. Most of our elementary teachers do not consider themselves very expert in math and science and many have shared their fear of these subjects. Al Shanker, leader of the American Federation of Teachers (AFT) has said that "the United States is the only industrial country in the world where most elementary teachers do not know science, and they'll tell you they don't. They feel very uncomfortable with it . . . There is no point in firing these teachers because there aren't a bunch of scientists and mathematicians lined up waiting to take their jobs."

We simply fear what we don't know or understand. Staff training, subject specialists, and improved curriculum and materials for the early grades would allow students to start off on the right foot. High school is much too late to kindle a student's interest in science as a career. If I were the President, I'd see that the Department of Education puts a major emphasis on math and science at the elementary grades, even at the possible expense of increased efforts at the high school and college level. Start at the right end of the pipeline, and it will be full and productive for years to come.

Inside Out

Science and scientists have to change the way they think and talk about science to the non-scientist. It must be demystified and made accessible to the lay person.

First, they must drop the jargon and the ponderous phrases. Bilingual education is fine, but one shouldn't be required to speak "Sciencese" to access an understanding of the wonders of our world. I realize the danger of an educator criticizing someone for the use of jargon, since it is a major disease of ours, but if science is to be improved and enriched it must become open to the multitudes.

Second, the way science is taught must be changed dramatically. We must find ways to put the wonder and excitement back into it. We have many children today who hate science but are intrigued with black holes, space travel and dinosaurs. So much of science teaching today consists of telling children what we know and making them remember that. We need more emphasis on what we don't know. The exploration of the unknown is the true adventure and it is there for everyone.

This is particularly true for urban education. James Shymansky and Ted Bredderman have done extensive research to determine the most effective ways of teaching science. One finding was clear—the benefits of hands-on science were greatest for disadvantaged students.

Third, we must remove the "fear of trying." Again, to the extent that teachers and students are afraid of science, progress will not be made. We must let them know that it is a much messier discipline than it is purported to be. We must make it clear that it is a discipline that is built on failure and that it is OK not to know everything. That is why we experiment.

Fourth, we must replace the esoterics with the universal. If we are to turn around the dismal statistics that show only a tiny percentage of women and minorities entering the field, we must start stressing the universals. This is the quickest way to open access. To the extent it is believed to be a closed priesthood, problems will continue. The divisions between the disciplines must be broken down. Subjects should be better integrated. A focus on the history of science allows students to understand how society is affected by science. Our art, economy, and even our living habits are all intertwined with it. Language arts and writing are crucial skills for the scientist. Liberal arts must underpin any aspiring inventor. We must make a real commitment to improving the teaching of critical thinking and problem solving.

In fact science can be a subject that breaks down some of the disparities that children bring to school. David Hawkins, a scientist and educator, has argued that, in fact, math and science can become the great equalizer in schools if taught properly. He asserts that curriculum based on topics familiar to all students can equalize opportunity. All children may not be able to go to museums, but every child can observe bugs or small animals. They can wonder about clouds or about the water that comes out of a tap.

Finally, the science community needs to go easy and hard on educators. The going easy consists of much of what I have already mentioned. Like the Berlin Wall, the barriers between the science community and the rest of us must fall. However, as we work with our teachers, I hope the science community will have some high expectations for us. They should invite high school teachers into the universities and businesses and treat them like peers. They have good sense and are very capable of understanding and participating in much of what is happening today. Have high expectations for educators.

School districts must make certain that elementary schools are focusing on these issues. Adequate inservice support must be there for all teachers so that they can feel comfortable in delivering their material. We must insist on a problem oriented, hands-on curriculum. And we must offer adequate administrative support. Providing a specialist to support teachers and administrators and to guide the inservice seems to be a minimum.

When I was Superintendent in Princeton, New Jersey we developed and implemented an Inventor in the Classroom program, much like the Artist in the Classroom program many districts enjoy. Children must be given access to the unique and interesting ways inventors and scientists think and it must happen early in their lives if it is to have an influence.

If schools and the science community can join hands to turn around our present situation by turning things upside down and inside out, perhaps the teaching of science and math in this country will stand as a fitting tribute to President Bush—his pharoaich pyramid.

"SO YOU WANT TO BE NUMBER ONE!! THE VIEW FROM THE SUBURBS"
Louis J. Hebert

Sputnik And The Bush Challenge

The long list of articles, reports and recommendations on the teaching of science in recent months - including the cover story in *Newsweek* (April 9, 1990) - has produced no response in the school community which is remotely comparable to the reactions to Sputnik. Simply announcing that the U.S. should lead the world in science and math education by the year 2000 captures none of the drama of the 60's. That challenge was elegant in its simplicity and in its honesty, i.e., the threat of Soviet superiority is real . . . what resources must we marshall to meet or beat that threat?

This was not a parade of rival forces—this was the real thing—a war to be fought in laboratories, classrooms and factories. Because the call to arms was real, the response was real. The resources of the nation were to be put at the disposal of new warriors whose targets were clearly defined: put a man on the moon and make U.S. science and technology second to none. The authenticity of the challenge and of the response were not in question. No one pretended that the job could be done easily or cheaply.

The impact on American schools was extraordinary. A different kind of science had to be taught and it had to be taught differently. The new science was to focus on the methodology of the scientist; to teach the student to think as the scientist thinks. To do this the student was to participate in "hands on" science lessons instead of listening passively to teacher dominated lectures and lessons. This kind of science required new laboratories, new equipment, supplies and specimens consumed by students not teacher demonstrators, more lab time and a teacher who knew how to work in this milieu. School districts across the country responded with eagerness and confidence that they could and would make a difference. That optimism was the hallmark of the new science.

Demographics + Federal Distancing = Deprivation

What has happened in the last 20 years? Why aren't school districts, including those in affluent suburban communities, responding as they did 20 years ago? Demography tells part of the story. It is no longer possible to assume that taxpayers will provide the best science programs available now that the words taxpayers and parents are no longer synonymous. With suburban parents of school-aged children making up less than 30% of the taxpaying public, the annual quest for higher taxes is meeting increasing resistance. For taxpayer groups, lab-centered science programs (which have high start-up costs, require smaller class sizes and have high maintenance costs) are not seen as solutions but as one of the problems of a first rate school district.

The summer institutes of the 60's provided an unparalleled level of teacher training for many of the science teachers now approaching retirement age. Without training in the hows and whys of the new science programs their impact would have been felt by far fewer students. The Federal Government's financial support for university and school district staff was a powerful affirmation of the rhetoric of national goals. The cessation of that support left intact a cadre of teachers well-trained in the science of the 60's, but there was no provision for renewal or for learning how the explosion of technology and scientific knowledge which followed should change the teaching of science.

Current contractions in local support, because there are fewer parents with school-aged children, are not being offset by increases in federal and state support. Instead, the state's share of district instructional expenses is reduced annually just as the state bemoans the reduction of federal support for its budgets. Deprived of access to needed state and federal assistance, assurance that science education is effective is left to individual teachers and their already beleaguered school districts.

Experienced science teachers - who are faced with the need to give up summer income and to pay high university tuition for science courses that do not guarantee new insights, new methodologies or even new content - often say no. Elementary teachers enrolled in graduate programs regularly elect less demanding education courses in lieu of science. Cynics say that teachers with the greatest need reject the option to learn but not the option to teach. Confronted with the reality of individual choices and the terms of union contracts, suburban school districts find that they can mandate participation in science inservice programs and pay teachers for their attendance or they can accept the alternative - less effective science instruction.

It should come as no surprise that a local school district cannot support training sessions to rival those of the National Science Foundation. The research base, the time and the money to establish new directions are simply unavailable to local educators. In light of the federal government's tepid response to a genuine international challenge, the absence of excitement among local educators should come as no surprise. The desire to protect a school district's investment in its staff by ensuring that knowledge and skills remain current appears to diminish with salary increases. The need remains. Unfortunately, reductions in support for teacher education result in a diminished capacity for the effective teaching of science, i.e., science which engages the student.

Accountability, Thinking And Time: Compatible??
The cost of offering activity-centered courses and the dearth of teacher training opportunities have certainly contributed to the growing dominance of text-based science courses, but no more than the move to greater accountability. Those with stated concerns for high standards, including the colleges and universities to which suburban school districts send their graduates, insist upon measurable evidence of program effectiveness. Unfortunately, what is most easily and most commonly measured is less often an effective assessment of scientific reasoning. Courses made up of bits and pieces of information about science, produce students who perform well on instruments designed to sort the best from the rest, but often fail to excite good students about the significant issues of science. Moreover, there are tests and there are tests. The performance of American students on international tests of scientific knowledge, laboratory and inquiry skills suggests that the test-driven science courses to which most students are exposed are poor science courses. Many suburban school districts

do well in science competitions; they will do better when more students believe that science is worth the time and the effort it demands.

If the school day is no shorter than it was 20 years ago why is there competiton for science students and for time within the school day? Why is it so difficult to provide double lab periods in the secondary schools and regular science experiments in the elementary schools? Because there are new competitors for the available time. Consider the impact of computer education, drug abuse education, AIDS education, the education of the gifted, the handicapped, the non-English speaking and a host of others. In the best of circumstances it is difficult to meet the needs of the full range of students in the hours and days allotted to the task. Sharp contractual limits to the use of teacher time during and beyond the school day impact the time available to set up experiments, gather and distribute materials and thereby create yet another barrier to activity based science courses.

Wanted: Well Educated And Well Trained Teachers

Given the inhibitors to effective science education identified above, the fact remains that there are many good, and some outstanding, science programs in secondary schools. Because this is the case, it is important to identify factors which make it possible for science programs to flourish. Unquestionably, the first of these is the presence of a well educated science staff. In Abington High School, 92% of the science teachers have a master's degree in science; at Abington Junior High School more than 70% have a master's degree in science. Most of these teachers have been in the school district for more than 20 years and have accumulated at least 30 credits beyond the master's degree. However, few credits beyond the master's are taken in science unless the teacher is involved in a doctoral program - an option most teachers believe to be of limited financial or instructional value.

A senior and stable faculty with substantial preparation in the field of science is a major asset but its potential would not be realized were it not for the support provided by the school district. That support includes excellent laboratory facilities, equipment, supplies, instructional materials and the class sizes which make effective instruction possible. Because science labs are configured for 24 students, that is the number assigned. As a result, teachers are not only assigned lab-centered science classes - they teach them. Should that comment strike the reader as odd, I would add that teachers in many school districts regularly skip the teaching of labs because they lack equipment, supplies, time, or even laboratories.

Who Takes Science And Why!

In Abington School District all students are required to take science for eleven years; 59% take science for each of 12 years. With a student population of 1176, Abington High School serves a prosperous middle class, a substantial blue collar community and a minority population which approaches 15% of the total. In 1989-90, 140 students were enrolled in Advanced Placement biology, chemistry and physics with 56 more in honors level chemistry or physics. There were 309 students enrolled in chemistry, 207 in biology, 153 in physics, 82 in aerospace science, 80 in physical science and 47 in science and technology.

Elementary Education: The Start Of The Beginning

In Abington, as in most school districts, elementary teachers came to us with little preparation in science or its methodology. Faced with ideas, experiments and equipment that were not "user friendly," many elementary teachers found that they just "didn't have the time" to teach science as anyting but a reading exercise. To deal with the need for a stronger content base, an elementary principal enlisted the help of a parent who is chairman of the science department at Beaver College. This professor and a classroom teacher with strong science background worked together to convert an existing classroom into a laboratory. Then they developed classroom activities that teachers could understand and students learn from.

This development occured at about the same time that the superintendent established K-12 curriculum committees in each subject area. The district science committee concluded that all elementary teachers needed to know more science and how to teach it. With the full support of the Board of School Directors, 30 hours of science inservice was approved for each teacher in grades 1-6. Instruction took place in 10 three hour sessions which alternated between A.M. and P.M. sessions (covered by substitute teachers) so that no teacher missed the same subjects more than five times. Dr. Rose was paid $9,000 for the 30 days he spend working with 115 teachers; the substitute teachers were paid $28,000.

The mandatory participation of teachers and the very strong suggestion that principals attend all sessions could have produced a negative response. Instead, the reaction to science inservice was the most positive the district had yet experienced. More importantly, this first half of a two year program made science education in the elementary schools real. Not only did teachers now spend the stipulated amount of time teaching science, most found ways to exceed the recommended 100-125 minutes per week in primary grades and 225 minutes in the upper elementary grades.

The elementary science program also benefitted from giving certain teachers responsibility for science/math and others for language arts/social studies in the fifth and sixth grades. In the lower grades, science continued to be taught in self contained classrooms, but by teachers with a new belief in their ability to teach science. When other staff members were given the added responsibility of ensuring that needed supplies and specimens arrived when needed, activity centered science became a reality. Was this two year training program worth the $80,000 it cost?

What happens to the secondary program when it encounters a better prepared elementary student? This was the question posed to the outgoing chair of the junior high school science department. After a thorough analysis of the new elementary program, the chair concluded that the sixth grade program was now more rigorous than the seventh grade program. The response was to eliminate duplications and to plan a total revamping and upgrading of the junior high school curriculum. Although it has created more work for the junior and senior high schools, the secondary schools are effusive in their praise of the work done. The newly trained elementary teachers have already had a demonstrable impact on student learning and attitude about science.

The Future vs The Budget

What will happen to strong science programs
- As pressures mount to hold costs constant?
- When class sizes exceed the number of lab stations?
- When funds for conference attendance are eliminated?
- When funds for curriculum and professional development are reduced or deleted?
- When the cost of activity-centered programs becomes a greater issue than the effectiveness of the programs?
- When staff members are no longer selected because they are the best available but because they are the best available at an entrance level salary??

What will happen if court decisions or legislative action dictate that no school district can spend more than the least affluent of its neighbors? The questions are real; the responses are frightening to contemplate. Unless there is a national strategy to ensure that the least effective schools become better, the most effective are likely to become mediocre.

What can be done? Local school boards can and must remain advocates for children rather than for taxpayer groups. The pressures to define desired outcomes in terms of dollars saved must give way to insistence upon the best education possible, because the best is barely good enough. The states must accept their constitutional mandate to educate and the fiscal mandate that requirement implies. The nation must restore its support for schools so they can assume the position of educational leadership once held.

One elementary school in Abington School District was recently cited for performing in the top 10% of schools involved in an international math competition. More of our schools need to make it to the top. I am persuaded that we can perform equally well in science. I am also convinced that we have little chance of improving our performance without strong teachers, strong and ongoing teacher training, an improved curriculum, more instructional time and the money required to transform the dream of better education into reality.

SUMMARY: Paul D. Houston

One of my favorite quotations is from the great American philosopher, Yogi Berra, who opined that "It ain't over 'til it's over." Clearly this country has far to go before it can claim preeminence in the field of science education K-12. To take that position will require thought and effort before it is over. Yogi is also quoted as saying "this feels like deja vu, again." Unfortunately, much of the effort now being mustered to become number one stands a good chance of being simply a pale rerun of what was done 30 years ago right after Sputnik. If we follow that particular trail, we will not become number one and will probably move further behind. As a nation and as a profession we must avoid the tendency to relive our history and try to avoid the "deja vu again" syndrome.

The challenge facing the improvement in science education in this country is no less grave or serious than it was in the early sixties. It will require imagination, and real commitment on the part of our leaders and will require real leadership from the imaginary and the committed. We cannot afford the "tepid response" cited by Hebert and we cannot leave things right side up and out as Houston suggests. One concern arises from our education President who has set the lofty goal of being number one. This is also the same President who has suggested that you can't solve problems by throwing money at them. Certainly that thought has merit if one considers a wild pitch notion of solving problems where the resources are unguided and unfocused. However, both Hebert and Houston have suggested that focused resources do matter. Staff development, preservice training of teachers, experiential science programs will all call for expenditures not currently available in strapped local distric budgets. The challenge at the federal level will be to identify resources that can be targeted to be the most effective. Again, both superintendents suggest heavy investment at the front end of the educational system—K-8.

This resource issue is particularly important in the staff training area. Hebert describes a program of training that was most effective for his teachers in improving science instruction. For a similar program to be developed in the Tucson system, nearly $400,000 would need to be made available only for that one inservice program. With most school system budgets having less than one percent available for staff training, the problem becomes immediately apparent.

Perhaps of even greater need from the federal government is a kind of leadership which has been sorely missing for a number of years. With the shifting demographics moving those with a direct interest in what happens in schools into a distinct minority and with a growing underclass whose day to day existence hangs tenuously in the balance, the vision and direction of what should be done and how it should be done, and perhaps most importantly WHY it should be done, must come from the federal level. We need a voice from the Secretary's office that goes beyond decrying the problem and provides a real sense of direction and useful information for school systems to build upon. This type of leadership was provided after Sputnik and the burst of activity and improvement was palpable, if short-lived. A sustained effort is now demanded.

Another Berra quote seems an appropriate way to end. This one comes from Dale, one of Yogi's sons, who responded to a friend of mine who had commented to Dale that he and his father had a lot of similarities. Dale replied to this observation with a quote that best summarizes the preceding article on the view from the cities and the suburbs—Dale said, "Yes, but our similarities are different." Funding, flexibility, and vision mark great differences between suburb and city. But the need to improve teachers, focus on experiential learning, and start early mark the similarities. If America is truly to become number one we must find ways to bridge the differences and build on the similarities.

Science Education in the United States: Issues, Crises and Priorities. Edited by S. K. Majumdar, L. M. Rosenfeld, P. A. Rubba, E. W. Miller and R. F. Schmalz. ©1991, The Pennsylvania Academy of Science.

Chapter Twenty-Five

ROLE OF STATE ACADEMIES IN SCIENCE EDUCATION

KURT C. SCHREIBER

Past-President
The Pennsylvania
Academy of Science
Professor Emeritus,
DuQuesne University
1812 Wightman Street
Pittsburgh, PA 15217

Kurt C. Schreiber
Former President, PAS

It has already been pointed out elsewhere that science education in the United States is currently in a crisis situation[1]. However, it needs to be remembered that this is not a new situation. Ever since the middle of the 1950's studies have shown that our science and mathematics education on the pre-college level has been deficient. F. L. Fitzpatrick reported that "the Science Manpower Project was conceived in recognition of the fact that in the early fifties the flow of prospective scientific personnel through the schools and colleges had not been adequate to meet the needs . . . and indicated that it was at the high school level that many students with potential scientific ability made decisions into other areas . . .[2]. The "A View from the National Science Board. The State of U.S. Science and Engineering", published in Feb. 1990, puts it "American education in science, engineering, and mathematics is a national problem that has not yet been effectively addressed"[3]. Section IV of this report outlines the "Necessary Actions" that should be taken to alleviate the situation. It may be instructive to look at some of these and see how state academies have worked in these directions and what other actions they should take. Before doing this, however, it may be instructive to give a very cursory review of the historical development of state academies.

Historical. The chartering of state academies can be divided into three eras. The first of these is the period prior to World War I. The majority of these were founded in the nineteenth century, some, like Maryland, Iowa, and Wisconsin[4], before the development of discipline related societies. The second period was between the two World wars, especially in the twenties, when science in th United States took a big step forward. Finally, a few of the state academies have been established since World War II, primarily in the fifties.

At all times the state academies have encouraged research and interdisciplinary communication. The initial primary thrust of the academies formed during the first period was the unity of science and providing a forum where research from different disciplines could be shared, thus, broadening the knowledge base of all. However, their governing documents did include concern for science education as one of their functions. This concern led in due time to the formation in most states of junior academies.

Junior Academies. The report mentioned above[3] in its "necessary actions" section urges "dramatically improved pre-college and undergraduate education", with special emphasis on

a. students being "required to take science and mathematics throughout high school",

b. the need for better science and mathematics teachers and more of them, and

c. "undergraduate science needs major improvement, especially in areas such as research opportunities".

The American Chemical Society testifying before Congress in November 1989 presented a blueprint entitled "Education Policies for National Survival"[5]. Among the many points made, one states "three years of mandatory laboratory-based science for all secondary school students."

State academies have always been advocates for the concept that science cannot be properly taught without hands-on laboratory experimentation.

It needs to be kept in mind that even when the high school science courses have a laboratory component, and, unfortunately, too many do not, these exercises are usually of the cookbook variety. In many cases the students can fill in the answers in the blanks provided without even doing the experiment. Such exercises, when performed, may provide the student with an understanding and experience in the skills necessary to conduct scientific investigations, but they do not expose them to the excitement experienced in doing science. Neither do they represent the kind of work in which scientists are usually involved.

Somewhat more stimulating are experiments which require students to obtain answers not readily obtainable from their textbooks or other references. Experiments in which the student has to report the results of an unknown belong to this category.

The state academies realized early that research even on the pre-college level can furnish the students with the essence of science in a manner not accomplished in most science classes as they have been and continue to be taught in our schools. The junior academies were formed by the state academies as a vehicle to encourage student research a the pre-college level.

In Pennsylvania the state academy was formed in 1924 and the Pennsylvania Junior Academy of Science (PJAS) started to operate ten years later. While the participation in PJAS was very small in the thirties and forties, with the onset of the space era in second half of the fifties, science became a much discussed topic and interest in science activities increased. Ever since the late fifties involvement in the junior academy has steadily increased. A numerical example of this can be given in terms of one of the ten regions which comprise the PJAS. Region seven includes Allegheny and Westmoreland counties (Pittsburgh and its surroundings). While at the 1957 regional meeting 45 papers were presented, the number had risen to 1344 by 1990[6]. On a state-wide basis probably more than 6000 students were engaged in this program in the 1989-1990 school year. Through the Junior Academy organization at their respective schools students are able to contact a sponsor, usually one of their science teachers, under whose supervision and guidance they will then pursue a project of their own choosing.

Doing research, learning about science in the laboratory and experiencing the excitement of obtaining results that have not been part of our knowledge, is one of the benefits students derive from the junior academy program. Each region sponsors annually a regional meeting at which students who have been involved in a project have the opportunity to share their results with others. There are several benefits the students derive from this. First of all they must prepare an oral presentation. This involves careful analysis of their results, drawing valid conclusions from the data, organizing the talk in a logical and succinct manner, and preparing the required visuals to accompany their oral presentation. The second benefit is the presentation in front of an audience which includes other students, teachers and scientists who serve as judges, ask questions and make comments and suggestions. The judging is conducted against a set standard. Thus, in a particular session there could be all "First Awards", or there may be none.

For those who receive a "First Award", there is the additional benefit to go to the state meeting of the PJAS, to present their project there, and to compare their achievements with students from the other parts of the state. Finally, there is the benefit of recognition and awards for outstanding projects in various categories both on the regional and state level. Again, this meeting aspect of the program gives the students a flavor of the environment in which a scientist works.

One of the unanswered questions is how effective is this approach in convincing students to choose careers in science, mathematics and engineering. This is a difficult question to answer. Of the students who continue their research through their senior year in high school, a very high percent will enter college with the thought of majoring in one of the above mentioned. However, the question immediately arises whether these students would have been in such a program even if they had not been involved in research. It is clear from speaking to individual students that some would have, but others would not. In either case the students indicated that their commitment to that career goal had been greatly increased by their research experience. Another indication of the effect is given in "Science & Engineering Indicators—1989". Talking about the academic persistence of high-ability high school minority students, it states, ". . . and science fair/independent research projects also seem to have been influential in their persistence in science and mathematics studies."[7]

While one might hope that everyone of these students would pursue careers in science, mathematics, or engineering, it is important to realize that even if they do not, they have received an insight in how research in science is conducted. Some of these students will become lawyers, politicians, judges, business and community leaders, and they may be the ones who will be asked to make decisions that will affect science. Through their exposure to research they will have a different outlook on the problems that face society, how these need to be addressed, and the role that research plays in solving them.

State academies, including the Pennsylvania Academy of Science, have provided funds for high school student research. While these funds are usually very limited, it gives students the opportunity to pursue research when no funds are available at their schools. In Pennsylvania, a student has to submit a short proposal which outlines the project and includes a budget on how the money is to be used. This proposal is then reviewed by a committee of college scientists before funding may be approved.

Another program operated by at least one region of PJAS since 1976 is the Science Workshop Series. This program is conducted in the Fall on Saturdays and is a joint undertaking between the region and the three universities in its area. The aim of the program is several-fold. First of all, through this hands-on program students have the opportunity to become familiar with instrumentation not available in their high schools and to learn where these techniques are applicable. Secondly, they get acquainted with the universities in the area and meet faculty members and graduate students. While the evaluations by the involved students each year indicate that the program is achieving its goals, the fact that the program is limited to about 100 students each year is most unfortunate. Furthermore, since the program is a Saturday program many students who would like to participate are unable because of other commitments. Single day workshops with some laboratory components are also common in many regions, and similar programs are conducted by other state academies.

Many state academies are involved in another similar type program, namely science fairs. The Alabama Academy of Science should be mentioned in regard to conducting an outstanding program of this nature. In this program, too, students are required to work on a research project; however, the results are presented in a display/exhibit fashion, including three dimensionals and electrically activated ones. In addition the student must prepare a 250 word abstract of the project. Judging is conducted by interview between the student and the judges. This provides a more personal contact between student and scientist. The program is especially well suited for engineering projects which are difficult to present in the PJAS program.

Another type of program that has developed in recent years, in which some states acdemies are involved, is the Science Olympiads. These might be classified under the heading of "It is fun to do science." This program is still in its developmental stage and it is difficult to characterize it in detail since it varies somewhat from state to state and from year to year.

Summer Institutes. The success of all the programs mentioned above and especially the research programs by pre-college students depend heavily on qualified science teachers in the high schools of the state who can give guidance and supervision to the students with their projects. Many such teachers have been identified over the years from their students' participation in the meetings of PJAS as well as in the activities of other states. However, despite the growth of high school research, there are still many school in Pennsylvania as in other states that are not involved and that do not have a single teacher who can help students get started in planning and executing projects.

Research has not been a part of the college curricula which prepare most science teachers. They have not received any instruction on how to involve students in research, how to guide them in selecting a project, how to proceed to set up experiments to give the unequivocal results, and on how to draw valid conclusions from the data obtained. Thus, they lack the knowledge and background to serve as sponsors.

In the early 1970's for three years, with Federal funding, the Pennsylvania Academy of Science was able to conduct one-week summer Science Leadership Conferences. Obviously, in one week it is impossible to prepare teachers to be research scientists. It is possible to demonstrate to them what students at various levels are able to accomplish research-wise, to introduce them to the basic steps in developing a research plan, and to expose them to research techniques that might be applicable in their high school laboratories. Most important, one can expose them to the enthusiasm that other teachers have who have successfully worked with students in the PJAS program. Furthermore, it brings them in close contact with university scientists to whom they can turn for help as questions arise in guiding students.

The teachers who participated in these Science Leadership Conferences have served as catalysts for research in their respective schools and as mentors and models for other teachers in their schools. Their impact has been felt in the increase of students involved in research and presenting their results at the PJAS meetings.

In recent years there have been no Federal funds available for the kind of high school teacher training program described above. Considering the importance of giving all students a better understanding of science and of attracting more students to choose careers in science and engineering especially in terms of female and minority groups, it is of critical concern to develop in each high school of our state at least one science teacher to the point that he/she will be able to start a student research program in that high school. With this in mind the Pennsylvania Academy of Science must in the years to come develop other sources of funding in order to hold these Leadership Conferences. One other point needs to be made here. The teachers who would be candidates for such conferences must be qualified science teachers, that is they must have appropriate undergraduate education in the sciences.

The argument has been made that there are programs available where high school students can be involved in research at university laboratories and, therefore, programs of research at the high school level are unnecessary and to develop teachers for this is a waste of funds. While programs in the university setting are highly commendable and should be continued, there are some points that make the above arguments questionable. First of all, the number of students involved is small in comparison to the number that is and could be served by the PJAS program; secondly, it caters primarily to students who are already turned on by science; thirdly, rarely will eighth or ninth graders be chosen; fourthly, such opportunities may not be readily available to students who live at a considerable distance from a university; and, finally, there may not be the opportunity to continue the project during the school year. There is one other item that should be mentioned. High school teachers receive a great deal of stimulation and satisfaction from successful student-research programs. This is absent when the research is conducted in the university setting.

In all fairness, it must be mentioned that there are some important positive factors involved when high school students pursue research at university laboratories. Among these are: the research will tend to be more sophisticated and carried out with equipment not available at the high school; the student will experience the college atmosphere and come in contact with university faculty and graduate students.

As already mentioned above the success of all programs depend on qualified teachers—teachers who have the proper background in science and mathematics, who are interested in keeping up-to-date on the professional level and who are concerned about the students in their classes. While there are many who fit this description, there are too few of them. With respect to course preparation the National Science Teachers Association recommends 50 semester hours of course work in science. According to a survey conducted by Weiss[8] only a small percentage of biology, chemistry, and physics teachers have that kind of preparation. During the fifties, sixties and early seventies, science and mathematics teachers were helped in keeping current and in removing course deficiences through a variety of summer programs funded by the National Sciene Foundation. NSF in 1957 described these summer institutes in the following terms;

"To stimulate interest in science and science teaching, to improve teaching methods and to provide a greater number of secondary school students with a deep appreciation of science and science careers"[9]. S. Schenberg[10] interviewed teachers in New York City who had attended summer institutes and wrote that "teachers reported that they were able to enrich their own teaching. They encouraged students to engage in research and to work on projects suitable for science fairs and other contests." The NSF pre-college science education funds reached a maximum in 1964 and slowly trailed off until they reached nearly zero in 1983 expressed in constant 1988 dollars[11]. Since 1983, especially through the very aggressive campaign of Bassam Shakhashiri, until recently NSF's Assistant Director for Science and Engineering Education, funds for pre-college science has again been increased.

State academies must utilize their prestige to convey to representatives and senators from their respective states the importance of supporting such programs both by NSF and the Department of Education. College faculties, who represent the major component of state academies membership, need to realize that efforts to improve high school science through both upgrading of teachers and reform in curricula, is in their own best interest. Increases in science, mathematics, and engineering enrollments can be the result of exciting students early in their schooling about science and mathematics.

Some state academies have been able to conduct a variety of summer institutes with private funding. While most of these are aimed at college faculties, some do have relevance to high school science teachers and provide, thus, up-to-date knowledge and experience to these teachers on a specific topic.

Science Talent Searches. Many state academies, including the Pennsylvania Academy, sponsor state science talent searches. These activities are coordinated through the Science Service in Washington, DC and are extensions of the national Westinghouse Science Talent Search. They serve several purposes. First of all, they serve to encourage promising students to participate in the national search. Secondly, they identify and give recognition to outstanding science students who for various reasons have not entered the national competition.

The Pennsylvania Science Talent Search (PSTS) operates in conjunction with the PA Junior Academy of Science (PJAS). Students who are juniors or seniors in high school are interviewed by a panel of two or three scientists and/or college admissions officers at the regional meetings of PJAS. The students must have pursued research, the results of which they are presenting at the meeting. Based on their high school record, SAT scores, high school teachers' recommendation and the interview, students may receive the appropriate recognition. The PSTS will also write letters of recommendation to the colleges the students apply to. At the state meeting the director of the PSTS meets with high school juniors to alert them about the Westinghouse Science Talent Search, encourage them to start working on a project and to enter the contest, explaining the benefits to be derived from entering, and clarifying questions about the procedures, deadlines and requirements of the competition.

Curricula. It has become evident that science instruction on the elementary school level is an important contributant to the developing behavior pattern of children. This is the age when they are still curious, intrigued by the world around them, and astute in discerning the attitudes of adults. Many of the elementary school teachers have little science background and, therefore, are uncomfortable in teaching science. The students quickly pick up that dislike and often model their behavior pattern in accordance with it.

The Pittsburgh Public School system has formed under the leadership of Doris Litman with the support from Superintendent Wallace, the Pittsburgh Science Institute devoted to bring excellence in science education to the system. "The Institute identified teachers from each elementary school, a total of 90, and provided 240 hours of special training in science instruction. It also supplied the schools with the necessary equipment and supplies."[12] A similar 75 hour training program for middle school teachers taught them the "hands-on" approach to teaching of science. Funding for these special activities has come largely from non-public school sources. Programs like this one should be operating in many other parts of the U.S. and should receive support, guidance, and help from state academies.

The science background of elementary teachers is one problem, but there is another. In order to provide an outstanding program of science instruction, it must be coordinated, integrated, and continuous from kindergarten through the senior year of high school.

The Am. Association for the Advancement of Science is in the process of developing such a program. It is entitled "Project 2061: Science for All Americans". Details about this are given in another chapter of this book. What are the responsibilities of state academies with respect to this or any other program of this nature? Above all, it must be understood that the academies must cooperate and work with other groups to accomplish the goals to be mentioned. Three words describe the tasks: dissemination, adoption, and implementation. The first aspect refers to the need that information about these programs must be conveyed to teachers, school board members, and departments of education. With this in mind the PA Academy of Science sponsored a symposium on "Project 2061" at the 1989 annual meeting of the PA Science Teachers Association. Much of this task can be accomplished on the national level through articles in appropriate journals and other descriptive mailing. The adoption of programs and their implementation is an activity on the state and local levels. It is in these aspects that the state academies must be active. If the programs referred to are properly designed, they will cut across the lines of disciplines and project the unity of science. The state academies are organizations whose membership are made up of scientists, mathematicians, teachers, and engineers from an abundance of different fields. Their voice will not be speaking as the special interest group for a particular discipline.

College Science Education. Little has been said in this chapter about the activities of state academies as they relate to the college level. This has been the main thrust of the academies throughout their existence. Nearly all state academies have grant programs for college research, even though in most cases the amounts are very limited. The journals published by the academies report the research activities primarily from colleges and universities. In recent years results of undergraduate research conducted at colleges without graduate programs have been reported in increasing numbers at the meeting of the PA Academy. The involvement with state legislatures and departments of education on various matters differs from state to state.

Suggestions for the Future. In various sections of this chapter recommendations for future actions have already been given. No attempt is being made to repeat these here in detail. This section serves to summarize and present general guidelines. It is apparent from the previous discussions that state academies need to

a. continue to support their junior academies, help them to expand their programs, and provide financial aid as much as possible,
b. work with other organizations to promote appropriate legislation with regard to science education and the operation of science on the Federal and their respective state levels,
c. use their influence in the development and adoption of coordinated K-12 science curricula on state and district levels, and
d. consult and act in concert with concerned groups in developing meaningful measures to have properly trained science personnel teaching at all levels of the school system.

REFERENCES

1. The National Commission on Excellence in Education, *A Nation at Risk: The Imperative for Educational Reform,* Department of Education, Washington, D.C., 1983.
2. Fitzpatrick, Frederick L. 1959. The Science Manpower Project. *Science Ed.* 43, 121.
3. National Science Board. 1990. *A View from the National Science Board. The State of U.S. Science and Engineering.* U.S. Government Printing Office, Washington, D.C.
4. Shyamal K. Majumdar, editor, The Proceedings, Directory and Handbook of the National Association of Academies of Science 1989-90.
5. American Chemical Society. 1989. *Education Policies for National Survival.* American Chemical Society, Washington, D.C.

6. Data taken from the program of the 1957 regional meeting held at Duquesne University on March 16, 1957 and the program of the 1990 regional meeting held at Shaler Area High School on February 3, 1990.

7. National Science Board. 1989. *Science & Engineering Indicators-1989*. U.S. Government Printing Office, Washington, D.C. pp. 32 (NSB 89-1)

8. Weiss, I.R. 1988. Course Background Preparation of Science Teachers in the U.S.: Some Policy Implications. In: A.B. Champagne (Ed.) *Science Teaching: Making the System Work*. American Association for the Advancement of Science, Washington, D.C.

9. National Science Foundation. 1957. *Sixth Annual Report*. U.S. Government Printing Office, Washington, D.C.

10. Schenberg, Samuel. 1959. An Evaluation of the 1958 Summer Institutes Attended by Science and Mathematics Teachers from New York City High Schools. *Science Ed*. 43, 114-120.

11. Worthy, Ward. 1989. Diverse, Innovative Programs Revive Precollege Science, Math Education. *Chem. & Eng. News*. Sept. 11, 7-12.

12. Atkin, J.M. & Atkin, A. 1989. *Improving Science Education Through Local Alliances*. Network Publications, Santa Cruz, CA, pp. 100-101.

Science Education in the United States: Issues, Crises and Priorities. Edited by S. K. Majumdar, L. M. Rosenfeld, P. A. Rubba, E. W. Miller and R. F. Schmalz. ©1991, The Pennsylvania Academy of Science.

Chapter Twenty-Six

THE SCHOOL BOARD'S ROLE IN THE DEVELOPMENT OF SCIENCE EDUCATION CURRICULA

JEREMIAH FLOYD*

National School Boards Association
1680 Duke Street
Alexandria, VA 22314

Determining exactly what the curriculum should be in public elementary and secondary schools is a matter of continuing debate. While a modicum of agreement does exist that local school boards, administrators, staffs, and various publics all must play a key role in the development of school board policies dealing with the curriculum, nowhere is this assertion more evident than in the field of science.

In a Nation at Risk the President's Commission defined its "New Basics" in the broadest of terms and among its recommendations was a call for three years of science, and six months of computer science, along with increased and stringent requirements in other disciplines. The President clearly stated where the responsibility for giving substance to these subjects lies when he called upon parents to demand these and other reforms in their local schools, and to hold their local officials accountable.

Judging by the earliest reactions to the Commission's report and the half dozen years that have elapsed since it was issued, there can be no question that school boards have reacted and continued to respond to the President's plea. Nor is there much doubt today that the new calls for educational excellence flowing from science and mathematics achievement goals enunciated recently by President Bush and the nation's Governors will require boards to continuously reexamine policies on the science curriculum.

---

*Dr. Floyd is Associate Executive Director of the National School Boards Association. He served as President of the Wilmette (Illinois) Board of Education and also as Vice President of the Montgomery County (Maryland) Board of Education.

The Commission also likened the "stuff" being taught in American schools today to a cafeteria line in which the appetizers and desserts can be mistaken for the main courses. By use of this analogy, one is compelled to ask: What is the nature and level of the scientific knowledge (appetizers, desserts and main courses) and skill acquisitions that today's youth require in order to be literate and scientifically proficient as adults in the 21st Century? School boards undertaking an assessment of the science curriculum that might be effective in answering this question must certainly seek a community-based definition of what basic science skills and knowledge students must master to effectively function in today's world and which will enable them to learn new skills to meet the changes that tomorrow's world will bring.

Boards also must be prepared to link this definition to standards and expectations of the school community. As the Commission noted, the weighting of the curriculum would reflect its relevancy to the desired outcomes of teaching. "In some schools," the Commission lamented, "the time spent learning to cook and drive counts as much toward a high school diploma as the time spent studying mathematics, English, chemistry, U.S. history, or biology."

The more immediate question is: Should the board start any assessment of excellence in the Curriculum with the presumption that curriculum is the most important of all school programs? Certainly. Curriculum in science or any other subject is what ultimately determines the quality of the education produced. It is the central program that most closely touches every student, parent, teacher, administrator, and school board member. Curriculum—that which is taught—is also the one program that determines the schools' contribution to society at all levels, local, state, national, and global. But since this is said of the total curriculum—all subjects, upon what basis can the science curriculum lay claim to being given top priority?

In Educating Americans for the 21st Century, the National Science Board Commission on Precollege Education in Mathematics, Science and Technology declared that we need to return to basics, which they described as sustained attention beginning in the elementary grades, to mathematics, science and technology. This Board also called for high school curricula to contain three years of science and one semester of computer science among other technological offerings. In addition, the Board strongly recommended that computers and other technologies be incorporated in the teaching of all mathematics and science courses as well as other subjects as appropriate.

Other genre of reform reports released during and since this era called upon school boards, administrators and teachers to develop curriculum which is effective, relevant, and which emanates from a common expectation that demands the very best efforts of all students, whether they are gifted or less able, affluent or not, college bound or not.

This is a tall order for any school board. Yet, if the Board is to fulfill its role in the promotion and development of excellence in the science curriculum, board members must accept this challenge as their first order of business.

Indeed, as we look to the current and future global scene and the likely increases in international competitiveness, many school leaders believe today that this effort can—and should—reflect a commitment to rearm and refinance education equal to our present commitment to restructure and reinforce our nation's defenses in the changing world.

What are Some of the Barriers which Affect the Science Curriculum and its Development?

Although curriculum content in general and science content in particular are important concerns of board members, such concerns traditionally have been major sources of frustration for them. Board members, administrators, and teachers have long recognized that Boards face several significant barriers to developing effective curriculum policies. Boards sometimes lose their control over or interest in curriculum development because of both physical and mental barriers.

Among physical barriers to effective curriculum policy-making are:

- State and federal mandates or judicial orders that predetermine elements of curriculum content.
- Lack of finances through budgetary restrictions, and limitations to the tax base because of citizen intervention (such as California's Proposition 13 adopted a few years ago).
- Lack of a clear plan or process through which the board can address curriculum policies.

What Can Boards Do to Improve the Process of Development of Science Education Curriculum?

Before going to the heart of this question, a brief retrospective look at federal and state roles and their impact on textbook content and classroom practices may be relevant here.

FEDERAL SUPPORT

The role of the federal government in curriculum development perhaps is best illustrated by the surge of popularity which the sciences enjoyed in classrooms across the U.S. in the late 1950s. The National Science Foundation (NSF), was established by the federal government in 1950 to promote basic research and education in the sciences. But it did not really get off the ground until the Soviet launching of Sputnik in 1957.

That year, then, NSF undertook the process of curriculum reform on a major scale. It is well known that extensive NSF support was given to develop a number of alternative science programs. Among these: Elementary Science Study (ESS), Science—A Process Approach (SAPA) and Science Curriculum Improvement Study (SCIS). NSF also developed several innovative textbook programs, such as Biological Science Curriculum Study (BSCS) and Introductory Physical Science (IPS).

Federal influence on curriculum in the late fifties and sixties extended beyond the National Science Foundation. In addition to NSF, the National Defense Education Act (NDEA) of 1958 and the Elementary and Secondary Eduation Act (ESEA), passed by Congress in 1965, provided considerable financial support for curriculum development, equipment purchases and teacher education. What actually happened as schools attempted the implementation phase of these programs in the nations classrooms will be discussed later in this paper.

The trend of increasing federal financial support for the sciences and for education generally peaked in the late sixties. Federal revenues reached the highest level of 8.8% of the school funds nationwide in 1967-68. That share dropped to, or below, 7.5% each year from 1968 to 1975—with the exception of 1971 to 1972. Afterwards, the researchers conducting a science study for NSF concluded in light of past patterns of state and federal funding it is not likely that many states will give science a high priority since federal legislation does not. The most recent decade has shown this conclusion to be correct. In fact, the percentage of federal support dropped drastically in the 1980's.

THE STATE ROLE

Some researchers hold that the statistics bear out this prophetic conclusion for states as well. The percentage they say, of state support for science education has remained virtually unchanged since 1955. In addition, while this may change in light of the nation's governors joining with President Bush in calling for a specific goal in science and mathematics achievement by the turn of this century, it is true that neither science nor mathematics were generally included in state education assessments in the 70s and 80s. In those states where needs assessments have questioned citizens about priorities for what students learn, "knowledge of basic skills" focusing on reading and writing and "application of skills to real-life problems" emerged high on the list of needs. In most states, relatively little attention was given to the history, status or needs of science education. Similarly, when science was included, the stated needs increasingly focused on concern for practical life and work skills in science.

However, while state *support* for science education has remained level, the *influence* of states on science education has increased markedly during the last fifteen years. Since the mid 1980s, the number of legislation and regulation items has increased markedly all over the country. While funds have been provided by a few states in some of these instances, in general funding has been left to local sources to be provided. It is therefore clear that passage of legislation or regulation items without funds can be and often is (an) action influencing the science curriculum.

Specific state requirements that affect local district curriculum decisions include such items as increasing the number of science courses for graduation requirements or the establishment of specific course requirements within subject areas, e.g., health and hygiene within the area of science. Whether or not state requirements are advantageous to a particular subject area depends, of course, on the importance placed on that subject by the state and the outcomes to be derived.

Some education critics believe that the growing influence of states in public school curriculum matters has had both positive and negative effects. On the positive side they say that state policies encouraging school district consolidation in the state which were common in the fifties, sixties and early 1970s helped foster larger schools which in turn were able to offer a wider variety of courses especially in science and mathematics and thereby provide greater educational opportunities for more students.

In the negative column, as state priorities have changed and moved away from curriculum concerns, so have state funds. State activity in larger societal issues— such as health, welfare, transportation, and others—accelerated rapidly in the middle 1960s and continues to this day. These requirements have, in many cases, set a high priority and demand for state funds, and thereby leaving education, especially in certain curricular areas with less of a priority claim.

Although state governments wield a growing say in how and where state education dollars are spent, state regulations concerning the curriculum provide relatively few flexibility options. And neither does the federal government. In the 1980s, as in the decades before then, federal aid to the public schools was categorical, and often the precise nature, scope and direction of curriculum reform is predetermined. That has not changed. Funds from the Elementary and Secondary Education Act of 1965 and subsequent amendments, for example, are channeled into "supplementary" programs (programs above and beyond what the school is providing) for certain groups of students. Local education agencies are compelled to adhere to specific curricular guidelines.

Promoting change and determining the direction of curriculum is not, then, simply a question of dollars. Most often, divergent interests of a variety of individuals and groups must be taken into account. However, funds—be they federal, state or local,—should be looked upon as an investment. And, as with any investment, information must be collected and used to ensure a successful outcome. In the case of the Science Education Curriculum information regarding practices in the schools must be gathered from all available sources and the information must then be effectively applied. Only then can the needs of all achievement levels of students be balanced. And only then will the payoff result in production change in the science curriculum.

How Can Restructured Internal Organization and Improved Science Curricula Affect Better Instructional Outcomes in The Classroom?

Recent national reports on education, usually referred to as "reform reports" contend that teaching-learning objectives in math, science and social studies did not change much over the decades of the 60s and 70s. Moreover, they contend that where changes were instituted during this period, they produced little lasting impact on classroom procedures and outcomes. Just as *what* students are learning has not changed greatly, so too has *how* they are being taught changed little since 1965, they say.

Educators have long searched for a better pattern of school and classroom organization. Accordingly, some new practices have been introduced in the past quarter-century. In 1955, the use of specialists—especially in math—was seen as the answer to poor preparation of elementary school teachers. During the early 1960s, various nongraded and multi-graded instructional approaches were unveiled and team-teaching was proposed as an alternative to departmentalizing. The "open classroom," often espoused in the late sixties as a way to make schools less rigid, does not show up very often today in research papers on teaching and learning. And although portions of other innovations linger today, the general pattern of school and classroom organization continues little-changed since the Sputnik era. The graded, self-contained elementary classroom and the fixed-period schedule at the secondary level still predominate in this the last decade of the 20th century.

WHAT IS THE ROLE OF THE TEXTBOOK IN SCIENCE CURRICULUM DEVELOPMENT?

Educational pundits and some practitioners also tend to agree that within the classroom, teaching procedures have changed little in thirty years. The textbook is still central in determining both instructional patterns and curriculum content and cannot be over-emphasized. It continues to play the primary instructional role in science as well as most other so called "academic" subjects. Teachers tend to adhere firmly to the idea of "covering the material" in the text. Moreover, they rely on covering it in the sequence as presented in the text.

Textbooks not only determine the science and most other curriculums in individual classrooms and schools, but also tend to create a core of subject matter for students in classrooms all over the nation. The basic components of the science curriculum thus becomes that which shows up in textbooks. The differences that exist tend to be largely in approach, design and amount of space allocated to scientific topics.

There is some evidence, albeit scant, that computerized technological science programs are finding their way into the nations science classes. These new instructional materials and techniques seem to have taken hold in a few American school classrooms. But, these are all too often the exception rather than the rule. Textbook—the almighty book—still reigns supreme.

Textbooks and other instructional materials obviously play a significant role in what children learn. It follows, then, that the process by which these instructional materials are selected is singularly important. Principals, superintendents and district curriculum coordinators responding to the NSF-sponsored study in the mid 1970s reported a patchwork process by which textbooks were selected. Each of the three groups agreed that school board members, students and parents were not, at the time, significantly involved in the selection of instructional materials. Well over half of the districts reported that school board members were not involved *at all* in textbook selection. At the same time, only half the curriculum supervisors and principals said they were heavily involved. Surprisingly, only twenty percent of the districts reported that the superintendent was heavily involved in choosing textbooks.

Who, then, is selecting textbooks and other learning materials? Responses from each group suggest that teacher groups and individual teachers were then wielding the most say in the textbook selection process. This finding, however, is at variance with an Educational Products Information Exchange (EPIE) study conducted about the same time in which teachers were asked about their involvement in textbook selection. Almost 45% of the teachers surveyed claimed they had *no* role in selecting instructional materials. Apparently, no group feels it has control over the textbook selection process.

WHAT USE IS MADE OF EQUIPMENT IN SCIENCE INSTRUCTION?

It is not surprising that the classroom teacher remains the primary conduit of instruction for most students. The NSF-supported studies referred to earlier found that although there has been an increase in student-centered learning activities, such as lab work, the lecture method is still predominant in most classrooms as the mode of instruction. Why is this so? One might surmise that it is due to a lack of science equipment available for instruction. But this same NSF-sponsored national survey conducted in the early 1970s found that while more than three-fourths of U.S. elementary schools have microscopes, only 28% of the grade K-3 science classes and 59% of the 4-6 grade glasses ever make use of them. Similarly, the availability of computer terminals in schools was more widespread than their use would lead one to believe. The availability and physical presence of computers in the classroom have both grown to a considerable degree during the decade of the 80s but their common use as a tool of instruction has yet to be realized in most classrooms.

Today's specialists in teaching methodology also say that classrooms have changed little over the past three decades. The predominant patterns in classrooms, they say, continue to be: instruction with total class groups; tell-and-show, followed by individual student work (in elementary schools); and lecture-homework at the secondary level.

HOW THEN CAN SCHOOL BOARD LEADERSHIP IN SCIENCE EDUCATION CURRICULUM DEVELOPMENT BE STRENGTHENED?

Above all else, the school board's responsibility is to set the philosophical framework—and to provide adequate resources—for science and other curriculum development to occur.

The first step toward achieving this reality is for school board members to see to it that curriculum considerations are not pushed aside by other board business. Too often the board meeting agenda is devoted exclusively to the 3 B's of business items—Building, Buses and Budgets—leaving no time for discussion of what is happening to students in the classroom. In addition to ensuring a greater focus on curriculum matters, the Board's leadership role must include an increase in their influence in federal and state legislation affecting curriculum.

But to return to consideration of the "How" of Science Education Curriculum Development, there is general agreement that science curriculum development is best viewed as a *process* rather than as a *product*. As a process, science curriculum concerns are ongoing. The school district should set annual goals, specifically in science education, and continually assess progress toward those goals. Further, the board should establish and adopt these goals as written policy, regularly review the science curriculum and communicate the results of their findings to the public. Subsequent decisions will thus become matters with which the public expects the board to deal. And, this process, by its very nature, identifies the science curriculum for priority board attention and facilitates systematic review, study and change as found appropriate.

BROAD-BASED PARTICIPATION IN SCIENCE EDUCATION CURRICULUM DEVELOPMENT IS A MUST

School board leadership in science education curriculum development must be designed to encourage broad-based participation from a wide variety of actors—school staff, students, parents and others in the community. Only in this way will the board be able to develop curricula that ensures the inclusion of "mainstream" opinion rather than "single issue" perspectives. While school boards are elected to deliver what the public wants, board members must not feel pressured to adopt every science instructional innovation which appears on the horizon.

Of course the board must first find out what the community wants its school system to provide in science education. This step usually involves conducting a needs assessment of the community and inviting parents, students, teachers, administrators, and other interested citizens to attend special board hearings and present their ideas, opinions, and needs. Many valuable resource persons can usually be found in the community to assist in the planning and improvement of the district's science instructional program. The school board should let the public know it is taking an in-depth look at the science curriculum to assure quality education in the district and that the public and administrative staff participation is welcomed. However, the board should make it clear that final decisions about the science instructional program cannot be delegated to others and will be made by the school board.

Questions and answers in the hearings between school board members and community members should be as concrete and concise as possible. They should be pragmatic discussions of what scientific knowlege, skills, attitudes, and values the district wants its students to acquire and exhibit. Some questions boards might consider asking their constituents are:

- What scientific knowledge, skills, and attitudes will our students need to have to be productive, contributing members of the community, nation, and world in 10, 15, and 20 years?
- How will scientific knowledge, skills, and attitudes better enable our students to interact with adults, elderly people, various nationalities and racial groups and each other, as they grow to adulthood?
- What scientific knowledge, skills, and attitudes do we want our students to have to better prepare them for further study, work, leisure time, and effective citizenship as adults?

Following these efforts the Board should next develop a written statement of the district's educational philosophy and goals in Science Education. The board must follow-up to ensure that the public perceives these goals to be more than mere words on paper. And aggressive public relations thrust is necessary so that everyone will at least have an opportunity to know what to expect from the schools.

The four most important questions that the school district's policies in science education must seek to address are:

- What are the educational purposes in science instruction to be achieved?

- What scientific educational experiences are appropriate to achieve the purposes?
- How should the experiences in science be organized for optimum learning?
- How can we evaluate whether the purposes have been achieved?

With appropriate written policies in place, it is the professional instructional staff, administators, curriculum specialists and teachers who must design the instructional program to achieve the goals. That is, they are the chief actors. They are responsible for deciding how the objectives of the district will be met in each grade in each school. Even though the district's professional educators will develop the specifics of the instructional program, school boards can become involved and show their concern in several ways.

For example, school boards can demonstrate their interest in what professionals are doing in the curriculum development process by requesting brief written reports from the superintendents. If boards are periodically briefed on these types of activities, board members will be in a better position to make subsequent decisions about their districts' educational programs. The reports, however, should be given to the entire board in a public meeting, not just presented to the board chairman. If the reports are presented and discussed in detail at board meetings, everyone on the board as well as any visitors at the meeting will be kept abreast of what is happening in the schools.

The actual instructional facet of this process is the area in which the learning program is further defined. In this area, teachers play the major role. They write lesson plans, evaluate students, and decide how to meet the goals and objectives laid out by the school board and the administrative staff. School boards can become involved at this stage primarily by assuring that qualified teachers are hired and by ensuring through board policy that teacher evaluations take place systematically to ensure that they are performing up to established standards.

How Does The Personal Experience of The Student Fit Into
The Science Curriculum Development Process?

The final phase in the science curriculum development process rest with the personal experience of each student, that is, what the students actually learn. Although students are the key people here, this phase provides the Board an opportunity to assess the effectiveness of the instructional program in science and to determine what is working well, what can be improved, and what should be discarded. A school board actively involved in judging its instructional program considers program evaluation a formative as well as a summative process. Answers to the following questions will help boards judge whether their schools are meeting the standards in science instruction set by the community, the board of education, the administrators, and the teachers:

- Are the students learning the scientific knowlege that was intended for them to learn?
- Are the students learning the scientific process so that they can apply it to new situtations and learn new scientific knowledge as it is created?
- What is the rate of student enrollment in the science courses?
- Are we employing qualified science teachers for our students?
- How are our graduates doing in post-secondary scientific study in work, in technical fields, as family members and as community members?
- What is the attitude of the parents and community members about our graduates and the science education that they reccived?
- What do the students say about the science program?
- Is our science program cost effective?
- Which of the school board's policies need to be altered to help improve the science program?

WHAT IS THE SPECIFIC LEADERSHIP ROLE OF THE SUPERINTENDENT IN SCIENCE CURRICULUM DEVELOPMENT?

Two of the most important jobs confronting school boards, are developing a curriculum policy that strikes a balance between the desires of parents and the public and what a consensus of the board views as educationally sound, and hiring a chief administrator who will then effectively implement that policy.

Establishing a cooperative relationship between the school board and the superintendent emerges as especially important. Board members should expect that their chief administrative officer will advise them on curriculum matters. Board members usually do not have the broad specific knowledge base and background to decide independently whether or not a given course of action in science edcuation is predictably educationally sound. And therefore the superintendent's evaluation of the situation and recommendations become significantly important.

While board members must trust and depend on the superintendent's advice on science and other curriculum matters, school boards should be careful not to relinquish and delegate their decision-making responsibilities. Boards must also see that teaching and administrative staff have the time and resources to plan, test and evaluabe curricular programs.

PLANNING FOR THE FUTURE

Finally, boards must plan science and other curriculum with an eye toward the future. Education may very well be subject to something like a 10-year half-life. Half of what students are now learning in school will be of no value ten years from now. Conversely, half of the scientific knowledge that will be needed in a decade or two does not even exist today. Boards therefore must be flexible, allowing for growth and change in science curricula, yet working to develop and build upon sound, established goals and objectives for the overall instructional programs in science.

In sum, there are several key tips that any serious school board member seeking to impact development of science education curricula might bear in mind. First, respect the views of specialists who come before the board, but be skeptical and ask questions to clarify major points. Also, be sensitive and attentive to interest groups that offer advice and use their input to assess how the science curriculum might best be balanced to reflect what they perceive as basic needs. Other factors to consider are:

- If administrators or professional staff propose science curriculum adjustments, do not simply "rubber stamp" and approve their plan. Instead, wisely disseminate these proposals and encourage public comment on them.
- Whenever the board initiates a science curriculum project through a policy adoption or proposed change, encourage the full participation of all constituencies.
- If board agendas are too crowded with other school matters, schedule special meetings or public hearings through which the board can receive public input on the science curriculum.

Second, research indicates that schools which produce positive learning outcomes operate best when their curriculum content reflects district-wide curriculum goals, and when the school staff is committed to achieving those district-wide goals. This is true generally across the range of subjects offered and is certainly likewise true for the science curriculum.

Obviously, giving some sense of "pride of authorship" to the professional staff who must teach and administer the science curriculum is a key ingredient in the district-wide commitment to excellence. The board should:

- Involve teachers and other staff in the same surveys, hearing, and other activities afforded to the public.
- Set out specific tasks that involve the professional community and support staffs in an assessment of the science instructional goals of the district. This assessment should determine the present condition of the school's science curriculum and instructional program in line with the science instruction goals of the board. It also should result in identification of areas in which additional public support, investment for laboratory facilities, or other program expenditures are indicated in order to meet the board's goals for the science curriculum.

Third, involve the administration and other professional staff. A district-wide commitment to excellence in the science curriculum portends implementing policies that are workable and manageable in the "real world" of an operating school district. To ensure that its policies are practical, the board should fully involve the superintendent, other key administrators, and teachers in policy-making and goal-setting, and in the analysis of input from the public and the professional community.

And fourth, determine your curriculum policies. Once the board has weighted the input from its three major constituencies (internal, external, and special)—and once the board has recognized special advice and opinion from its constituencies—it then should decide how formally its policy language should define its goals. To move toward excellence in science instruction and achievement the board must also:

- Determine how progress towards the goals and objectives will be monitored.
- Set a schedule for periodic evaluation of the district's instructional program in science education.
- Either through a policy or management directives, guide the administration on how the instructional program in science education is to be implemented. These policies or directives *must* indicate clearly how administrative responsibilities will be aligned to:
  — Carry the assesment process forward through the schools and, ultimately, to the patrons of the school system.
  — Determine the time and resources necessary to make the instructional program in science work at the building level.
  — Provide for continuous monitoring and evaluation of progress in implementing the science curriculum on a district-wide level.
  — Provide for "quality control" up and down the personnel ladder that links the community to the schools, the schools to the administration, and the administration to the school board.

Science Education in the United States: Issues, Crises and Priorities. Edited by S. K. Majumdar, L. M. Rosenfeld, P. A. Rubba, E. W. Miller and R. F. Schmalz. ©1991, The Pennsylvania Academy of Science.

Chapter Twenty-Seven

HELPING HANDS: WHAT INDUSTRY IS DOING TO IMPROVE SCIENCE EDUCATION

P. ROY VAGELOS
Chairman and Chief Executive Officer
Merck & Co., Inc.
P.O. Box 2000
Rahway, NJ 07065

INTRODUCTION

In his first State of the Union Address, President Bush set some lofty goals for American education. Among other objectives, he stated that, "By the year 2000, U.S. students must be the first in the world in math and science achievement." Realizing this goal will be no easy task, considering the fact that during the past few years, U.S. students have consistently scored embarrassingly low in these areas.

This situation is perplexing, since children begin life with an inherent curiosity about the world around them. And that interest is unabated even in the early school years. It has been my experience that science has enormous attraction for young people.

The Appeal of Science

Three years ago, I received a letter from a class of third graders that demonstrated how interested children are in studying and understanding the world—and how determined they are to make a contribution to society. The children wrote the letter to express their appreciation for Merck's decision to donate supplies of ivermectin for use in poor, tropical countries. Ivermectin is a drug that has proved to be extraordinarily safe and effective in treating a disease called onchocerciasis, or river blindness.

River blindness is a parasitic disease that afflicts 18 million people in sub-Saharan Africa and other tropical countries. An additional 85 million people are at risk of getting the disease, which has blinded more than 340,000 people now living. River blindness is transmitted from person to person by black flies that breed in fast-flowing rivers. The flies carry the *onchocerca volvulus* worm, the parasite that causes the disease.

Effects of the disease include terrible itching, dermatitis, soft-tissue nodules filled with worms, and—worst of all—scarring of the eye and ultimately complete blindness. The victims, usually elderly people, often have to be led around by children whose eyesight is not yet impaired.

Ivermectin, discovered in 1975, has been effective in controlling a wide variety of parasitic worms in livestock and dogs. For many years Merck scientists have been looking for a way to use the drug in humans. One of these scientists, a physician who had seen the misery of river blindness while working in Africa, suggested that ivermectin be tested against this disease.

Clinical trials in Africa yielded dramatic results. The drug proved so effective that *one dose per year* provided the necessary protection. But we knew that the people who most needed the drug could not afford to pay for it, nor could their governments. So, we decided to charge nothing for the drug.

Our decision resulted in favorable comments from many people in all walks of life. But the letter from the third grade class was particularly poignant. They wrote:

We are Central City, Iowa Third Graders. We read the article about River Blindness in the Cedar Rapids Gazette.

We feel very bad that some people in West Africa and Latin America are getting blind from the worm larvae that the flies spread causing itching, weight loss, ugly skin bumps and finally blindness.

We think that you are very nice because you are giving the medicine away free to the poor people in poor countries.

We would like to help pay for the medicine so we brought some of our allowances to school.

We also told the other classrooms about River Blindness and about the Merck Drug Company giving the medicine away free. Some of the boys and girls in the other rooms donated some money too.

We are sending a check for $38.32 to help pay for the medicine and for shipping it to the poor people.

Thank you for making and sending the medicine so that people will stop getting River Blindness.

The fundamental interest and concern that these children exhibited, not just in the welfare of people in faraway lands, but in how one can reach out to others through science, is a precious commodity. But their letter is both satisfying and disturbing. It is pleasing to know that these young students appreciated the tremendous benefits that science can bestow upon humanity. But it is distressing to realize that for many—perhaps most—of these students, the allure of science has faded in the intervening three years.

What has happened and why? These are crucial questions still not adequately answered. It is clear, however, that as a society, we are failing to keep our children interested in science. Numerous studies have demonstrated the shocking results of that failure.

The Failure of Science Education

In 1988, the International Association for the Evaluation of Educational Achievement (IEA) conducted a study of students of comparable ages from 17 countries. The results of that study were alarming. American 14-year-olds scored 14th in overall science achievement. For more advanced U.S. students in the 12th grade who were completing their second year of biology, chemistry, or physics, the results were equally unimpressive:

— 13th (last place) in biology
— 11th of 13 in chemistry and
— 9th of 13 in physics

The report's authors summarized the performance of U.S. students with a superlative understatement. They observed that, "For a technologically advanced country, it would appear that a reexamination of how science is represented and studied is required."

But even before the IEA study, we had fair warning that science education in the United States was sputtering. In 1985, a commission of the National Science Foundation summarized our failure in science education with these words:

Alarming numbers of young Americans are ill-equipped to work in, contribute to, profit from, and enjoy our increasingly technological society . . . At a time when America's national security, economic well-being, and world leadership increasingly depend on mathematics, science, and technology, the nation faces serious declines in skills and understanding in these areas among all our youth.

Unfortunately, the overall situation has not improved much, in spite of the fact that more than 300 reports on American education have been issued since the publication of *A Nation at Risk* in 1983.

Not only does the declining interest in science and mathematics mean that the scientists and engineers we need for industry and research are not being trained; but also tomorrow's leaders in business, politics, education, and scores of other fields will not be scientifically literate. In a world beset by highly complex scientific questions concerning the environment, public health, industrial development, military armament, and so forth, scientific literacy is an essential element. F.H. Westheimer, Morris Loeb Professor of Chemistry, Emeritus, at Harvard University, observed that today's high school graduates are tomorrow's decision makers. He said:

> They will have to exert judgments on problems concerning the safety of nuclear energy plants, the desirable and undesirable effects of specific chemicals in our society, the dangers and advantages of genetic engineering . . . and dozens of other questions . . . It will serve our students and the nation well if they know enough science to provide them with a background for future learning, and if they can at least listen intelligently to the arguments of experts; perhaps they will even be able to distinguish them from purported experts. Lack of knowledge will not prevent them from having opinions; it will only prevent them from having informed opinions.

More than 30 years after a pulsing Sputnik sounded a scientific wake-up call for this country, we are again challenged by the growing sophistication of science education in other nations, a declining interest in science in our own, and international business competitors who understand—and effectively use—an ever-increasing armament of high-technology breakthroughs.

What, then, can be done by industry, government, and educational institutions? Rarely does a day go by that someone does not suggest in some national newspaper, magazine, or on television or radio, an answer to that question.

The Role of Industry

It is indeed encouraging that great efforts are being made in restructuring the way science and mathematics are taught in this country and considerable discussion is taking place about how to capture the interest of our young people in studying these subjects. In these efforts American industry is playing an important role.

For example, in its charitable giving Merck places first priority on the need for the revival and enhancement of science education. Since 1957, when The Merck Company Foundation was established, the Foundation has made cash contributions of more than $72 million—most of it in grants to higher education in medicine and health-related sciences—through programs initiated by the Foundation. In 1989, the Company and the Foundation made educational grants totaling nearly $9.8 million—up from $8.8 million a year earlier.

In 1988, American businesses donated $2.1 billion to educational causes. Approximately 90% of that total went to universities and vocational schools; 10% to public schools. Public schools, however, are receiving an increasing share of the pie, with companies such as IBM, Coca-Cola, RJR, GE, Mobil, and Merck earmarking more and more of their corporate giving for precollege public education.

By necessity, industry has accepted a supporting role in science and mathematics education. But we cannot be the primary patron of American education's restructuring. The task is too large and our combined resources too inadequate. Industry is, nonetheless, an active partner and a generous supporter of countless education-related programs on the national, state, and local levels.

National Programs

The immensity of our failure in math and science education is compounded by the size of our educational system. The total educational enterprise in the United States employs more than three million people and spends nearly $200 billion a year. Elementary and secondary schools alone serve 50 million students located in more than 80,000 schools in 50 states. Obviously, the main responsibility for bringing about change in an enterprise this extensive must fall on the shoulders of government officials and education professionals—not executives from private industry.

While the government has been slow to take the reins of leadership, to their credit, government officials have recognized the need for cooperative ventures with industry and educational institutions. In 1986, a report of the White House Science Council Panel on the Health of U.S. Colleges and Universities urged the establishment of multidisciplinary centers:

The strength of the nation in trade, defense and health has been directly related to past investments in science and technology. Our future position in global markets will similarly depend on our willingness to respond to opportunity and to mobilize our strengths today. To this end, we must promote a broad interdisciplinary approach to problem-solving by focusing on university-based centers that will improve cooperative linkages between scientists, engineers and industry.

The purpose of these university-based complexes is to bring together government, industrial, and university expertise to solve problems related to broad national needs and to supplement other university-based programs.

The report called for renewal of the government-university-industry partnership as essential to our national interest:

We are certainly not alone in recognizing that science and technology are critical to our future. Nations everywhere are investing in these capabilities. We conclude that we must rethink and, in many ways, rebuild the critically important interactions between universities, government and industry that have served this Nation so well in the past. The Federal government-university relationship is too fundamental to the maintenance of our national science and technology base to be taken for granted, and the industry-university partnership is emerging as critical to exploiting that base in order to compete in the world marketplace.

Even before the members of the White House Science Council voiced these words, industry had been working with universities, colleges, and institutes to sponsor innovative research projects, to reequip science classrooms and laboratories, to endow professorial chairs, and to provide fellowships for promising scientists.

In addition, industry recognized that extraordinary steps must be taken to attract women and minorities to scientific careers, a traditional domain of white male Americans. Merck, for example, has identified 16 colleges and universities with large minority studentbodies and exceptional science, math, or engineering programs and has made a special effort to recruit new scientists and engineers from these institutions. But we certainly are not unique in using this strategy.

As the White House Science Council Panel pointed out, the interrelationship among government, industry, and universities has been an important reason for America's preeminence in many highly technical and innovative industries. The implicit challenge in the Panel's findings, however, is to make this interrelationship effective in restructuring America's system of science and mathematics education, not only at the university level, but also in the primary and secondary grades as well.

Much, if not all, of that effort must take place on the state and community level. The work on the national level has generally focused on marshaling government resources and encouraging government leadership in educational reform. To that end, industry has called national attention to the seriousness of the problem and has lobbied for effective government action and much-needed restructuring of science and math curricula.

Several years ago, people were excited about the major motion picture, *Stand and Deliver,* which demonstrated that disadvantaged students could excel under the dedicated tutelage of a talented teacher. Few knew that the film was underwritten by Atlantic Richfield, one of the nation's major oil companies. This motion picture had tremendous impact—not just because of the quality of the script. The subject treatment was factual; the plot, believable; the characters, warm and human; and the problem of student underachievement, a heartbreaking reality.

More recently, many public television stations have shown a documentary entitled: "Crisis: Who Will Do Science?" This film examines the problems faced by minority students in the United States who choose to expand their interest in science. For many of our minority students, there are simply no role models to follow in pursuing a career in science and mathematics. This Merck-sponsored documentary tries to fill that void by featuring minority scientists who have built a career in the sciences in spite of the obstacles. It examines, as well, the more troubling question of how we can interest all of our young people in studying science.

Documentary films and major motion pictures are vehicles for attracting public attention to the problem of waning interest in science. They are not, however, the limit of industry involvement. Joining or sponsoring alliances with other concerned groups and working with leaders in education, American industry is playing an important role in addressing the shortcomings of our educational system. The Triangle Coalition for Science and Technology Education, for example, is closely associated with the National Science Teachers Association. This coalition has assembled a diverse group of leaders in business, industry, education, labor, and other organizations who are dedicated to improving science and technology education.

The mission of the Triangle Coalition is to work with local action groups that have formed in communities and states around the country with the aim of supporting better education. As part of that mission, the Coalition has participated in a number of activities, including serving as a clearinghouse for information on new proposals and successful math and science education projects and helping to organize and train participants in the National School Volunteer Project in Science, Mathematics, and Technology Education.

Developing new education curricula is part and parcel of any restructuring of our system of education. Here, also, industry—as consumers of the human product of our schools—is offering its own unique perspective on what skills are needed for career success. Amoco, for example, is a primary sponsor of the University of Chicago School of Mathematics Project, which is developing a new math curriculum with supporting materials for students and teachers in the primary and secondary grades. Taking another approach, Xerox founded the Institute for Research on Learning to study how children learn, with the ultimate goal of using this knowledge to develop more creative and effective teaching techniques.

Industry is also interested in the quality of education and training being provided by colleges and universities. Merck's Ambassador Program is one example of the growing number of industry-university liaisons that benefit both industry and education.

The Ambassador Program enables scientists from the Merck Sharp & Dohme Research Laboratories to develop and maintain close contact with our nation's top research universities. Each year, Merck researchers visit more than 30 major universities across the country. These visits provide our scientists—who are, in many cases, alumni of the schools they visit—with opportunities to discuss areas of mutual research interest with faculty members and graduate students and to describe career opportunities in industry.

In addition, we have many other programs that encourage extensive interactions with academia, ranging from grants for basic research to postdoctoral fellowships, consultantships, and research collaborations. The goal of these programs is to improve the quality and expand the size of the talent pool from which industry draws new employees. In short, we hope to inspire—so that we can eventually hire—gifted men and women.

Efforts on the State Level

The corporate approach to educational restructuring on the national level admittedly tends toward the general and theoretical. On the state level, however, industry efforts become more concrete. In their efforts to bring about change in educational institutions within the states, leaders in business and industry begin to see familiar faces in the classrooms—the faces of their own children and the children of their employees. Looking at the problem at the state level, the stakes suddenly become clearer and dearer. Not surprisingly, industry's commitment of resources to restructuring state educational systems also becomes greater.

One of the most important activities being undertaken by a corporate consortium is the Business Roundtable Task Force on Education. More than 200 business leaders have been asked to select a state governor to work with during the next ten years in a unified effort to restructure education in the 50 states. The goal is not just to begin a dialog with the nation's governors, but to work with them and their commissioners of education in developing a plan of action for educational reforms that will benefit industry and society.

Talking with governors and state education administrators, while important in setting the course for new education initiatives, is not, by itself, enough. To bring about change, to make a course correction, requires involvement with educators, schools, and groups that are willing to sponsor workshops, training seminars, and symposia that focus on the need for change and how to bring it about.

Bell Atlantic, for example, has paired up with the American Association for the Advancement of Science to sponsor two-week skills enrichment programs for teachers from the Middle Atlantic region. On the West Coast, Pacific Telesis Group, a telecommunications company in San Francisco, has committed $2 million to a program called Education for the Future. The six California schools involved in this experiment will propose a model for the future of education, complete with curricula.

Merck has also contributed to state and local efforts to improve education. In New Jersey, where our headquarters are located, the governor and commissioner of education have drawn upon the resources of Merck and other corporations in developing a number of innovative policies in science education. Merck also supports programs for science teachers in urban districts, and was instrumental in organizing a special commission to examine how corporations and others can assist in the educational process. Science teachers work with

Merck scientists during the summer to revise curricula, and scientists visit the teachers' classrooms during the year to help in instructing students.

Merck is also involved in a number of other state projects to enhance science education. For example, we support the New Jersey Center for Advanced Biotechnology and Medicine, a center for basic research in molecular biology as well as for clinical research. The Center is funded by the state and Federal government together with industry.

As chairman of the Center's Scientific Advisory Board, I am committed to ensuring that industry continues to work closely with staff members from the Center so that the Center's two objectives will be met: to provide new basic information on biological systems and to prepare graduates and postgraduates to work in industry.

Merck is also working with the New Jersey Business/Industry/Science Education Consortium, an assembly of interested representatives from government, industry, and education. The goals of this group are: (1) to help upgrade the quality of the teaching and learning of science, mathematics, and computer science in elementary, junior, and senior high schools in New Jersey; (2) to provide career information about science, mathematics, and technology to teachers, so that they can be more knowledgable and enthusiastic in encouraging student interest in these fields; (3) to sponsor workshops and other programs for improving scientific and technological literacy; and (4) to strengthen the ties between education, business, and industry.

In addition to consortium participation, many companies also sponsor state science fairs, which bring talented students together to compete for cash awards and individual recognition. Other corporations have funded state academies that aim to improve science and mathematics education by updating teaching and management skills. Students and teachers are not the only beneficiaries of these programs; administrators receive training as well.

As with national and community efforts, it is impossible to catalog everything industry is doing, in cooperation with state government and education leaders, to bring about positive changes in science and mathematics education on the state level. It should come as no surprise, however, that the bulk of industry efforts are aimed at the communities where companies have plants, offices, or other facilities.

Community Programs

Most industry leaders want to have their companies regarded as good neighbors, so many have taken a long, in-depth look at the problems of their host cities—large and small—and have initiated programs that address some of the most pressing social and educational problems facing these communities.

Citibank recently announced a $20-million, 10-year commitment to improving urban schools in various ways, including developing new teaching approaches and promoting classroom technology. Six schools in New York, Chicago, Houston, and Los Angeles are receiving Citibank support, which involves money, tutoring help, adult mentors, and job training.

In many parts of the country, corporations are creating partnerships with local schools and are providing support to students with demonstated talent and promise. Cigna has joined with five Philadelphia schools, committing $1.5 million in funding and supplying equipment and employee volunteers to assist teachers and students. Cigna is also giving school administrators management training—something industry is uniquely qualified to do.

Other companies are helping teachers develop innovative and creative lesson plans. Since 1987, Rohm and Haas, a Philadelphia chemical and plastics company, has offered Delaware Valley science teachers an opportunity to work in the company's research laboratories with company scientists. As a result of their experience, a group of teachers and scientists prepared a series of lesson plans suitable for classrooms of varying levels. Called Project L.A.B.S. (Learning About Basic Science), it is just one of many projects that industry has sponsored to create an exciting classroom environment for teaching science and mathematics.

Although it should not be the case, making science exciting for students has been a particularly difficult challenge in the elementary grades. Many teachers in these grades have little training in the sciences or mathematics; hence, they are reluctant to undertake programs in these areas. Compounding this problem is the fact that few elementary schools have the equipment or facilities needed for scientific experimentation.

Our own experience with elementary teachers from the Plainfield (N.J.) School District is a case in point. About four years ago, Plainfield's elementary teachers attended a Merck-sponsored program at the Center for Elementary Science at Fairleigh Dickinson University. Upon returning to their classrooms, these teachers requested and received the first science equipment the district had ever purchased for its elementary grades. Since then, this equipment and training has been used to make science an important part of elementary education in the school district.

But providing money for equipment and sponsoring summer internships, workshops, and seminars for teachers—important as all of these are—are not industry's greatest contribution to improving science and math education. The greatest contribution is the personal time that industry representatives spend with teachers and students.

Establishing mentor programs is one way for industry to provide the individual attention that many students—including disadvantaged minority students—so desperately need. A classic example of such a program is one sponsored by the Federal National Mortgage Association (Fannie Mae) in Washington, D.C. Fannie Mae's program has two phases. In the first phase, mentors focus on the most talented and promising students in a targeted inner-city minority high school. Students who get all As and Bs are assigned a mentor, who works with them much as a mentor in the business setting would, giving them encouragement and advice. The purpose of this phase is not just to help students establish career goals; mentors may also have to deal with a host of social problems facing these young people.

Phase two of the Fannie Mae program helps these promising students find summer jobs. Those who spend the summer with Fannie Mae find themselves back in school, at least part of the time. They have morning classes in writing, goal setting, and career development. In the afternoons, they work at a variety of jobs—all for attractive salaries. Each program participant also receives a company-paid stipend that goes into a savings account for college tuition.

At Merck, we have had success with a similar program started by one of our employees who is not only a concerned parent but also the president of a local parent-teacher organization. Merck's program is also multiphased. One objective is to bring about meaningful change in the school district's science curriculum by introducing science education in grades K through 12. Several of our employees, working on their own initiative, have met with administrators, teachers, and students to set up summer internships that each year bring into our laboratories several science teachers to work side-by-side with our scientists. One year, on an experimental basis, two high school sophomores spent the summer working with our scientists.

The effect of this program has been encouraging, with teachers returning to share their experiences with fellow teachers and students. As for the student interns, their summer experience has served to solidify their interest in science as a career.

But not every student will want to be a scientist, nor will everyone go to college. What can we do for those students who do not want to attend college or those who, because of family finances, must start working after graduation from high school?

To help these young people, Merck provides mentors from among our crafts people to talk about their careers in the trades: carpenters; mechanics; heating, ventilation, and air conditioning specialists; pipefitters; electricians—to name just a few. These mentors go out to the classrooms to meet with eighth and ninth grade students. We also bring students to our facilities to show them how our crafts people apply their trades in the workplace.

We have found—as have other companies—that succesful mentors must deal not only with mapping out career goals and giving advice on educational planning; they also have to counsel students on such societal problems as drug and alcohol abuse and teen pregnancy. The role of a mentor is not easy, nor is working with young students a race for the short-winded. Those who hope to bring about substantive changes in education must be committed to the long haul.

Conclusion

The program variations on the theme of improving science and mathemetics education are endless. But whatever form these improvement projects have taken, they all involve a substantial industry commitment of employee time and company resources. The most valuable component of successful programs, however, is the one-on-one attention that comes from dedicated adults working with students. Monetary considerations, while significant, are secondary.

Lack of money may prevent a good education program from getting better; but adequate funding—by itself—will not guarantee the success of any project. Many proposals to improve education or requests for support for new outreach programs look good on paper. But, when the essential element of motivated and dedicated people is wanting, they are doomed to fail.

To assure success, one of the finest things our public schools, colleges, and universities can do is to nourish an interest in, and appreciation for, the sciences. Only by acting now will we be able to instill that interest in today's children— children who will become tomorrow's teachers, business executives, government leaders, and parents.

For the sake of coming generations, teachers and school administrators cannot permit interest in the sciences to languish. Those students who grow up unaware of the irresistible excitement—the captivating allure—of science suffer a tragic loss. But that loss is more than personal; it is shared by all of us.

A change in current thinking about science education certainly will not happen overnight. It will take years, perhaps decades, to bring about a complete restructuring of our approach to science and mathematics education. Industry, of course, can do more, on the national, state, and local level. More money needs to be spent, more personal attention given. But at best, industry must remain a junior partner in this effort. Governments, both federal and state, need to take the lead; and school administrators, teachers, and professors will have to come forward with well-conceived plans to implement the necessary changes.

The key element, I believe, is concerned and motivated people, at all levels: parents, teachers, university professors and administrators, community and government leaders, and business and industry executives. Without everyone working together, nothing meaningful, nothing lasting can be achieved.

It is true that the failures of our educational system are analyzed in terms of the poor performance of our children. The real blame for system shortcomings, however, belongs with America's adults. We are failing our chidren because we do not demand—and provide funding for—well-equipped classrooms and schools, better teacher training, and more interesting and relevant curricula. We are failing our children because most adult Americans, themselves, do not understand the importance and usefulness of science; they lack the zeal of scientific converts.

Whether the steps government, education, and industry are not taking will bring about the needed changes is the subject of much optimistic speculation. Time, of course, will be the ultimate judge of our efforts. Unfortunately, we have little time to waste, for even now we risk squandering our nation's most precious resource: the minds of our children.

REFERENCES

1. Bush, George. 1990. State of the Union Address. *New York Times,* 2/1/90, p. D22.
2. International Association for the Evaluation of Education Achievement. 1988. *Science Achievement in Seventeen Countries: A Preliminary Report.* Pergamon, New York.
3. National Science Board Commission on Precollege Education in Mathematics, Science, and Technology. 1985. *Educating Americans for the 21st Century.* National Science Foundation, Washington, D.C.
4. Morrison, Ann M. 1990. Saving Our Schools. *Fortune,* special issue on education, p. 8.
5. Westheimer, F.H. 1987. Are Our Universities Rotten at the 'Core'? *Science,* 5 June 1987, pp. 1165-1166.
6. Dumaine, Brian. 1990. Making Education Work. *Fortune,* special issue on education, pp. 12-14, 18, 22.
7. American Association for the Advancement of Science. 1989. *Science for All Americans: a Project 2061 report on literacy goals in science, mathematics, and technology.* American Association for the Advancement of Science, Washington, D.C.
8. Task Force on Science and Technology, House of Representatives, 99th Congress, Second Session. 1986. *Report of the White House Science Council Panel on the Health of U.S. Colleges and Universities.* Science Policy Study Hearings, vol. 23. U.S. Government Printing Office, Washington, D.C.
9. Kuhn, Susan E. 1990. How Business Helps Schools. *Fortune,* special issue on education, pp. 91-106.
10. Triangle Coalition for Science and Technology Education. 1988. "The Present Opportunity in Education", a position paper on the current state of science and technology education in the United States. The Triangle Coalition, College Park, MD.
11. Brozan, Nadine. 1990. From Citibank, Millions for Schools. *New York Times,* May 16, 1990.
12. Rohm and Haas Company. 1989. Project L.A.B.S.: Learning About Basic Science. Philadelphia, PA.

Science Education in the United States: Issues, Crises and Priorities. Edited by S. K. Majumdar, L. M. Rosenfeld, P. A. Rubba, E. W. Miller and R. F. Schmalz. ©1991, The Pennsylvania Academy of Science.

Chapter Twenty-Eight
DIRECTIONS IN SCIENCE EDUCATION: A TEACHER'S PERSPECTIVE

THOMAS C. ARNOLD
Earth Science Department
State College Area High School
650 Westerly Parkway
State College, PA 16801

INTRODUCTION

As we prepare to enter the 21st century education, specifically science education, is reeling from the report by the National Commission on Excellence in Education. This report, *A Nation At Risk,* reveals grave concerns about the competency of today's educational system and its ability to meet the expectations of an increasingly technological society. Implicit in the report is that neither the teacher nor the curriculum is addressing the demands of today's society. In spite of the fact that many schools are not doing an adequate job in preparing our nation's youth, the quality of excellence reflected by our National Merit scholars, science competition participants, and technical tradespersons indicates that some schools are developing highly competent students capable of succeeding in the next century. Often, what separates a successful program from those that are non-productive is the quality of teaching. What are the traits of the successful teachers, why are they likely to succeed when others do not, and more importantly, which traits will be required in the 21st century?

student hears; not what the instructor means to convey, but the impression the student receives. The teacher and student together play important roles in the learning process. As a result of this partnership, science educators have developed a shift in emphasis within the classroom. In order to instill new and deeper appreciation for the structure and strategies of science, the science teacher emphasizes the ability to learn new knowledge rather than rote learning of isolated facts and principles. This major emphasis results in the structuring of new curricula that have developed in the learner the ability to face new learning situations in much the same way as the professional scientist does via the inquiry method.

A LEARNING MODEL

Learning theorists recognize inquiry as a learning model involving a disciplined movement from a starting point via pathway to an appropriate end. They suggest that dissonance or a discrepant event initiates the learning process because it is highly motivating. In the learning process, whatever one senses becomes a part of what the student has already learned if there is a place to anchor the new knowledge. Facts that refuse to fit provide precisely the irritants that prevent closure of systems of conception. This indicates that there must be a responsive environment that includes a store of ideas and a conceptual model to organize data. Science instruction must involve the learner in the processes of discovering problems in the environment of society, observing data concerning facts or events, hypothesizing solutions, and testing the hypotheses in an attempt to reach reasonable conclusions.

The inquiry process then is a thinking process instigated by some discrepant event causing internal conflict between perceived knowledge and existing cognitive structures. The inherent drive of curiosity will encourage the learner to become cognizant of this discrepant event, and the internal conflict will result as the individual is thwarted in the attempt to seek closure. This implies that in order to resolve the conflict, the individual will have to become active in the learning process. Thus any learning task can be considered "problem solving" or following the scientific method if it requires the learner to discover the correct response either from his repository of previously learned material or from new knowledge specified from the task. From this discussion, we note that different learners react to new learning situations in different ways. The implication is that in a structured learning environment, the teacher must first determine if the student has the propensities that are required for learning to occur and then for ensuring that the individual reacts to the new situation in the required manner. This action would dictate that students entering an activity with differing levels of aptitude and attitude (readiness) cannot be expected to function in the same rigid fashion. That is, different studies should be able to address the new learning situation at a level commensurate with the readiness levels.

IMPLICATIONS FOR TEACHING AND LEARNING

In summarizing this information, we can evolve a model for effective teaching. As teachers, we must always be cognizant of the fact that only a small segment of our students will pursue a science career but that all students are curious and can learn the fundamental precepts of science required of a literate citizenry. Our goal must be to emphasize the processes of science and continually integrate these processes with other social/emotional skills. In so doing, we should not be surprised to learn that considerable knowledge and experiences that were once thought limited to higher education can in fact be assimilated by students at a much earlier age. As the body of scientific knowledge doubles, the expectations of the teacher also increase. Science teaching cannot be treated as separate from the science process. The effective teacher must be reasonably proficient as a scientist, and a practitioner of the scientific method, and conversant in the language of science-mathematics.

A MODEL FOR PROFESSIONAL DEVELOPMENT

Most teacher-education programs are designed to prepare prospective science teachers to teach in one or possibly two areas of study. However, as many urban schools experience enrollment reductions, or as many rural schools experience the frustrations of limited enrollments, many science teachers are called upon to assume teaching assignments quite unrelated to their fields or their levels of competency. These projections coupled with teachers with limited science background being assigned to general science classes are forcing teachers to reshape their roles in the classroom. They may no longer be the initiators of the curricula in which the teacher is a dominate presenter of information expecting passivity in his or her students. Instead, the teacher is expected to become a demonstrator of tools and media, a resource person who assists the learner in seeking information employing the new tools of the society, and a facilitator who directs the learner in using an inquiring mode of instruction. To become proficient in these new expectations, the teacher must be competent in the subject area. Implicit in competency is not only to be well versed in the knowledge of the content area, but also to be cognizant of the scientific processes employed to derive that knowledge. If the teacher is to present discrepant events to motivate student learning, he or she will have to be sufficiently educated to recognize those events that will stimulate the learning environment. This presupposes that as teachers are assigned to new responsibilities, that they will engage in activities that will enrich their own education. Inservice activities can involve participation in professional organizations, subscriptions to professional journals, or continuing education. In the past, active programs sponsored by the National Science Foundation have upgraded the science competency of many of our nation's teachers. If our teachers of science are to improve their knowledge bases and become familiar with the changes in the tools of science and the processes of scientific research, then universities, industry and the NSF will again have to pursue aggresively the funding for this action. Excellence in teaching begins with sound preparation in the content areas.

EDUCATION

A problem facing any curriculum experiencing rapid changes in information and technology is the identification of the essential material germane to the discipline. Remembering that the student is only enrolled in a specific content area for approximately 135 hours of instruction and considering the number of lost hours to assemblies, field trips and other school functions, the science teacher must understand that learning needs to focus on the 120 hours that the average student will spend with the teacher. In short, the average student will engage in a provocative learning environment for less than 150 days during the year. Obviously this means that the teacher must be very selective in the number of activities and challenges that can be offered in a specific discipline. These numbers further accentuate the problems associated with emphasizing information over processes and rote learning over meaningful learning. Teachers not familiar with the changes in the content and technology of the discipline risk teaching 19th century knowledge verified by crude 20th century technology to potential member of the 21st century whose knowledge and technology base have yet to be defined. The best we can do in any discipline is to introduce students to the newly identified "essential" concepts so that we might engage them in the processes of discovering new knowledge and the possible directions this knowledge portends for society. Thus, the task of the teacher is not so much to teach the knowledge of science but to encourage the student to want to learn science. This places the greatest burden on teachers. They must constantly be renewing their knowledge base, evaluating past practices that must be discarded, and seeking methods of integrating new discoveries and technologies into the dynamic curriculum of the future. If the teacher of the future earns a Master's degree in a subject area, most of the knowledge, processes and technology brought to the classroom will be obsolete in eight years. However, a teacher trained in the processes of science and knowing the computer and mathematical algorithms of his or her discipline should be able to engage in the same pursuit of life-long learning that we will expect of our students. The major hurdle of the future is one that we experience now. New technology is expensive, and without federal and state support, the majority of our teachers and students will enter new learning environments technologically illiterate. This will not only impede academic growth and curriculum reform, but also will entrench the already archaic and inefficient teaching practices found in today's schools.

PEDAGOGY

The methods and techniques of instruction employed by the teacher are critical in establishing a positive learning environment. As was established earlier, it is imperative that teachers be cognizant of the theories of learning. Knowledge of how the student learns is essential to develop or implement a theory of instruction. Such a theory is a guide to what to do in order to achieve the objectives of learning. A functional theory should concern itself with the factors that predispose the student to learn effectively, provide for the optimal structuring of knowledge and processes, identify the appropriate sequences required for learning, and establish consequences for learning such that the student exits the instruction equipped with the necessary skills for self-initiated learning. Recent changes in the preparation of science teachers from many of our colleges and universities have provided for dramatic improvement in this domain. As colleges have redesigned their teacher preparation programs, students have participated in new courses of study developed to increase their awareness of theories of learning and instruction. Today, more than ever, we see young men and women entering the classroom with the observational and diagnostic skills previously limited to "veteran teachers". The new teacher is more aware of the psychology of the student, more adroit in initiating changes in teaching to accommodate individual differences in the learner, and more familiar with the myriad of science curricula than were teachers from past generations. However, many of these improvements have been achieved at a high cost to the content preparation of the teacher. The number of credit hours required for graduation has not significantly changed in the past twenty-five years. In order to complete these new courses of study, the science teacher often has been forced to limit the number of pure science credits that have been usually selected for enrichment. The result is a teacher with a narrow development in science content and more enriched awareness on the pedagogical theories. Unfortunately, many of these entering teachers lack sufficient subject competencies to develop challenging, innovative curricula and must rely on the expertise of the mature teacher and the limitations of the more traditional science curricula. The solution might be to require future science teachers to complete their graduate education requirements in specific content areas with a trend to becoming proficient in at least two areas of study. This requirement coupled with the necessity to become involved in a renewal process that emphasizes involvement with the evolving processes and technology associated with their subject areas should provide the experiences required for them to continue to motivate students with new and invigorating curricula. In short, unless new teachers make an overt effort to supplement educational deficiencies associated with a already crowded college curriculum, American education continues the risk of preparing future students with outdated educational models and inadequate subject-matter competence that are only successful for students of past generations.

TECHNOLOGY

For many educators the term "technology" is often limited to the discussion of the use of computers in the classroom. However, in science more than in any other discipline the discussion of technology must include the tools of the trade. Electronic balances, magnetic stirrers, advances in microscope technology, video-microscopy, video disk technology, lasers, as well as plotters, scanners and computer interfaces continue to evolve while most teachers interact with early 20th century technology. For most of the 20th century, science instruction has been more concerned with preparing students to enter a non-technological society or a post high school learning environment that is only slightly more technologically sophisticated. As our society has become more sophisticated, and our students have become more reliant on the electronic media, their personal familiarity and facility with the changing technology have exceeded the skills of their teachers. This surprising event coupled with the documented aging of our nation's teachers has presented another dilemma. That is, if most of our tenured teachers are in their 40's and 50's, and they are not now technologically literate, what motivation will be required to correct this dilemma such that we adequately prepare this generation of students for the future? As we exit the 20th century, we leave behind the policies of isolationism and enter into a new era of the world community. Our students will not just be competing with his or her neighborhood friends but also with the best students in the rest of the world. If these students are to be prepared, then their teachers will have to become continually involved with changes in technology. If teachers do not use the technology of science in their classrooms, then by default the students will be denied effective instruction in the processes of science and most probably will enter post-secondary instructional programs with sufficient deficiencies as to hamper their potential for success. When students from schools experiencing financial limitations do not have access to the tools and technology of science, we effectively eliminate them from realizing careers in science and technology.

It is imperative to realize that we will not automatically improve science teaching and learning through the introduction of technology into the science curriculum alone. However, if we are to be consistent with the goals reviewed earlier in this chapter, then curriculum review and learning objectives must be continually evaluated. Efforts must be made to identify needed changes in the technological tools of the discipline. As dated and obsolete equipment is replaced, teachers cannot be guided by what is comfortable, but by the changes in the research methods of their discipline so that they may prepare their students to enter society and higher education at a readiness level that will encourage success. This requires that the successful teacher of science will also be a a practioner of science. Students entering the teaching profession must bring with them the knowledge of innovative changes associated with their discipline.

Veteran teachers have the responsibility to learn the new technology and the understanding and skill to include this technology in the dynamically changing curriculum. In the teaching of science, technology is disguised as a double-edged sword. Not only must the teacher allocate limited funds for the upgrading of the tools of science, but also must provide for the technological advances that enhance instruction. This will require continuous evaluation of past practices and the discarding of comfortable teaching techniques. An example might be to eliminate costly field trips and to replace these activities with teacher-designed videotapes of these experiences or the inclusion of interactive videodisk technology. In a time of decreasing financial commitment from federal and state agencies, limited funds might be better invested in capital for new technology rather than in consumable activities and/or materials. Instead of six or ten mechanical balances, the curriculum might be better served by fewer electronic balances. As schools upgrade their computer facilities, science departments can adapt the older computers as interface terminals to replace thermometers, pH meters, photometers and other highly specific yet costly equipment. Teachers within a department must be aware that if information is increasing exponentially, so is technology. Thus, the most efficient method of dealing with these changes is for different members of the department to assume the responsibility of change within specific technologies and to provide the in-service training to others. This is consistent with the goal of being committed to continuous education concerning the knowledge and processes associated with teaching a specific discipline.

INNOVATION

Innovation in education is often associated with creativity and the ability to stretch the department funds such that the program survives another year. However, in many instances the creative and/or innovative label is applied to the successful teacher who continually motivates, encourages, and engages students in active and dynamic learning experiences. These are the teachers students identify as "making a difference in their lives". The student outcomes may not result in a plethora of college science majors, but rather a student population confident in its ability to succeed and adapt to a changing society. These students having been instructed in the processes of science enter society comfortably literate in the knowledge of science and secure in their belief that they can deal with the changes that time will bring. It seems only appropriate to investigate some of the additional parameters or guidelines that these teachers incorporate into their instructional environment.

CLASSROOM ENVIRONMENT

One of the observations most often made by students of these teachers is that their science class was fun, not in the sense that they consistently had free time, but rather that the instructional atmosphere was such that it was non-threatening to be in class. Teachers were constantly presenting mind probing questions, every students opinion was valued, and thinking toward the correct answer was more valued than arriving at the correct answer. In short, if science is the fine art of guessing with the aid of inductive reasoning, then science could be fun if teaching techniques employing the processes of science were fun. In these classes a greater proportion of time is spent exploring the new directions that knowledge opens rather than the mastery of knowledge. In these classes emphasis is placed on the significance of learned materials and the implications of this new knowledge for growth in later courses or in its utility in a changing world. This does not imply that the students are not required to master minimal numbers of facts, concepts, and laws associated with specific discipline. In fact in many of these classes the students attempt to do more because the teacher is confident in his/her own knowledge base and proficient in the processes and technology of the discipline. In addition the teacher is able to empathize with the students anxiety and through compassion and humor introduce them to a love of science.

DOING SCIENCE

Perhaps nothing enlightens students more than being able to engage in the processes of science. For most students this is an illusion presented as a "laboratory experience". In too many instances, "laboratories" are instructional activities carefully contrived to allow the student to experience or discover a specific concept or law. This form of learning is a valuable asset to the science curriculum. However, the identification of the problem, the procedure, and in most cases the conclusion are already derived for the learner. Students who actively engage in self-initiated, self-directed scientific study experience a greater appreciation for the processes of science. Different teachers and/or schools enable students to enter this activity through different channels. In many districts, students have the opportunity to participate in science fairs. Here students interested in the processes of science are encouraged to engage in the scientific method in an attempt to derive an original solution to a self-initiated problem. In other schools teachers encourage students to participate in science clubs and competitions such as science olympiads. Here students learn to research information and learn to use inquiry and inductive reasoning. A third endeavor is to involve students in field experiences. Here students become involved in the skills of observations, data collection and analysis, and the utilization of the tools of the trade. For these activities to succeed it requires a teacher who not only is committed to extracurricular activities but also has been trained and involved in these forms of scientific endeavors. For most successful teachers, these are skills learned through involvement with professional organizations and graduate education commitments.

CREATIVE ACTIVITIES

In the category of creativity resides that teacher who possesses the rare talent to discover the novel demonstration, laboratory activity or unit of study that is recognized via magazine publications and or convention presentations. It should be noted that these activities are not discovered in a vacuum but are the cultivated result of actively participating in professional societies, observing others, and receptive to changes in the profession and discipline. In most instances these are veteran teachers who have experienced the frustrations of teaching poorly motivated students, receiving insufficient funding, and working in overcrowded classrooms. Many of their contributions evolve because of their commitment to science education, science, and their desire to instill into their students a self-appreciated love of science and its associated processes.

SUMMARY

Attempts to visualize the future of science education are fraught with numerous "black holes". Evaluation of the literature with specific emphasis on the philosophers, learning theorists, curriculum specialists, and more importantly practicing teachers provides reasonable direction and expectations. It would appear that the great changes proposed and implemented since the early 1960's will continue to influence the direction of science teaching well into the 21st century. The practice of rote learning, classroom demonstrations, authoritative teaching styles, and non-participatory science students have become outmoded with the beginning of the "information age". As science educators practice innovation, investigate new technology, and become more cognizant of pedagogical recommendations, they will continue to improve in preparing students who are literate and comfortable with science and technology. However, as noted earlier, it will be necessary for support services to be provided to the classroom teacher to enable change. Within the four areas identified in the model for professional development, change will only occur if the following circumstances evolve.

1. Education
   Entering teachers must be able to participate in undergraduate programs that will strengthen their science education. If the current curricular changes brought about by the addition of courses in pedagogy remain, then new

teachers should be expected to fulfill their post-graduate studies concentrating on science courses with the goal toward certification in two fields of science.

Veteran teachers must be able to identify and participate in meaningful science courses designed to improve both their sciences content and research experience. This goal has proven to be elusive for at least a decade in that few colleges offer summer programs in the sciences beyond the introductory freshman science courses. Unless support similar to the PA STEP programs or former NSF summer institutes that help supplement the expense of initiating these inservice instructions emerge, there is little hope for change. As the situation now stands, teachers in communities some distance from quality higher education institutes have no opportunity of enriching their education during the school year and less chance during the summer months. For the past decade, the burden of continuing education has been assumed by the efforts of state and regional professional organizations which have neither the resources nor expertise to assume this responsibility adequately.

2. Pedagogy

Recently, teachers entering the profession have demonstrated a pedagogical sophistication seldom observed in the past. This proficiency in pedagogical practice is the result of changes initiated by teacher preparatory institutes. As a result of this in depth preparation, these teachers should be expected to eschew inservice programs and graduate studies that are designed to enhance proficiency in their content area. For those teachers entering the profession without the pedagogical sophistication, it is imperative that inservice programs be developed to meet their needs. Recent changes in state requirements for providing inservice for all teachers help address this need. More importantly, successful models like the PA STEP programs are now available to provide free or low cost credits in diverse locations for all teachers regardless of their locations.

Veteran teachers most often review changes in pedagogical practices through involvement in professional organizations, participation in district inservice programs and more recently in the new programs like PA STEP. Recent efforts by professional associations such as the Pennsylvania Science Teachers Association have surveyed teachers to identify their inservice needs. These efforts have resulted in the development of a series of programs that provide meaningful experiences, while eliminating the often distasteful experience many teachers have encountered in past.

3. Technology

The problems associated with integrating technology into the science curriculum involve lack of familiarity with new technology, inadequate funding to procure the technology, and limited educational opportunities for proper instruction in the uses of the new technology. In too many instances teachers have experienced frustrations in trying to "make do with the old" knowing that few of their students will ever again engage this antiquated technology. If science educators are to evaluate efficiently and implement change in existing curricula, most of the technology associated with the curriculum should "roll over" every five years. Often the new technology, because of its sophistication, lacks the durability characteristic of the less complex technology associated with earlier curriculum materials. This implies not only that funds will need to become available for purchasing the technology but also that sufficient reserves will need to be made accessible for the maintenance and repair of the technology. Experience has demonstrated that with the use of this new technology student motivation increases, time on task is enhanced, and learning is improved, but that vast amounts of teacher service time are required to ensure the successful operation and effective use of the technology. This is the area in which wealthier school districts have advantages over the rural and urban schools. The prohibitive costs of the technology for curricula funded under 1960 guidelines, coupled with the decreased support for public education from both federal and state sources, has inhibited the acquisition of this technology. Unless the federal government reestablishes support similar to the NDEA and NSF programs of the 1960's, there is little hope of addressing this dilemma in the near future. Public schools and state supported college science departments can enact few improvements with capital budget increases that seldom approach 10% of the cost of one personal computer. In these instances, we are only fooling ourselves if we believe that this generation of students is entering that competitive would market on par with, let alone ahead of, other industrial nations with regard to science and technology.

4. Innovation

The innovative teacher is identified as one who engages students in active and dynamic learning experiences. It has been implied that this goal is only achieved by the teacher's actively seeking means of improving the classroom environment, by getting students involved in doing science and by developing new and unusual activities. Too often the science teacher is restricted in his/her own school environment and unaware of the contributions of others. Membership in professional organizations can be cost prohibitive, and released time from teaching responsibilities to attend professional meeting is often only a dream. If school districts want to encourage "renewal" and innovative teaching strategies in veteran teachers, then more efforts will have to be extended to allow teachers to achieve these goals. Too often teachers must decline to participate in these activities because their limited budgets are needed to address the needs of the curriculum or because administrators will not allow released time and funds for these activities. Under optimal conditions, each teacher within a department should have the opportunity to attend a professional meeting at least every five years. By alternating years with other members of the department, all teachers can potentially be "recharged" by the participating member. As each member of the professional team develops, the development of the entire team progresses. In the absence of adequate and appropriate content enrichment and inservice opportunities, this remains one of the most cost effective means of upgrading and improving science education in the state and nation.

The demise of effective and productive science education in our state's and nation's schools is recognized, and it is hoped that this perception of problems and potential solutions will provoke and initiate action toward providing for excellence in science education for the future.

SELECTED BIBLIOGRAPHY

Ausubel, D.P. "learning Theory and Classroom Practice," Bulletin No. 1 The Ontario Institute for Studies in Education, Toronto 5, Ontario, Canada, 1967.

Ausubel, David P. *Educational Psychology: A Cognitive View.* New York: Holt, Rinehart and Winston, Inc. 1968.

Baughman, Dale. *Teaching Early Adolescents to Think.* Junior High School Association of Illinois, Urbana Illinois, 1964. ERIC #011867.

Berlyne, D.E. *Structure and Direction in Thinking.* New York: Wiley, 1965.

Brandwein, Paul F., *Substance, Structure, and Style in the Teaching of Science.* New York: Harcourt, Brace & World, New York, 1965.

Bruder, Isabelle. "Future Teachers: Are They Prepared?," *Electronic Learning.* 1989, *8,* pp.32-39.

Bruner, Jerome S. *The Process of Education.* Cambridge, Massachusetts: Harvard University Press, 1960.

Bruner, Jerome S. *Toward a Theory of Instruction.* Cambridge, Massachusetts: Harvard University Press, 1960.

Chan, Lyn, et. al. "Christa McAullife Educators Reach for the Future," *Electronic Learning,* 1988, *8,* pp.10-14.

Cornish, Edward "The Science Teacher as a Futurist," *The Science Teacher,* 1969, *36,* pp. 21-24.

D'Ignazio, Fred. "An Inquiry-Centered Classroom of the Future," *The Computing Teacher,* 1990, *17,* pp. 16-19.

Dewey, John. "Science As Subject Matter and as a Method," *Science,* January 1910, pp.121-127.

Dewey, John. *Democracy and Education: An Introduction to the Philosophy of Education:* New York, The Free Press, 1944.

Dowden, Edward and B. Smith, "The Practical Interface: PA STEP and Pennsylvania Teachers," *Technological Horizons in Education,* 1989, *17,* pp. 69-71.

Fowler, H. Seymour. *Secondary School Science Teaching Practices.* New York: The Center for Applied Research in Education Inc., 1964.

Gagne, Robert M. *The Conditions of Learning.* New York: Holt, Rinehart and Winston, Inc., 1970.

Hill, Winfred E. *Learning: A Survey of Psychological Interpretations.* San Francisco: Chandler Publishing Company, 1964.

Holte, Jeff. "Technology and the Science Class: Going Beyond the Walls of the Disk Drive," *Electronic Learning,* 1989, 9, pp.38-42.

Hurd, Paul DeHart. *Science Teaching for a Changing World.* A Scott, Foresman Monograph on Education, 1963.

Hurd, Paul DeHart and Philip G. Johnson. "Problems and Issues in Science Education," *Rethinking Science Education* Fifth-Ninth Yearbook of the National Society for the Study of Education. Chicago: University of Chicago Press, 1960.

Lacey, Archie L. *Guide to Science Teaching in Secondary Schools.* Belmont, California: Wadsworth, 1966.

'Not Just for Nerds: How to Teach Science to Our Kids," *NewsWeek,* 1990, 95, pp. 52-64.

Piaget, J. "Principal Factors Determining Intellectual Evolution from Childhood to Adult," In Edward Douglas Adrian, et. al. (eds.) *Factors Determining Human Behavior.* Harvard University Press, 1933, pp.33-48.

"Planning for Excellence in High School Science," National Science Teachers Association, Washington, DC. 1961.

"Power On! New Tools for Teaching and Learning. Report of Office of Technology Assessment, Washington, D.C. 1988.

Science for All Americans. American Association for the Advancement of Science, Inc., Washington, D.C. 1989.

Shulman, L.S. and E.R. Keislar. *Learning by Discovery: A Critical Appraisal.* Chicago: Rand McNally, 1966.

Suchman, Richard J. *Inquiry Development Program.* Science Research Associates, Inc., Chicago: 1966.

Suchman, Richard J. "Inquiry: In the Pursuit of Meaning," *The Instructor,* 1965, 75.

Suchman, Richard J. "A Model for the Analysis of Inquiry", In Herbert J. Klausmaier and Chester Harris (ed.s) *An Analysis of Concept Learning.* New York: Academic Press, pp. 177-189. 1966.

Tyler, Ralph W., "The Interrelationships of Knowledge," *The National Elementary Principal, 43,* 1964.

Young, Virgil M. "Inquiry Teaching in Perspective," *Educational Technology,* 1969, pp.36-39.

Science Education in the United States: Issues, Crises and Priorities. Edited by S. K. Majumdar, L. M. Rosenfeld, P. A. Rubba, E. W. Miller and R. F. Schmalz. ©1991, The Pennsylvania Academy of Science.

Chapter Twenty-Nine

SCIENCE AS A LIBERAL ART

HEINZ G. PFEIFFER

Consultant, Electric Utility Problems/Energy and Environment
Barnes Lane, Allentown, PA 18103

The concept of a liberal arts education goes back to an era when there was a clear separation between the small class of the educated, those who could read and write, and the rest of the population. A liberal arts education meant acquiring the knowledge of man's history, thought and accomplishments, and some individuals as recently as Francis Bacon could claim to have a comprehensive mastery. Increase in public literacy, impact of foreign cultures, anthropology and the scientific revolution during the past two hundred years has made a comprehensive mastery of all human knowledge impossible for an individual. During this period the original aim of a liberal education was perverted to limit it to the basic curriculum suitable to a medieval education, i.e., language, philosophy, history, literature and abstract science. C.P. Snow's[1] essays on the two cultures and the subsequent tempests they provoked among the literati illustrate how strongly the classical view of liberal arts was still held in certain circles. At the same time that the narrow definition of liberal arts reduced it to an area that many considered irrelevant to the mainstream of human effort, the demands of career oriented curricula crowded out broader studies in the humanities. Much of the turmoil in education in the period 1960 to 1990 has been the effort to reconcile the dilemma between the love of knowledge as an end in itself, with disregard for its practical relevance and the narrow demands of a specific career. What appears to have been lost sight of during this period is a definition of liberal arts as the basic knowledge necessary to carry out an individual's role as a private citizen. Neither the classical liberal arts nor the career oriented curricula provides this basis. Of the knowledge necessary to understand the modern world, scientific knowledge is perhaps the most critical.

The basic education should be just that, an education to provide a basis for the future. Major career changes are becoming the norm rather than the exception; the success of these changes is dependent on a broad base of knowledge and many of these career changes involve technical and scientific understanding. The challenge of science education is to provide an understanding of the basic principles without the detail necessary to become a specialist, to enable one to understand the arguments and develop an informed opinion on scientific issues of the day and, if necessary, provide a base for more scientific studies that can advance a new career.

Critical to any liberal arts education is the method by which our knowledge is derived. As the scientific method and the available tools have been refined, all fields of study have been affected. The Shroud of Turin and the dating of the Dead Sea scrolls are examples that are affecting religion, computerized analysis of literary works and scientific examination of works of art are among the many examples that have affected other fields. An understanding of quantitative methods is essential for any educated individual. Unfortunately many in science education have denigrated any knowledge that is not measured or weighed and have alienated people in disciplines where qualitative and connotative aspects are dominant. The challenge in science education is to provide an understanding of the quantitative methodology and an understanding of the basic scientific principles without detailed quantitative exploration of the principles themselves. An appreciation of non-quantitative knowledge must be maintained.

During the next generation national policy decisions will have to be made on such diverse subjects as information networks, resource development, strategic defense development, energy policy, environmental problems and waste disposal. In all these areas the citizen will be bombarded on one side with arguments that ignore the long term consequences of the actions and on the other side with non-viable alternative solutions to real problems. An understanding of the technical problems involved is essential to arrive at solutions that are neither economic nor ecological disasters.

One of the challenges of science in the liberal arts curriculum is the appreciation of science and technology as basic knowledge and the role of the practitioner of the scientific and mechanical arts. As recently as the nineteenth century, the inventor and the scientist were seldom the same individual. The prototype of the inventor so well developed by Mark Twain in "The Connecticut Yankee in King Arthur's Court", or in life by Thomas A. Edison was the antithesis of the image of the scientist as embodied in folklore as the absent minded professor, or in life by Albert Einstein. The development of the large academic and industrial laboratories in the twentieth century has practically eliminated the historical distinction between the scientific and the inventive geniuses. It was only during the last half of the twentieth century that the National Academy of Engineering was formed to recognize the equivalence.

One of the greatly neglected areas in science education is the study of the impact of science on life today. This is most striking in the areas of communication, medicine, public health and the availability of energy. In the industrialized world there has been a doubling of the life expectancy, and consequently a more than doubling of the population at a constant birthrate. For the first time in recorded history the female life expectancy exceeded that of the male in the industrialized countries, certainly a major contributor to the great social changes of the recent past. The explosion of population and improvement of communication have both created problems in the whole world as more people have become restless with the knowledge of the tremendous differences in opportunity and quality of life that exist. These problems are serious threats for the future of the west. The thesis that McNeill proposes in "The Rise of the West,"[2] that the industrial, scientific and technical revolution characteristic of the western countries will dominate the future does not guarantee that those countries that initiated the revolution will be the most successful practitioners in the future.

We have the materials available to provide sound science education as part of the general curriculum. The basic structure to provide the background in science to non-science majors was developed during the 1950's and early 1960's both for public schools and colleges. At the same time very fine audio-visual material was produced that provided the instructor with the means of demonstrating basic physical principles to the non-specialist without violating accuracy. The curriculum revolutions in the middle sixties and seventies directed the emphasis away from science education and less of this material seems to reach the classroom today than twenty years ago. A great deal of fine material has been developed in the interim for television and for popular demonstrations at places such as the Exploratorium in San Francisco and the Franklin Institute in Philadelphia.

Both the television presentations and the working demonstrations have content that is so often lost in classroom courses in the basic sciences. They are prepared for an intelligent audience with a minimal background in the specific field. They do not insult the specialist in the field by being too "cute", nor do they get so involved in detail that they bore the non-specialist. Although the material is available, it has generally not been integrated into specific coursework and often the student can avoid exposure to science in his education even when it is available. There are new initiatives underway under the auspices of the National Science Foundation and the National Academy of Sciences but it is not certain that these will use the tremendous amount of material already available but that they will start a whole new education approach to the problem as though nothing had been done, or is presently available.

At the university level a further problem exists in providing an adequate background in science as part of a liberal education. That is the compartmentalization of the natural sciences into specific fields and where a beginning course in any one of them will provide education in the use of the scientific method they do not teach the unity of the principles underlying all of the sciences. A survey of the sciences still tends to have so many lectures in Physics, so many in Chemistry, etc. What is needed in an interdisciplinary approach that emphasizes the unity rather than the individuality of the sciences.

The formal educational curriculum sets the stage for life-long continuing education, which for adults is motivated by personal interest and career necessity. The early years of education have content that is determined by the educational institution and reflects the perceived needs of society at large. The education beyond the formal years is determined by the individual and is often limited by the person's previous background. Provision of a fundamental understanding of basic scientific principles and scientific method are important both for career development and personal involvement in the changes taking place so rapidly in the world. Liberal arts education has always had the goals of intellectual development and participation in the affairs of society. Science is an essential component of a liberal education if these goals are to be met.

REFERENCES

1. Snow, C.P. Two Cultures and a Second Look. Cambridge University Press.
2. McNeill, W.H. 1962. The Rise of the West. University of Chicago Press.

Science Education in the United States: Issues, Crises and Priorities. Edited by S. K. Majumdar, L. M. Rosenfeld, P. A. Rubba, E. W. Miller and R. F. Schmalz. ©1991, The Pennsylvania Academy of Science.

Chapter Thirty

SCIENCE EDUCATION AND THE ELECTRONIC MEDIA: THE IMPORTANCE OF COMPETENT SCIENCE REPORTING ON RADIO AND TELEVISION

MARTIN WEISBERG

1015 Chestnut Street
Suite 620
Philadelphia, PA 19107

INTRODUCTION

Seatbelts may be hazardous to your health!

Pap smears are a waste of money!

Most operations are unnecessary!

Don't worry about your cholesterol!

DPT immunizations may kill your child!

These stories and more at eleven.

All of these news reports aired in the Philadelphia area during 1989. Each was heavily promoted and advertised. Each was presented in a multi-part series. Each showed a crying patient or loved one, and each carried with it that unwritten endorsement of truth, "if it's on the news, it must be true."

Of course, each of these stories did contain some truths. A seatbelt can be a cause of a ruptured spleen, or might trap someone in a burning car; some Pap smears are improperly interpreted; not all operations are necessary; many people do not need to worry about elevated cholesterol; and yes, there have been deaths from DPT immunizations.

The benefits from seatbelts, Pap smears, surgery, a prudent diet, and immunizations are well known and proven, but they are not hot news.

Although newspapers and magazines often present well researched, well documented, complete stories about scientific breakthroughs, the electronic media, radio and television, educate by headlines and emotional impressions. Usually, buried deep in the report is a ten to fifteen second statement by a competent scientist or physician. The expert is often seen in a white coat, in a lab or office. His or her statement is usually unaccompanied by graphic pictures or crying relatives, and is often taken out of context. the expert is rarely experienced in talking to a camera or microphone, and often appears nervous and fidgety. The result is that the person who knows most about the subject often has the least impact and leaves the least important impressions.

But radio and television are not going anywhere and they will continue to play a major role in mass education. Therefore, scientists and experts in every field need to learn how to work with this small portion of the electromagnetic spectrum to insure that the information presented is accurate and helpful, or at least not harmful.

THE PROBLEMS WITH HEALTH REPORTING TODAY

Most science and health reports are accurate, but there are good reasons for some of the bad reporting we see. They usually have to do with the immediacy of the electronic media. Reporters and producers rarely have the time to research these stories the way they should be researched. A story sometimes comes off the wire just minutes before it is aired. If there is time to call an expert for comment, there is rarely time to allow that expert to research the question.

Experts are often asked to comment on a newly released journal article before even receiving the journal. But experts are experts, and the news is the news, and people believe both of them.

Money plays a major role in what gets on the air. I do not mean that there is graft or payola, but the purpose of all of those lights, and expensive cameras, and special effects, and all those perfect teeth and the plastic hair is to get people to watch the newscast. If people watch the newscast, it will have high ratings. High ratings translate to more advertising and more money.

Yes, reporters want to get and report an accurate story, and news directors want the highest quality newscast, but as important as quality is, they also want people to watch and listen, and if it takes sensational headlines, partially nude bodies or graphic footage to get those people to watch, they will stretch (but usually not cross) the boundaries of good taste and responsible reporting to get there.

There was a time when a respected scientist or physician would not consider appearing on television or radio. It was considered unethical and some professional societies forbade it. Now, many feel that it is unethical not to grant interviews. If the scientific community does not participate in the education of the people, someone else will do it, and probably not do it as accurately. Lay reporters may be able to interview and write well, but they all lack the perspective and the background that only an expert in a particular field could have.

The truth is that there is no such thing as scientific "news." By the time the "new research" hits the wire services, it has likely been presented and discussed at numerous meetings, and preliminarily discussed within the field for several years. The experts have been aware that the work has been going on and have likely heard about preliminary results.

It is unlikely that I need to convince readers of this book that scientific research is important or that science education is a good thing. The remainder of this chapter deals with health and science education by way of the electronic media. The information is based on the authors thirteen years of experience with radio and television news and talk shows, both as a guest and a host. As a physician, never formally trained in journalism or communications, I offer only the fruit of my own personal experiences.

THE SCIENTIST AND THE MEDIA

Scientists and other professionals interact with the media in many ways. Some have become well known, like Dr. Art Ulene, the doctor on NBC's Today Show, Dr. Timothy Johnson, medical editor for ABC, and Dr. Frank Field, a long time science reporter.

Others have a few minutes in the spotlight. You might not remember their names, but who can forget the physician who addressed the press when John Kennedy was shot, or the meteorologist from the National Weather Center who appears when there is a devastating hurricane, or the geologist who showed us why there was an earthquake in San Francisco.

More and more, the electronic media news is relying on experts to embellish, authenticate or add credibility to their reports. Most television stations have a health and science reporter. Some have both, and many have physicians and scientists filling these rolls. But even then, experts from the community are frequently called upon for interviews. Medical reporters rely on these experts to explain, comment on or refute these news stories.

Guests, expert or not, are invited to be on radio and television for a variety of reasons, and in order to be good guest and to get your point across, its helps to know why you have been invited.

Experts are often asked to be interviewed as part of a report on current news. Obviously, the newsmaker is the first choice, but if that person is unavailable, someone else in the field will suffice. If you invent a perpetual motion machine, you will no doubt be contacted to explain *who* you are, *what* the machine does, *where* can you buy one, *when* it will be available, *why* it is important, and *how* you got the idea. These, of course, are the five w's (and the h) of journalism. If you are not available, the chairman of the physics department of the local university will be asked to comment. These interviews are usually relatively benign provided you know the subject. That, by the way is the most important rule in dealing with the media. If you do not know an answer, television is not the place to fake it.

The interview might take from fifteen minutes to longer than an hour. You might present a flawless well structured coherent lesson on perpetual motion only to find that your appearance in the news report lasts for ten seconds, and for six of them your eyes are closed, the lights are reflectng off of your bald spot like a thousand watt laser and the most important thing that you want to say was left out.

Expert guests are often invited to present opposite points of view. Controversy makes for good television and radio. The louder and less polite, the better the interview. That is fine for the show, but it may not be the way you choose to be seen and heard. If the topic is abortion, animal rights, or euthanasia, you can probably expect an opponent, but they can also show up when you least expect them. It is always a good idea to ask who else will be interviewed, and will the report or show have a controversial element.

Sometimes an expert is invited to a news studio or talk show simply to be a resource for the host or reporter. In these cases it is perfectly acceptable to inquire, in depth, as to what topics will be discussed. This will allow you to brush up on any information that you may need, and to make sure that you are the appropriate guest.

Although most media appearances will be for one of the above purposes, occasionally you are invited only to be attacked. Sometimes people are ambushed in their own offices. We have all seen the 60 Minutes, 20/20, approach. My best advice is never to respond to anyone who ambushes. Do not say "no comment." Do not scream and yell. Do not confirm or deny anything. Just shut the door if you can, or advise them that you require interview requests in writing.

Ideally, a reporter calls the most knowledgeable expert in the community. Ideally the expert answers the phone on the first ring. Ideally the expert has read the journal in which the hot breakthrough news has been published, is intimately familiar with the topic, and has already formulated a well thought out, well researched opinion. Ideally the expert then takes all the time necessary to educate the reporter enough so that he is able to write a balanced piece, explaining the particular field of science as it was before the breakthrough, exactly what the breakthrough is and how it may change the future of the field. Ideally, the reporter will stress the limitations of the new study and present it in perspective with reference to the existing body of scientific knowledge.

In reality, however, the reporter makes several calls until he finds someone who has the time to talk. In reality, the reporter has received the journal a day before the expert. In reality, the expert knows something about the breakthrough but has not had time to study it. In reality, the interview is not comprehensive, and in reality, the news story may run anywhere from five seconds to two minutes.

For these reasons, it is important for the scientific community to understand the workings of the media, and learn how to get the important points across.

It also must be remembered that it is not always the scientific breakthroughs that get the ratings. Falsified research, misspent funds, unexpected complications, bad results, accidents and catastrophies, either natural or man made, also make for good ratings.

It may also be the expert who requests to be interviewed. Television and radio are good ways to market an agency or a practice, to promote a book or a research project, to respond to a previous report or guest, or simply to get your message out.

Stations are required by law to allow responses to editorial material, and some shows and newscasts welcome the solicitations. You must, of course, remember that the ulterior motive of every producer is to put together a show that people will watch or listen, so that the ratings will be good, so that the advertisers will spend more money. Sad, but true.

DO PEOPLE WANT TO HEAR THIS?

Critics of on-air experts have said that the public does not want to hear about complicated scientific things. They say that graphic medical footage is offensive, and that scientific concepts are too difficult for the lay public to understand. Those critics have not seen the research that, time after time, find that people are interested in these things.

One of the fastest growing cable networks is Lifetime Cable Television. During the week, the newtork airs a variety of shows about a variety of things, but on Sunday, Lifetime devotes a good part of its day to health and medical programming. These shows are aimed at doctors. They make no attempt to speak to lay people. Medical terms are not defined, actual operations are shown, questions are asked by doctors and answered by doctors. Yet about three out of four viewers of these programs are not in the medical field. They are lay people who are just curious, or maybe they think that they can get some inside secret information. Whatever their motivation, they watch and they learn.

HOW TO DO IT?

It is my true belief that the electronic media can be an effective teacher of health and science information, and that we, the scientists and medical experts are the ones best qualified to impart that information responsibly, in an accurate manner, and in perspective. To do it well, is another matter.

The remainder of this chapter will discuss some of the things that I have found helpful in dealing with the media.

A scientist would never start an experiment without knowing what he was looking for. It would be wise to keep that in mind when you are asked to be on television or radio. As discussed elsewhere in this chapter, every guest is invited for a reason. It may be that they want you to follow up on a hot news item, or present an opposing point of view, or just to serve as a resource. Whatever their goals, you need to set your own.

For example, many times an expert will be called in to comment on a new diet. It is my feeling that regardless of how he feels about the diet, the goal of that expert should be to leave the viewers or listeners with good nutrition information.

It is important to go into an interview with the goal strongly formulated and to communicate it as often and as clearly as possible. In advertising this is known as making your copy points. If you feel that one of your important points will not be brought up by the interviewer, bring it up yourself. There are several ways to do this. The bridge technique almost always works. In this maneuver you answer a question and then "bridge" to your point. An example may be: "Yes, you will lose weight on this diet, but it is important to understand the importance of good nutrition."

Television studios can be very distracting. There are lights and cameras and microphones and wires and monitors, and people milling around. It is very easy to become distracted. It is essential that a guest concentrates one hundred percent on the interviewer. It is also helpful to forget that there may be hundreds of thousands of people watching, and to talk one on one to the interviewer, not to the audience. It is very important for a guest on any television or radio show to know the show. It helps to know in advance, the pace of the show, the demeanor of the host, and the format of the interview. If it is not possible to see or listen to the show in advance, it is perfectly acceptable to ask for a tape.

In addition to knowing about the show, it is essential that the guest is clear about his role on the show and the type of material that will be discussed.

WHEN THEY'RE OUT TO GET YOU

As scientists, we sometimes do things that the lay public may not understand. Animal experimentation, food additives, anything with the word nuclear in it can all create fear or outrage among the uniformed. An important part of science education is to explain to the public the importance of some of the things that they might find distasteful, scary, or even merely pessimistic.

On a purely logical basis, we may be able to win our case most of the time, but on the electronic media, we are likely to look like the bad guys. Crusaders and zealots are often experts in oratory and stage presence. They have usually presented their cases hundreds of times before and most importantly, they usually either lie or extrapolate inappropriately. We as scientists tend to tell truth and state facts that we can absolutely support.

A good example of this occurred on a medical talk show I was hosting on a call in talk radio station in Philadelphia. My producer booked as a guest a man who had written a book on the miraculous powers of cod liver oil. He claimed that, in addition to reducing cholesterol and heart disease, it also lubricated the joints and cured arthritis.

It is true that some components of some fish oils can have beneficial effects on the lipid levels in some individuals, and some fish oils have prostaglandin inhibiting qualities, but no fish oil can cure arthritis, and the doses this "expert" recommended could have only the most minimal effect on cholesterol.

I insisted on having a medical expert on the show with the author, and we chose a board certified cardiologist from the faculty of a local medical school. In addition, this particular physician was a member of the Food and Drug Administration's committee to evaluate fish oil. He was well published and had the respect of the medical community.

The author gave a five minute sermon about the miraculous benefits of cod liver oil. My expert gave a ten minute scientifically accurate, articulate logical refutation of most of what the author said. Then we started taking calls.

Every call had to do with how much cod liver oil to take, what kind to buy, when it should be taken, and does it also work on . . .

This story, one of many I could relate, strengthens the argument that we should become good at getting our messages out because we have the disadvantage of being scientists and stating only that which we know is true.

We do have some advantages though. Because we speak the truth, our statements are likely to stand the test of time and also be supported by the mainstream professionals.

There are some rules that can be followed to help assure that your points are made. The key is to make every question work for you. Just watch a politician to learn how to do this.

One technique, already discussed is known as "bridging." Bridging involves answering a question and then shifting the focus to a point that you need to make. For example, if an interviewer asks, "Isn't it true that your laboratory murders monkeys by bashing in their skulls and all this is funded by taxpayers money?"

Your answer might be, "Unfortunately, in the process of trying to save thousands of human lives, often an animal needs to be sacrificed. By the way, were you aware [this is the bridge] that last year alone, fifty children with head injuries in this very city could have been saved by the techniques that were developed in our laboratories?"

Another technique is called "hooking." Hooking is answering a question in such a way that the interviewer has no other choice but to ask you the question that you want to ask. For example, "Yes we do animal research, but that is only the second most important thing we do."

It is doubtful that any interviewer could resist asking. "What's the first?" Thus, the term "hook".

It is also important to state your case as soon as possible and as often as possible. You never know when the interview will end.

CONCLUSION

The electronic broadcast media, radio and television are as much a part of today's educational system as are schools and books. They offer instant access to all ages, all levels of sophistication and all socioeconomic segments of society. (There are more televisions in this country than there are toilets!)

We as scientists need to become involved in this rich resource. Four decades ago, Don Herbert, also known as Mr. Wizard, discovered that television offered a wonderful opportunity to teach science. With a balloon and a piece of wool he taught me, and probably many of the readers of this chapter, about static electricity. He is still on the air today, and I continue to learn from him.

But we need more like Mr. Wizard. Every six o'clock news should have a science and health reporter who is more concerned about teaching and helping than getting ratings.

With the advent of interactive cable and talk radio, our students are now able to ask questions and participate in discussions, and new technology consisting of multi channel programs will allow a viewer to actually experiment and see the results of his decisions.

We, the scientists and engineers, developed these technologies. We, especially, should exploit them in the interest of science education.

fort, namely, Arkansas, Maine, Montana, North Dakota, South Carolina, South Dakota, and West Virginia. These seven states were encouraged to compete for five government awards of roughly $3MM each, over a five-year program, and five of these states, namely, Arkansas, Maine, Montana, South Carolina, and West Virginia were successful in being chosen for this first experimental program to try and bring greater balance in research capability between the highest and lowest states. It is my understanding that many of the states' academies played an important role in preparing their proposals.

This first program was deemed sufficiently successful so a new round of awards was initiated, with a second tier of low technology states and Puerto Rico selected for competition. The following potential EPSCOR states were selected, namely, Alabama, Idaho, Louisiana, Mississippi, Nevada, North Dakota, Oklahoma, Puerto Rico, South Dakota, Vermont, and Wyoming. Originally, Kentucky was not chosen as one of the states to compete for this second round.

However, at the urging of Professor Charles Kupchella, who had been Chairman of the 1978 study, Kentucky was then added to this list, largely on the basis of the evidence which had been developed in this earlier study previously described.

Through this series of events, and through the efforts of a number of energetic presidents and academy boards, the Academy began to focus on this need and how to make an impact and contribution to achieving this goal of a more competitive state scientific capability, and eventually to help in also bringing more advanced technology-based companies and industrial activities into the state.

It is the intent of this chapter to review, from the point of view of industrial involvement, the series of events and efforts which have ensued, leading to the beginning of a better interaction between industry and academia science within the Kentucky Academy with the thought that other academies that have similar goals may be interested in learning what is happening in Kentucky as we seek to bring about this interaction with industry. Perhaps they can build on this learning experience in their own state.

At the same time, the Kentucky Academy of Science looks forward to learning from other academies of their successes, and thereby, enabling us to also apply their learning experiences to our efforts. Our work is not finished, rather we believe it has just begun, and any help from our fellow academies will be most welcome.

Because of the difficulty of contacting many of the Academies in the short time allotted to preparing this report, it is not possible to summarize the present status of all of the state academies. However, the following questionnaire was prepared and enough contacts made to confirm that there is a wide variation in the interaction of academies with industrial scientists, with industry, and the day-to-day interaction and cooperation between a given academy and industry.

QUESTIONNAIRE

1. Do you have an industrial science section?
2. Do you have an engineering section?
3. What percent, roughly, of your members are industrial scientists, if any?
4. Do you have financial support from industry?
5. Do you have financial support from the government?
6. Are any of your officers or members of the board of directors from industry?
7. Do you have a permanent headquarter?
8. Has the interaction between the academy and industry led to any opportunities for industrial spin-off from academic research?

EARLY HISTORY OF THE KENTUCKY ACADEMY— INDUSTRIAL INTERACTION

The Kentucky Academy of Science observed its 75th year anniversary last fall at the site of its founding, the University of Kentucky. Going back in the records as best I could, and trying to determine or describe the early interests of the Academy, especially as it relates to industrial interaction, I learned that sometime, perhaps in the 30's, an effort had already begun to achieve better interaction with industry. This is only conjecture because the early record is not clear. All that I was able to learn was that as of the end of World War II, the Academy had already established a company membership category with two members being authorized for each company. These members represented their companies in the State Academy, and although I've not been able to confirm this, it is very likely that our founder of Ashland Oil, Inc., Mr. Paul Blazer, because of his many interests in technology, had played a part in the establishment of this corporate membership.

It certainly is apparent, however, even at this early date, that the Academy was seeking to achieve a productive and meaningful interaction with industry. At that time there were apparently not many industrial corporations with significant research facilities located in the state, and therefore, it is difficult to determine just how many company memberships existed. In trying to learn more, I discussed recollection of this period with Mr. Oliver J. Zandona, a retired executive, who was named one of Ashland's two corporate members in 1952. He recalled that Ashland Oil had apparently appointed members of the Research Department as representatives as far back as the end of World War II.

One has the impression that it was considered a perk and that the selected industrial representatives, instead of helping to improve interaction, kept it as a somewhat confidential arrangement.

By the late 60's, Mr. Zandona became the main company representative to the Academy and began to take a more active part in the affairs of the Academy by serving on the board in the years 1974-1978. A second industrial representative, Dr. John G. Spanyer from Brown and Forman, also went on the Board. It is also apparent that at that time the concept of company membership had become lost in the transition from one president to the next, and today many of the older members having a long time familiarity with the Academy, have no knowledge of this early company membership category.

He and Dr. Spanyer, along with other members did, however, direct efforts to get better industrial involvement, but apparently this did not meet with much success. A lapse of three years occurred until 1981 when Dr. William Baker from the G.E. Laboratories at Louisville, Kentucky, was encouraged to become a member of the board. Nevertheless, the trend of the Academy's effort to get greater industrial involvement does extend back to the 30's and perhaps to its founding. Records prior to 1933 are not available. It is interesting to note, however, that a spring meeting of the Academy was held in Ashland in 1953, but whether it was held under the sponsorship of Ashland Oil is also not clear.

FEDERAL FUNDING OF BASIC RESEARCH STUDY

Another element in this development and interaction with industry occurred in 1973 when a Task Force on Science and Technology, supported by the Kentucky Academy of Science and a $6,000 grant from the Department of Commerce, was authorized to make a study to determine how much federal support was available for research. This Task Force was headed up by Professor William Lloyd from Western Kentucky University and based its study on three criteria: the dollars of federal money awarded the state in terms of research grants and/or federally supported technology, such as, for example, the Oak Ridge national Laboratory in Tennessee, the Aerospace industry at Huntsville, Alabama, or the Air Force Laboratory in Ohio.

This evaluation compared federal support related to (1) population; (2) federal tax revenue received from the state; (3) average income of citizens, and showed that Kentucky was 48th out of 50 states in terms of federal support. A report of this study was prepared and printed and widely circulated in the state in the spring of 1974, but to our knowledge was never published in the *Transactions* or any other Journal.

Nevertheless, because it was focused on population and tax revenue and also involved money awarded to companies producing goods and services for the government, it became apparent that Kentucky was lagging badly both in research activities, as well as industrial performance.

This study was discussed in some detail in the following years by the board of the Academy and culminated in a request for state funding support to make a more detailed study of the situation. This urging of further support was spearheaded by Professor Charles E. Kupchella, who was then on the faculty of the

University of Louisville and President of the Academy. It culminated in Resolution-33, passed by the General Assembly in 1978, that a more detailed study be made of this disparity in federal research and technology expenditures.

This study was completed in 1978-79 and four reports, approved by the General Assembly and supported by the Kentucky Academy of Science, The Council on Higher Education, and the Legislative Research Commission, confirmed the findings of the earlier study. The findings of this study were published in a series of articles in the *Transactions* of the Kentucky Academy of Science in 1979-81.[1,2,3,4] These reports were widely circulated and became the foundation upon which more vigorous action was germinated.

The next stage of interaction with industry came about, as things usually do, by a series of unexpected events and from several different directions almost simultaneously.

INTENSIFIED EFFORT FOR ACADEMIC AND INDUSTRIAL SUPPORT

In the period subsequent to the 1978 study, a number of presidents began to work on this problem of interaction, not only with industry, but also with the universities and colleges in the state. As early as 1971-72, Professor Louis Krumholz, University of Louisville, at that time, President of the Academy, petitioned the University of Louisville to become the first academic affiliate supporter of the Academy. As a result, over the years, the Academy gradually began to accumulate university and college supporters.

Under the urgings of a number of excellent presidents, this effort increased in intensity, and finally, during the years 1980-85, under the urgings of Presidents John Philley, Ted George, J.G. Rodriguez, Gary Boggess, and finally, Joe Winstead, a further intense effort was made to enlist support of all colleges and universities in the state. By the end of Winstead's presidency, it had grown to 21 colleges and universities.

During this same period, an effort was initiated by Presidents Philley, George, and more intensely, by J.G. Rodriguez, to also seek similar support from industry. The effort initially met with little success until I.B.M. came on board in 1984 and since then, the number of industrial affiliate supporters has gradually increased in number and amount, and will be discussed in more detail later.

THE EPSCOR EXPERIENCE

During this period, the National Science Foundation also began its EPSCOR program, as previously mentioned, and Professor Kupchella, who had led the 1978 task force, happened to note in an issue of Science Magazine, a listing of the first EPSCOR states and the reasons for being selected. He also noted that a second round was in the offing and on the spur of the moment during a visit to Washington on other business, contacted Dr. Joe Danek, who was heading up and pioneering the program for EPSCOR. Being aware of our study which had shown us to be 48th, he called this situation to Dr. Danek's attention, and indicated surprise, in view of our findings, that Kentucky had not been included in the second round.

Much of this discussion occurred just as Kentucky was changing administrations. Previously, the Academy reportedly had appealed to Governor John Y. Brown for help in focusing on our state's poor technology performance problem and the EPSCOR opportunity, but without success. As a new administration came into office in 1984, it was suggested by Mr. Herb Leopold, at that time the Director of our Kentucky Junior Academy of Science, that the new Lieutenant Governor, Steve Beshear, might be interested in lending help to correct this situation, as it meshed well with his concerns for the future of the state, and a new program he envisioned, entitled "Kentucky Tomorrow". Members of the Academy under the leadership of then President Gary Boggess became members, along with others, of an ad hoc Science and Technology Committee, as part of the Kentucky Tomorrow Program, and this Committee then also took up a vigorous pursuit of the EPSCOR opportunity.

With a grant from the Governor's office, and with encouragement from NSF, members of this committee put together a proposal emphasizing Kentucky's interest in being considered for the EPSCOR program. The proposal to be included in this second round was successful, and based on NSF recommendation, an EPSCOR committee consisting of three representatives from government, three from academia, and three from industry, was formed to address the problem of preparing a competitive five-year plan for Kentucky to meet the goals of the EPSCOR program.

Based on guidelines issued for the EPSCOR program, the committee put together a project proposal plan that encompassed 15 projects and showing financial support from the universities, government, and industry. It was one of eight state EPSCOR plans out of the original twelve that received approval. This program in Kentucky is now in its fourth year, and promises to achieve all of the goals and objectives originally proposed.

In relation to the subject of this chapter, it should be noted that NSF also encouraged us to try and build into the program, projects that had spin-off potential to industry. Where possible, this was done, and several of these projects now hold promise for this aspect as well. Obviously, this aspect of the program was well received by state government officials, and the universities, but received only lip service approval from industry. In spite of its impressive impact and scientific accomplishments, industrial interaction and support has been slow in coming. Nevertheless, the Academy can be proud of the role it has played in this endeavor.

OTHER SPIN OFFS—THE KENTUCKY SCIENCE AND TECHNOLOGY COUNCIL, INC.

The success of the EPSCOR program, but the lack of significant industrial support and recognition, has in an indirect way, triggered a much more significant development. It had become apparent to many in the state that if a greater momentum in terms of science and mathematics education at all levels was to be achieved, and if more high technical developments and industrial activity was to be realized, then other approaches must be taken.

An interim committee which was formed during the previous administration, namely the Governors' Council on Science and Technology, has now been formalized and incorporated as the Kentucky Science and Technology Council. This Council, consisting of 50 members with experience in academic research, industrial research and/or technology-based industry, has now been formed with a Constitution and with an elected chairman, vice chairman, secretary-treasurer, and executive director. The chairman and vice chairman are from high technology firms and the secretary-treasurer represents academia. The executive director of this Council is Mr. Kris W. Kimel, formerly Administrative Assistant to the Lt. Governor. The Council is rapidly generating financial and industrial support and promises to become a significant force for high technology advancement in the state. Significantly, many of the members of the Council are also members of the Academy, including six Past Presidents. So in a rather round about manner, the Academy has served as a catalyst to achieve one of its long standing goals of increased industrial interaction.

INDUSTRIAL SUPPORT AND PARTICIPATION

Other efforts on the part of the Academy to achieve greater participation in the affairs of the Academy by industrial scientists have also begun to payoff. Although a member since arriving in Kentucky in 1977, as early as 1984 I was invited to become a member of the board. After spending several years in internship, I was elected a Vice President in 1986, President-elect in 1987, and President in 1988. I was deeply honored by this invitation to play an active role in the affairs of the Academy as this was both a new experience for me as well as for the Academy in having an industrial scientist as President. Therefore, as one of my major goals, I chose to focus on the challenge of achieving greater industrial affiliate support for the Academy. The Academy had, of course, as previously mentioned, been working on this objective for some time, but the first objective had been to get university and college affiliate support. This program was started in the late 70's and by 1984, as mentioned, 21 university and college affiliates were supporting the Academy.

Under the leadership of Presidents Joe Winstead in 1984, Charles Covell in 1985, and Larry Giesmann in 1986, the Academy began to also focus on industrial affiliate support and letters of invitation to join were sent out to a number of firms. This effort was slow in producing results. In 1985, the Academy had one industrial supporter. In 1986, the industrial affiliate support rose to three, and by 1987, it had risen to five corporations, with a continuing 21 educational affiliates. An intensive letter campaign was begun in 1987, followed up with further campaigns in 1988 and 1989. As a result, by the end of 1988, eight corporations had begun to participate. The campaign was further intensified in 1989 and by the end of 1989, member participation rose to ten corporations. Now in 1990, a fourth intensive membership campaign has been made and affiliate support is now beginning to catch on, and is rising rapidly. As of the date of this writing, it has risen to 21 corporations, with promise of continuing to rise in the years ahead. At least 125 corporations have been identified as being potential supporters.

PERMANENT HEADQUARTERS

For many years, it has been one of the goals of our Academy to establish a permanent headquarters and with a permanent administrative staff. But income has been such that the Academy has been unable to do so. Recently a greater effort has been made to achieve this goal by working hard on increasing membership, and especially including industrial scientists, and raising the dues sufficiently so as to make significant impact on income. Income has also increased from the annual meetings. This year the Academy also had impressive income from the participation of 21 industrial exhibitors. Income is also gained from the printing of the *Transactions* and this together with the income from the institutional and industrial affiliates, gives encouragement that our Academy soon will be able to realize the goal of a permanent headquarters.

With establishment of a permanent headquarters and with greater visibility, the Academy believes it will then be able to achieve one of its major goals of making a more significant impact on science, science education, and industrial growth through the means of high technology, with an accompanying increase in the standard of living of our people. Of equal importance, the Academy believes that with participation of affiliates, both industrial and academic, that it will serve to create a cohesive spirit among all sectors of the community of science, which should accelerate technical productivity now that a critical mass is beginning to form.

OTHER PROGRAMS TO ACHIEVE ACADEMIC INDUSTRIAL
INTERACTION INDUSTRIAL SCIENTIST MEMBERSHIP AND
OUTSTANDING SCIENTIST OF THE YEAR AWARD

INDUSTRIAL SCIENTIST MEMBERSHIP

The Academy has recently undertaken several other programs which should be mentioned, and which hopefully will also increase interaction between industrial and academic scientists. Previously, the Academy membership consisted mainly of academic scientists. With the industrial-focused campaign of the past few years, the number of industrial scientists in the Academy is gradually increasing. At first, industrial scientist membership increased slowly and without much enthusiasm, but as the number of industrial scientists has grown, there has become a greater awareness of the Academy, and that they are welcome as equals and encouraged to interact with academic scientists.

NEW ACADEMY OF SCIENCE, SCIENCE SECTIONS ENGINEERING AND INDUSTRIAL SCIENCE

For some time the Academy had an inactive Engineering Section. But the absence of engineers in the Academy, especially in these days of high technology, left much to be desired. Therefore, members of the engineering faculties were contacted to see if an Engineering Section could be reactivated. A very positive response was received and an Engineering Section has now been formed with much activity and enthusiasm. This seemingly small step has been an important development, however, as far as industry is concerned. Following this success, the Academy also approved formation of an Industrial Science Section. As a result, last year for the first time at an annual meeting, two full days of science papers were presented by industrial scientists. This is a significant new contribution to the Academy and if it can be maintained and caused to grow, will create and represent an increasingly strong interaction with industry.

SCIENTIFIC RECOGNITION TO INDUSTRIAL SCIENTISTS

In the past, the Academy each year has given an award recognizing an outstanding Scientist of the Year and similar awards for Outstanding Teachers. It readily became apparent, however, that industrial scientists were not being considered for the Scientist of the Year Award. In order to make some distinction between academic and industrial professionals, the Academy has, therefore, also approved the formation of an Industrial Scientist of the Year Award. It is believed that this award can serve to catalyze and encourage greater participation by industrial scientists. The program to increase membership by industrial scientists is also doing quite well. It has now risen to approximately 50 members with indications that it will continue to rise as more of these industrial scientists spread the word about the Academy.

CONCLUSION

With regard to interaction and cooperation of state academies with industry, there does appear to be a variation in this interaction and cooperation from state to state. While a complete study of all academies was not undertaken, it is apparent that some academies have excellent interaction and cooperation and are doing a fine job in this regard. Others apparently have difficulty in developing an interaction and relationship, and of course, some are somewhere in between.

In Kentucky it is quite apparent that the Academy has made a significant effort over the years to interact with industrial scientists in a state in which academic research is not well supported. Because not much high technology research exists or is required by Kentucky industry, the development and interaction has been very slow in coming. It is quite apparent, however, that the Academy has made much effort over a 50 year period in achieving this interaction, although only in the past 10-15 years has it begun to realize any success. The lesson learned here is that if an Academy is to be successful and if it is to make positive contributions to industrial development of the state, persistence in this effort at interaction can make it a reality. There are, obviously, many things that an academy can and should do to help increase this interaction.

Within the university this cultural interface must also be developed and encouraged. Many academic members have almost no familiarity with the activities, goals and operations of industrial research. It does not seem to be that they are not interested in industrial research, but rather that they have had little opportunity to become acquainted with it.

Several years ago (1982), Ashland Oil agreed to be the site of an annual meeting. In the course of that meeting, Academy members were given an an opportunity to visit our laboratories and plants and become better acquainted with what we industrial scientists do. Although this annual meeting occurred eight years ago, I still get compliments and expressions of how much they enjoyed this experience. So we see that perhaps another way to increase interaction is to schedule more annual meetings at industrial plant sites and/or for the Academy to schedule excursions to these industrial laboratories and manufacturing sites.

Encouraging project leaders in such as the EPSCOR projects or similar centers of team research which have a potential for industrial impact, to visit with industrial scientists, is another way that better interaction can be achieved.

Professor H. Ted Davis, University of Minnesota, gave a paper at the ACS meeting in Boston, Massachusetts, April 22-27, 1990, emphasizing how interaction with an industrial problem can actually stimulate fundamental research. It may come as a pleasant surprise to academic scientists to uncover such fascinating basic research problems through this interaction with industry.

While there are many ways to intensify technology activities and industrial success in the 21st Century, certainly a much stronger interaction between state government and industrial and academic scientist members of the Academy can be expected to make a major contribution, as the United States of America intensifies its effort to remain competitive in the community and the world.

REFERENCES

1.Kupchella, C.E., R. Simes, M.L. Collins and K. Walker. 1979. Federal funding for research and development in Kentucky. I. TKAS 40(3-4):149-153.

2.Kupchella, C.E., R. Simes, M.L. Collins and K. Walker. 1980a. Federal funding for research and development in Kentucky. II. Kentucky in comparison with other states. TKAS 41(1-2):1-11.

3.Kupchella, C.E., R. Simes, M.L. Collins and K. Walker. 1980b. Federal funding for research and development in Kentucky. III. Characteristics of colleges and universities with high levels of supports. TKAS 41(3-4):150-155.

4.Kupchella, C.E., R. Simes, M.L. Collins and K. Walker. 1981. Federal funding for research and development in Kentucky. IV. Economic impact of federal research and development funding in Kentucky. TKAS 42(1-2):29-32.

Science Education in the United States: Issues, Crises and Priorities. Edited by S. K. Majumdar, L. M. Rosenfeld, P. A. Rubba, E. W. Miller and R. F. Schmalz. ©1991, The Pennsylvania Academy of Science.

Chapter Thirty-Two

SCIENCE AND TECHNOLOGY EDUCATION: AN ENGINEER'S PERSPECTIVE

WILLIAM A. FREDERICK

Manager, Research
Pennsylvania Power
and Light Company
2 North 9th Street
Allentown, PA 18101-1179

William A. Frederick
Manager, Research, PP&L

INTRODUCTION

Scientists and engineers are the backbone of industry. Without their talents major industries could not stay in business. It is therefore necessary for industries to plan ahead to make sure they have an adequate, continuous supply of scientists and engineers in the future.

What is the outlook for scientists and engineers (S & Es) in the near term and in the long term future? Colleges and universities are reporting that there is a reduction in the number of applicants and a reduction of the quality of these applicants. Industries' needs for S & Es are growing, and most projections point to an increasing need over the next ten years. With an increasing need and a decreasing supply, we have a problem. This problem cannot be solved overnight.

The USA has been the greatest industrial nation in the world. Following the Arab oil boycott in 1972, and the resulting energy crisis, the USA lost its competitive edge in major industries, such as, steel and large machinery. The electric utilities purchased almost one third of the equipment manufactured in the USA on an annual basis prior to 1972. Following 1972, the economy slowed down and the demand for electricity decreased. The electric utilities found themselves with an excess of generating capacity that was planned in the 60's and 70's and came on line in the mid to late 70's. Today very little generating capacity is on order, and as a result, the manufacturers of the large utility equipment have reduced or sold their manufacturing facilities. However, as the nation grows, this manufacturing capacity must be rejuvenated, producing requirements for S & Es that will be unparalleled in our nation's history.

It is obvious that if the industrial base of the nation is going to grow, there will be a continuing need for quality scientists and engineers. Projections based on a U.S. Government Report (U.S. Department of Education, Center for Education Statistics) show a decline in BS awards in science and engineering from 1990 to the year 2000. A continuation of these trends could be a disaster to the economic viability of this country.

This chapter will deal, not with the statistics of what is expected to happen, but with the apparent causes of the decline in the quality and quantity of BS degrees in S & Es and explore the avenues necessary to produce a change in direction of the trend to improve both the quality and the quantity of S & Es in the future.

BACKGROUND

The USA has been a leader in world technology. The inventive genius of our scientists and engineers in combination with our free economic system has kept us at the top of the industrial nations in the world. In recent years, however, our position has been challenged, and we are losing ground in the industrial sector. Our production of scientific knowledge and new inventions is still far ahead of the rest of the world, but countries such as Japan are transferring our new technologies into marketable products in less time and at less cost than we are achieving. In many products we are losing our competitive edge. For example, the transistor and the integrated circuit were invented in the USA, but only one US television manufacturer still remains in business.

We can find many reasons for the USA getting behind in the world marketplace. Labor is cheaper in developing countries. They work longer hours. They are more dedicated than we are to quality and quantity, etc. Environmental laws in many countries are less strict than in the USA. Some countries subsidize their exports or provide trade barriers for imported products. They also provide less barriers than the USA in transferring new technology to a manufactured product. All these and many more reasons lead to the fact that the USA is losing its competitive edge in the world marketplace.

Let's look at some specifics. Twenty to thirty years ago US manufacturers such as General Electric, Westinghouse, and Allis Chalmers led the large machinery and electrical equipment market. Each company did its own research and continually surveyed the marketplace to find out what areas needed research to meet new equipment needs. In the last ten to twenty years foreign competition entered the US electrical equipment market. The US companies had to be more competitive to meet the price competition, and they did so by reducing expenditures on research. In the long run this hurt the US companies and they eventually lost both their technological advantage and their price advantage. In short, US companies have essentially withdrawn from the large electrical generating equipment market.

Another example, and for other reasons, is the demise of the nuclear industry in the US. Political, public opinion, licensing, financial, legal, and safety issues all add to the complication of building new nuclear electrical generating units in the US. The obstacles are so great that no US utility has ordered a new nuclear generating unit in over ten years.

Who is doing the planning in the US?—Politicians, with foresight equal to the length of their terms in office! Who should be planning the industrial future of the US?—Certainly the scientists and engineers should have a major role in the planning. The S & Es should also have a major role in seeing that these plans and goals are carried out.

If we are to retain the lead in industries where we still maintain a lead, and if we are to recapture the lead where we have fallen behind, we will need far more scientists and engineers than present forecasts predict.

What can be done about the shortage of S & Es? First, let us look at the steps in the process.

It is obvious by figure (1) that if we are going to have an increase in the number of scientists and engineers in the future we must influence students to have the desire to become S & Es at or before the sophomore level. We must influence them in time to take the algebra trail to get into the pipeline.

NEEDS

In a recent discussion with a college professor, I asked him what he thought would be done about the shortage of student applicants in the S & E college curriculum. He felt the colleges and universities would be forced to drop the entrance standards to attract enough students. Of course this is only a stop-gap measure and can lead to poorer quality graduates in the future. We need to address the problem from grades K to 12. We need to address the college curriculum, and then, of course, the continuing education throughout the productive life of the individual.

Let's start with defining what industry wants to see in a scientist or engineer. What capabilities and attributes would be considered ideal? The S & E should have the following:

1. An in-depth knowledge in his specific field.
2. A general knowledge of associated fields, i.e., if he/she has an electrical engineering degree, he/she should also have a general knowledge in other engineering disciplines such as mechanical, civil, etc. engineering disciplines.
3. Good communication skills, which include:
   a. Adequate speaking ability to communicate with peers, and to present ideas to higher management and to the public.
   b. Good writing skills for documentation and communication of ideas.

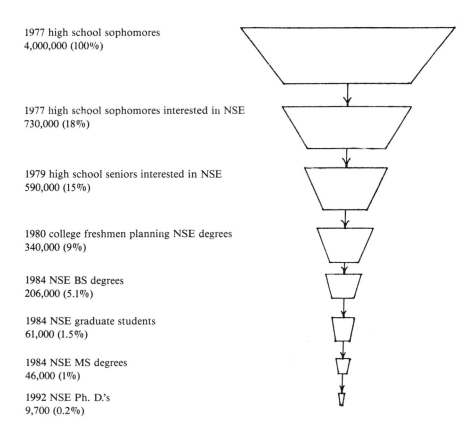

1977 high school sophomores
4,000,000 (100%)

1977 high school sophomores interested in NSE
730,000 (18%)

1979 high school seniors interested in NSE
590,000 (15%)

1980 college freshmen planning NSE degrees
340,000 (9%)

1984 NSE BS degrees
206,000 (5.1%)

1984 NSE graduate students
61,000 (1.5%)

1984 NSE MS degrees
46,000 (1%)

1992 NSE Ph. D.'s
9,700 (0.2%)

FIGURE 1. Natural Science/Engineering Pipeline (NSE): following a class from high school through graduate school (From "Educating Scientists and Engineers - Grade School to Grad School", Congress of the U.S. Office of Technology Assessment).

    c. Suitable social skills so that the individual as well as his/her ideas are accepted.

    d. Good reading skills to keep up with new technology and broaden one's horizons.

4. Planning and scheduling skills to make good use of time and develop personal as well as job-related goals and objectives.

5. Environmental skills, both for personal and job related work.

This is not a complete list but is good enough to develop the process for the present.

When we review the above it becomes obvious that all of the educational skills cannot be left for the college years. The entire education period must be carefully planned and carried out. This will be addressed later.

PROBLEMS

Let us now consider the problems and obstacles that a student faces in getting his/her education. These include:

1. Too many students in a class—not enough individual instruction.
2. Not enough encouragement at home—many one parent families or families where both parents work.
3. Too much television at home.
4. Not enough emphasis on memorization skills—such as multiplication tables, etc.
5. Not enough teaching aid equipment to go around.
6. Not enough attention to reading skills.
7. Not enough attention to comprehension skills.
8. Low pay of teachers leads to less teachers in the profession and poorer quality teachers.
9. Not enough attention to weeding out poor teachers.
10. Advisors are not adequately trained in many cases.
11. Poor planning for individual student's curriculum.
12. Not enough homework.

This list is not all inclusive and not all of the items apply to a particular school system. However, it is necessary to develop a comprehensive list for an audit guide on our existing system if we are to achieve improvement.

The next step is to review the needs and the problems, and then set out the task of developing solutions. If we are to achieve any major improvement in the educational process over the next ten years, we need a quantum jump in improvement in that process.

It is unlikely to expect this quantum jump to emerge without external help. The internal system has been working with these problems for a long time, and has stretched the existing ideas to the limit.

A scientist or engineer would look to technology transfer from another field when a new idea or quantum jump is needed. Where can we look for new ideas—changes that may work? Can we afford them? Where can we get the resources? Can we afford not to look for new ideas?

Other service industries where people's needs are met could be explored and compared. Caring for the elderly is a good example. In a nearby home for the elderly about 450 residents are cared for in all stages of their waning years. Some can physically take care of themselves with only frequent reminders about pill schedules, activity schedules, etc. Others may need complete care 24 hours a day and must be hand fed, bathed, etc. Some have no memory nor communication capabilities remaining. How many people are employed in this institution? About 450—just about one employee for each resident. Of course, this includes janitor staff, management, bookkeeping, food preparation, etc. The annual per capita cost for this service is about $20,000 to $30,000 depending on the individual care requirements.

How much time and effort do we spend on preparing an individual for a career compared to post career care? In the case of the elderly, society says it must be done, and circumstances dictate the cost. In the case of the young, society also says a minimum education must be provided, primarily at a minimum cost.

The elderly, on the average, have a one-to-one ratio of patient to support person. The student has about a 30-to-one student to teacher ratio. With all the clerical staff, supervisors, janitors, etc., this ratio may net out to about 20 to one. In addition, the student is in school less than a third of the day. If we provided as much attention, effort, and money on the front end of life as we do on the back end, we would have the resources to achieve a quantum jump in education.

SOLUTIONS

Assuming there is a source of money available, how can we improve the existing system?

Again, let's take technology from another area and examine it carefully. Look at a specific product. Can the product stay in competition in the marketplace without being improved? Does the product need to be modified to meet the needs of changing market requirements? Can production procedures be improved to reduce the manufacturing costs? Can raw material costs be reduced?

The automobile, for example, was available before the turn of the century. To a small degree it replaced the horse, but it didn't compete seriously with rail passenger transportation until Henry Ford set up a production line and reduced the cost to within the reach of the general public. Competition from other car manufacturers was always present, pressuring the car builders to continually improve their product in order to stay in business. The builders used test tracks to test new ideas before they were incorporated into the product in order to uncover component inadequacies and identify areas that needed improvement. This is a continuous process and a necessary part of the economic system.

Can we use ideas and philosophy from the industrial process to improve the educational process? I think we can. Industry puts more money and effort into a product that has already achieved a good market position, in order to maintain its market position, than it does in a poor product. Of course we cannot neglect the slow learners in the educational system, but we should also focus our attention on the best in the class, for there lies the greatest potential for growth, the greatest potential for creativity, the greatest potential for future achievement.

Let us take our best onto the test track and make sure all components are ready and able to withstand the rigors of the marketplace.

Some years ago I registered for a high school evening course in speed reading and comprehension. I had been out of college for about ten years and felt I needed some self renewal. In the entrance speed reading test I measured about 150 words per minute with a comprehension rate of about 70%. At the conclusion of the course, most of us in the class could read at between 600 and 800 words per minute with a comprehension rate of over 80%. Many achieved a scanning rate of 2000 to 3000 words per minute with about 30% comprehension. However, teaching scanning was not the primary focus of the course, and therefore, little effort was placed on it. The course cost: $1.65. What a bargain!!

The principles of the speed reading and comprehension course were simple:
1. Teach the student to bypass the mental pronunciation process and transfer the word image directly to the brain.
2. At this increased reading speed the mind is not idle for the pronunciation process and therefore does not drift and lose focus. Comprehension automatically increases with the increased focus of attention.

Speed reading and comprehension courses available at that time used strip film projectors with an adjustable motor drive so that the student could regulate the words-per-minute desired. A test was available with each film strip to measure comprehension. The student, thus, could operate the device at the desired rate and measure his/her own progress. A large library of film strip stories was available at all grade levels.

What a perfect match for a personal computer!! I have checked and found that this system is now available for the personal computer. I hope this teaching tool will be applied liberally throughout the educational system.

It has been said that if you give a man a fish, you will feed him for a day, but, if you teach him how to fish, you have fed him for his lifetime. Let us transfer that idea to the educational system. If you teach a student, he/she will learn for the day. If you teach the student how to learn, the learning process will continue for a lifetime.

Does the present system teach the student how to learn? Do we measure the student's learning skills? How do some students get through high school—

through the whole educational system—without being able to read? Some do. Why did they get so far without the system knowing and correcting the problem? What can we do to prevent this from happening?

Up to now I have emphasized the drawbacks, the apparent failures in the system. Of course the problems must be addressed if we are to develop solutions to improve the educational process. Solutions are not only available from outside the system but also may have been tried within the system and are not widely used.

Recently I saw on TV that a school in Texas used retired volunteers to help tutor students. Hospitals use volunteers. Homes for the elderly use volunteers. Perhaps therein lies a resource we could tap.

Earlier I mentioned the high cost of caring for the elderly and how little money we apply to education in comparison. If we want good teachers we will have to pay them a decent wage. If we want the best for our children, we must pay a wage that will compete with the best in industry. This country's most valuable assets are the lives and minds of its children. We cannot afford not to give our children the best education we know how to provide. We must provide the money to attract and to retain the best teachers.

We must also provide an incentive for teachers to continue their education. If a graduate with a BS degree is hired by industry, his/her education becomes obsolete in about ten years if he/she does not continue some phase of educational activity. Many industries, therefore, will provide some means of continuing education for the scientist or engineer such as an advanced degree or professional courses. A teacher has the whole summer to advance his/her education to become a better teacher. Many would welcome the opportunity for educational improvement if they could afford it. We should make this possible and could make it a job requirement!

The task of improving the educational system is not an easy one. I am by no means an expert in education. I have not pointed out or defined all the problems and obstacles to be overcome in the system. Nor have I by any stretch of the imagination provided solutions to all the problems. I hope that I might inspire those who can bring about change, to take a fresh approach with open minds to new ideas and methods that may improve the system.

I feel that a task force should be appointed to address the problems in education. This task force should include not only educators but people from many other disciplines. They should take a fresh approach, exploring all avenues of existing technology. They should study the transfer of technology from any area that is useful and should not be bound by bureaucracy or tradition. Their boundaries should be limitless. They should strive to improve the quality of education for every stage of life. In the end we will all benefit.

Let us take the poorer students by the hand and teach them to be good, let us inspire the good students to be better, and let us provide the tools and the guidance to allow the best students to become the geniuses of tomorrow.

Science Education in the United States: Issues, Crises and Priorities. Edited by S. K. Majumdar, L. M. Rosenfeld, P. A. Rubba, E. W. Miller and R. F. Schmalz. ©1991, The Pennsylvania Academy of Science.

Chapter Thirty-Three

SCIENCE EDUCATION: A HUMANIST PERSPECTIVE[1]

CARL N. SCHMALZ, JR.

Professor of Fine Arts
Amherst College
Amherst, MA 01002

On a stuningly bright blue and green day in late June, the last day of my junior year in High School, I was asked to stay after class. I was a bit alarmed. It developed, however, that my diminuitive, soft-spoken, spinster teacher wished only to have a word with me, to wit: "My boy, if I were you I would never take any mathematics again." The advice relieved me. I had struggled through six terms, alternately of geometry (*A*s) and then of algebra (*B*s), netting, of course, a final *B* for math, a grade I found disappointing. I later concluded that, though well meant and welcome at the time, this was rather poor counsel. Had my teacher been less circumscribed by regulations and procedures, had she seen that, for me, the way to math was through measuring *things*, perhaps I would not have spent nearly half a century trying to make up for that early loss. In the event, rather than math, I turned to drawing. That is, I turned to approximate measures rather than exact ones: for drawing is nothing more than estimating proportional relationships. For the visual artist (the poet, dancer and musician as well), whether representational or non-figurative, these proportions may be varied at will to suit the particular expressive purpose, but those variations are nonetheless proportions, measures. All artists are constantly and deeply involved in matters that are metaphorically mathematical.

[1]Though none is responsible for such errors or absurdities as may appear here, I would like to thank the many colleagues and students who have shared observations and ideas with me. I am especially grateful to Professors Duane Bailey and Joel Upton, and Mr. Tom Looker, all of Amherst College, and Professor Frank D'Isa of Youngstown State University. I am greatly indebted to Professor Robert F. Schmalz of Pennsylvania State University for many valuable years of pleasureful and fruitful conversation about these matters.

Besides this fundamental analogy to mathematical thinking, acquaintance with a spectrum of natural sciences is essential for both the practicing artist and the art historian. Anatomy and physical geology, for example, are fairly obvious. Others may come less immediately to mind—meteorology, physical anthropology, minerology, various aspects of physics and, of course, chemistry. The needs of other humanists, though varying, are similar, as are those of the ordinary citizen. By describing in greater detail the sorts of science that artists and art historians need I will try to suggest a model for the kind of knowledge required by those more general groups. Finally, I will offer some thoughts about science education, and education broadly, as seen from my own humanist perspective.

Science Needs of Art Historians and Artists

Art history is a profoundly inclusive study. Depending upon their particular interests, art historians find themselves necessarily concerned with literature, religion, philosophy, politics, economics, anthropology, sociology, music, and so forth. But because the greater proportion of works of art (which constitute their primary data) are to one or another degree representational, they must also be concerned with what things look or looked like. Likewise, because the objects they study exist physically, they frequently need to know about the materials and techniques of and by which those objects were made. Together these two facts thrust the art historian squarely into the worlds of the natural sciences.

Much of the sort of knowledge we require as art historians is empirical or pragmatic. Extensive study of zoology is unnecessary for the observer who notes that the leg muscles of Assyrian lions are conventionalized. On the other hand, some real knowledge of human anatomy will help a viewer perceive that Classic Greek male nudes are—in a different way—equally conventionalized. Similarly, the crystal character and hardness of stone can frequently affect the final look of a carved object, so the typically cubical appearance of ancient Egyptian sculptures may in part result from the difficulty of working diorite. A familiarity with basic physics is helpful to the scholar analyzing architecture: without a clear appreciation of how thrusts work, buttressing systems make little sense. It is often necessary for art historians to know something about the *history* of science, including the applied science of medicine. A good deal of the Mediaeval and Renaissance art of the West alludes in one or another way to alchemy, to the four humors, and to astronomy/astrology; and during the seventeenth and, especially, the eighteenth centuries many popular paintings and prints dealt with medical and dental quackery.

Today some art historians are working in concert with conservators, using highly sophisticated techniques to analyze the structure of objects of art with an eye not merely to learning about their physical compositions, but to discover how artists actually worked out their ideas, what sorts of changes they may have made as they progressed toward the formation of the final image, what materials they employed, and in what order, to understand more fully what effects they appear to have been pursuing. Such scholars have obviously to learn about the instruments and procedures required by their research. Occasionally an art historian must be (or become) acquainted with scientific theory. The formation of analytical cubism, for one example, during the years 1906–08, has been linked to Einstein's 1905 publication on Relativity. Clearly one must have more than a passing understanding of the Theory of Relativity in order convincingly to demonstrate such a connection.

The range of scientific knowledge required by the artist varies enormously. A "traditional" oil painter needs a good deal of basic information about both physics and chemistry, much of it handed down within the workshop system in earlier times and now—at least in the best schools—a part of the curriculum. I refer here to understanding the causes of warping in panels and stretchers, the reasons for working "lean to fat" as paint layers are built up, some fundamental grasp of paint chemistry to prevent unstable mixtures, and so on. Depending upon medium, knowledge of papers, drawing media, brush construction and quality, solvents, inks and acids, waxes, and emulsions are either necessary or desirable. In many instances the artist's own safety may depend upon his/her appreciation of these aspects of materials: the permanence of the work nearly always depends on such understanding. What is true for the painter and printmaker may be even more the case for those working with more complex techniques—sculptors, bronze casters, welders, ceramics artists—not to mention recent technologies including plastics and fiber glass, photography, video, and computer generated images.

As with art historians, artists who continue to work within the representational tradition also need a firm grounding in the "look" of things. This involves understanding structural actuality and principles, at least so the West has believed since the fifteenth century. Anatomical study provides this basic knowledge for creatures that move as botanical observation provides it for flora and physical/historical geology provides it for landscape forms. A fairly sophisticated familiarity with optics and the properties and behavior of light, and an understanding of the ways in which colored light behaves as opposed to colored paint, and the inter-relations between the two, are also important. Some knowledge of the chemistry, neuro-biology and psychology of visual perception may not be really necessary, but the fundamental devices of the artist, whether working figuratively or non-figuratively, are closely tied to the perceptual possibilities and proclivities of the human eye and mind.

In special circumstances artists may need to understand theoretical branches of natural science, or they may be inspired by the visual possibilities implicit in the notion of, say, black holes or recombinant DNA, and hence want to acquaint themselves with the astro-physical or bio-chemical hypotheses that underlie those phenomena.—There is little that may not offer grist for the artist's mill.

Importance of General Science Education

From the brief summary above, it should be evident that artists' and art historians "scientific" needs are usually of a practical sort; that is, one might better describe them as "technological" rather than "scientific". It is applications rather than theoretical constructs that they most frequently require. To a greater or lesser degree this is true for other humanists, and it is chiefly what Joe and Jane Citizen need as well: implications for the *teaching* of science ensue. But it is also desirable for artists, art historians, other humanists, and Joe and Jane (where possible) to understand enough theory and to possess skills that would enable each of them to pursue a given area of interest with a realistic expectation of actually broadening and deepening their knowledge. And this may suggest alternative teaching strategies.

Scientists will doubtless shake their heads as I attempt to pursue this distinction between "pure" natural science—theoretical formulation and research—and applied science or technology, arguing that the distinction is impossible owing to the invisible gradations between the two and the inevitability of interaction between them. I recognize the problem. Nevertheless, I believe that it is useful (and perhaps essential) to separate them, at least hypothetically, because the kind of science that most people deal with most of the time is of the latter sort—applied science. And while few thoughtful people would question pure science's disinterested investigation of nature, technological applications today almost always involve far-reaching issues of practicality, morality, or both. Clearly, one can back this up to ask whether, say, genetic research was and is *itself* good: for the purposes of this note, I should like to evade that query and assume that it is. The point is that we are still left with a mass of thorny questions about how, how much, where, when, knowledge so gained should be applied, and to what ends. Analogous considerations extend to much more mundane levels. *e.g.,* should personal radios contain built-in volume limits, or automobiles a top-speed control? These matters, of course, belong to the province of the law, and they are normally effected by political and legislative means. But in a democracy decisions like these are made ultimately by the citizens, and to act wisely on questions of this sort—as well as on the vastly more significant technological issues that today face the country and the world, such as those relating to defense, energy, medicine, environment, and so forth—the citizenry must be *informed*. That means that as many individuals as possible must have a sufficient background in the sciences to apprehend the fundamental issues, understand the arguments, be capable of detecting the biases of special interest groups, and so on. Otherwise they cannot rationally and responsibly grapple with the profound practical and moral problems that are becoming so vital to life's future.—But how to provide such understanding to as many people as possible?

Probably the most vexed—and vexing—question relative to natural science instruction concerns the need for mathematical skills, most notable the calculus. Plainly, anyone wishing to pursue research requires an easy familiarity with calculus, and some grasp of it is surely necessary to those who would teach themselves more about aspects of science that interest them. While acknowledging that use of the calculus is a skill *desirable* for all citizens, must and should we insist that it is absolutely indispensible? Much of scientific thought appears to be available in "translation", by analogy and through visual models. Many books, as well as the science articles and programs of the popular media— news magazines, newspapers, public radio, and television—present even complex theories and ideas in formats understandable by a reasonably intelligent and attentive audience, usually with little or no mathematics. Although providing citizens with basic "numeracy" is essential, schools, colleges and universities ought also to be capable of devising solidly informative science courses that depend as minimally on mathematical skills. This is by no means to deny the need for reforms in math teaching. Quite the contrary; and I understand that discussion and exploration of such reforms have for some years been seriously underway in many parts. Even assuming, however, that we devise new math curricula that engage students more fully and fervently, considerable time may elapse before such curricula are effectively in place. Meanwhile, we have in this country an enormous population of mathematically "illiterate" citizens, many of whom will be with us for the next several decades.

It seems to me that the ideal toward which we should aim is a thorough reintegration of the fields of knowledge, from the first grade more or less through high school and college. I say "more or less" because some students will continue to need special technical training if they are not going to college or a vocational school, while those going on to graduate study ought, for the most part, to continue to learn in a more integrated way than is usual at present. Two approaches to attaining greater integration are implicit in the remarks above: one is to improve math instruction so that students follow the arguments of science in its own "language", and the other is to incorporate applicable aspects of the natural sciences into the teaching of the humanities. Further, because of that large body of scientifically ill-educated folk, serious and persistent effort should be made to increase *greatly* the quantity and quality of non-mathematical science eduction publication and programming; and it should envision not only programs *about* science, but the inclusion of scientific references, explanations, reasoning and history within as much general programing as is feasible. This ambitious undertaking would deserve and doubtless require government funding, though corporate and other private sector backing should be vigorously sought.

Since improvement in mathematics instruction appears to be underway, and in the hands of competent people, and since I have not the space here to expand upon ways of tackling the matter of creating greater everyday public learning possibilities, let me address the remaining of my three suggestions: incorporating aspects of science into humanities instruction.

Educating Educators, and Related Thoughts

An initial difficulty is that many teachers of the humanities are among the scientifically knowledgeable poor, so a first step might be to set up adjunct learning programs through which they could better inform themselves. Such programs might best be directed toward particular disciplines, e.g. for historians, teachers of literature, etc. The process of creating such programs would certainly yield new ideas about effective teaching, encouraging depth, versatility, and range of thought.

Inducing teachers to attend such programs, let alone adapt them enthusiastically and imaginatively to their own classroom practice, presents another and different problem. Especially among teachers at the college level, we here collide with entrenched and accepted institutional practices. These are familiar: more or less rigid specialization and narrowness of professional focus, political intransigencies and concomitant departmental defense perimeters, jealousy of research time, plus the common, generalized resistance to change. In addition, reluctance to engage in educational innovation may be expected from some administrators who look apprehensively ahead to financial and demographic uncertainties during the decade of the nineties.

What is needed, I believe, to inspire and quicken our educational establishment is a new vision of the educated person, a metaphor based on a humane, or at least ecological, rather than a mechanical model. The notion of the "well-rounded" student that tacitly still underlies many of our educational assumptions, assembles the person from separate parts. It expects that the person who knows a little about a lot of things (some "required", some not) and a lot about one or two, will assimilate those bits of knowledge into his/her own *persona,* utilizing mathematical, logical, analytical and synthesizing tools, as well as analogical and metaphorical ones, as appropriate. And it further expects this assemblage to be performed by the unitary individual, much on the model of the "self-made" successful *men* of nearly a century ago. But knowledge has become so fragmented, and so encapsulated, and our students so heterogeneous, that basic ways of thinking are today much less easily transferred from one area of study to another than they were when the "well-rounded" ideal was formed. Societal changes, particularly the breakdown of the generally supportive family, have rendered the self-made individual both less attractive and less feasible than in earlier times. For these and other reasons students have much greater difficulty in "making themselves" from the methods and data derived from the specialized courses that they mostly encounter. To offer them a better chance of attaining their fullest capacities we need to imagine the educated person as one whose own knowledge (whatever it may comprise) is consciously interactive and interdependent, wherein the "parts" no longer appear as separate entities to be assembled, but interrelate reflexively as in a poem or, indeed, a mathematical proof. We need also to imagine the educated person as understanding his/her learning to be linked interactively with that of others. Most significantly, we need to present at least some knowledge *to* students in these ways.

Though this may appear a major shift in point of view, ample evidence suggests that many teachers—and some institutions—are already pursuing such a vision. Examples exist in so-called "interdisciplinary" offerings. They also appear in the "Studies" format—Asian, Black, European, Women's, for instance—as well as in a variety of the newer "Core" curricula and special Freshman programs. Characteristic of most of these approaches is examination of a broad question or an area of inquiry from the points of view of a number of relevant disciplines, frequently including those of the natural sciences. They may often in addition consider philosophical issues—moral, ethical and human concerns. At present, many courses of this sort are necessarily team taught, and perhaps for some time this will be the best way to take advantage of individual teachers' specialized training. For efficiency's sake such courses will be rather large: for learning's sake they should be broken into sections, laboratories, or discussion groups for at least some of the meetings. In many cases appropriate computer programs, video and other contemporary instructional techniques can be effectively employed, but care needs to be taken to avoid too extensive dependence on these allies.

I should perhaps briefly expand upon this *caveat*. To my mind, one of the most pernicious aspects of present educational practice is an excessive emphasis on what might be termed "intangible" learning, that is, transmission of ideas by verbal and abstract conceptual means. Throughout our lives, including our older years, we learn best not only by "doing", but by seeing, hearing, feeling, smelling, and even tasting. Our senses fundamentally underlie our experience of testing, rechecking, discarding and accepting. Only direct encounter with active minds and actual materials can nourish this basic form of learning. A professor of engineering lately lamented that his students, while fully apprehending the proper techniques, skills and formulae, neverthless had so little concrete understanding that they could not recognize their most manifestly wrong answers to problems. Another professor informs me that, in these days of ever-present calculators, the math class in which students work out their proofs on the blackboard will teach more effectively than one in which this does not happen. Other colleagues recently had students actually make glass and build vaults from blocks as part of their learning about the complex art of a Gothic cathedral. I believe that art historians, especially those not encountering originals face to face, who are innocent of personal experience in making art cannot bring either to their study or to their appreciation a necessary depth of understanding (any more than young artists can expect to accomplish worthwhile work without knowing the arts of the past). Concrete experience, properly conceived and employed, will nearly always reinforce conceptual ideas: it should not be dismissed as mere "sandbox" pedagogy suitable only to nursery schools. Similarly, shared learning, as teams of five or six students working together on a single project, helps to inculcate appreciation for intellectual diversity, the value of cooperation, and recognition that knowledge—however uniquely shaped by particular minds—is both the creation and the possession of all humanity together.

The concrete and pragmatic is surely not out of place in natural science teaching. But student advisees of mine have often mentioned the dreariness of much laboratory work. They are bored by repeating for the zillionth time in educational history an experiment of which they know the result. One wonders whether the experimental canon could expand to include fixing a flat iron or baking a pancake, or other means of linking basic scientific principles to our everyday lives. Perhaps, as well, teachers of science might allude more frequently and feelingly to its aesthetic dimensions, and even its abundance of metaphorical applications.

Finally, speaking as a humanist, what practically—given present educational structures and paradigms—can individual teachers in their individual courses do the help students understand better and more fully grasp the importance of the natural sciences? Chiefly, I think, we in the humanities must more conscientiously incorporate into our courses those ways in which scientific inquiry and discovery impinges, or has impinged, upon the matter of our teaching. While we certainly can benefit from the sort of supplementary study mentioned above, we can also begin immediately in much the same way that we have begun to include fuller coverage of, in my field, the artistic accomplishments of women and minorities. Let me provide some examples from my areas of interest.

The teaching of art history offers innumerable possibilities for inclusion of scientific information and information about the natural sciences (as, of course, do literature, history, *et al.*). For instance, in discussing Renaissance frescoes a teacher might explain how the chemical reactions between the pigment and setting plaster fix the color in place. Possibilities abound in relating stylistic peculiarities to the limitations of the artist's medium and/or craft. The hardness of diorite was earlier mentioned: discourse on the hardnesses of stone and the various properties of metals can (and, in my opinion, should) be a part of the study of both sculpture and architecture. Likewise, encounters with representational oddities should encourage geological, anatomical, botanical and zoological information. Virtually endless opportunities for mention of important aspects of the history of science inform all of the visual arts, from Pythgoras and Euclid, through Galileo and Copernicus, Lavoisier and Darwin, to Einstein and Crick. Jacques-Louis David's superb portait of Lavoisier and his wife (1788) enables one to speak about the definition of elements and its importance to the development of modern chemistry. Many other portraits offer like possibilities: *e.g.,* Tischbein's *Goethe in the Roman Campagna* and that great man's theorizing about color. George Stubbs' paintings of prize rams, pigs, etc., provide evidence of progress in selective breeding of domestic animals during the eighteenth century in England. The list goes on.

In teaching the practice of art analogous opportunities exist, though they may be more directly related to applied physics, chemistry and even biology. Optics, color theory, stress; pigment compounds, solvents and media; molds and mildews—all directly affect the artist and his/her product. The variety of ways in which clouds appear leads to simple meteorological considerations, and reflections in water prompt discussion of the equality of angles of incidence and reflection as well as the distinction between reflection and refraction. Properly presented, few aspects of the making of art can fail to elicit awareness of the profound significance of the natural sciences, or to encourage students to learn more about them.

CONCLUSION

It is evident that the foregoing examples do not represent scenarios in which pure science is being addressed. Nor, for that matter, is one teaching *art* when one speaks about Goethe's color speculation. Clearly some courses of study must continue to deal directly with the methods and matters of the various traditional disciplines. I think that it might be well, however, to push all such narrowly focussed work as far as possible toward the end of the educational cycle (the upper classes in college, and graduate school) in order first to concentrate on developing integrated minds. By this I mean minds constantly aware of, and alert to, the interconnectedness of knowledge because that is the way in which they have learned; minds understanding of the fundamental identity of human thinking no matter what its object, sympathetic to the common dilemmas facing all who pursue knowledge, stimulated by those leaps of imagination that characterize our successes, and persistent in the drudgery that lies between them.

Most would agree that there are some basic mental tools that we all should acquire and be able to employ appropriately. Logical reasoning and analysis, and the elements of mathematical thought are indisputably two. We should recognize the difference between proving a proposition and demonstrating its likelihood. We should recognize the weatherman's error when he tells us that there is a 100% possibility of rain. We should also recognize those situations or purposes which are served by metaphorical thinking. I would argue that, in a world wherein so much information is visually conveyed, we should as well understand something of how "ideas" are organized in visual forms. Above all, we need to be self-conscious about these alternate ways of thinking in order that we know them when we see them and use them for proper ends. Students should practice knowing and using them from their intitial educational years. In this way, to parapharase a colleague's observation, the geological phenomenon *volcano* can inform the historical phenomenon Vesuvius and *vice versa*, while both can be enriched by literary and pictorial interpretations.

I trust that it is clear that I have been speaking of integrated minds, not homogenized knowledge: distinctions among our ways of thinking are as important as similarities. Graduate education, primarily, must continue to bear responsibility for refining those special ways of thinking, although one would hope, as indicated above, that the rudiments of such differences will have been inculcated long before the graduate level.

I have never been sympathetic to the "Two Cultures" metaphor, whatever its superficial attraction. I believe that we have allowed its ghost to enfeeble us by encysting the various areas of knowledge, falsely accentuating distinctions, and promoting competition above cooperation. For, as I have tried to indicate in my opening paragraphs, I see human thought processes as fundamentally identical. Science and art may be considered as two ends of a spectrum of means by which human beings attempt to make sense of their individually minuscule, but magically and tragically self-aware transit through the immense Life of this planet. The social sciences and the other humanities, represent further excursions in the same search. Though science and art may appear thoroughly discontinuous, at their best they are—like the other branches of knowledge— rooted in creative thought; and all human creativity is born of essentially the same thought processes—only the "languages" differ. At various times in humankind's brief history one or another of these languages has seemed ascendant: today it is the language of science. But those complex and critical judgments that we shall have to make in the coming decades cannot be made soundly on the basis of scientific understanding alone: we shall have to acknowledge the falsity—indeed the *folly*—of the "Two Cultures" notion, and consciously yoke all our learning to those fundamental traits of the creative imagination that characterize the human mind. Paramount now must be neither the sciences nor the humanities: we cannot do with less than their coupled strength.

Finally, then, my plea is for much greater understanding and generosity of spirit among educators in all disciplines so that integrative cooperation can proceed. By and large, in my experience, it has been the humanists' embarassing ignorance that has inhibited such cooperation. This is a major reason for my recommendation that adjunct learning programs for humanities teachers be attempted, as well as for my proposal regarding general science publication and programming. Scientific modes of thinking have become deeply ingrained in our culture, so deeply that in more or less bowdlerized forms they have been adopted and adapted by many humanists. While this may have value, one unfortunate result has been a broad and precipitous decline in understanding of and respect for those more ancient, deeply human ways of making sense of our common experience in the world—literature, history, philosophy, all the arts,

Project 2061 has a three-phase plan of purposeful and sustained action that will contribute to the critically needed reform of education in science, mathematics, and technology.

Phase I focused on the substance of scientific literacy. *Science for All Americans* and the reports of the scientific panels constitute the chief products of that phase. The purpose of Phase I was to establish a conceptual base for reform by spelling out the knowledge, skills, and attitudes all students should acquire as a consequence of their total school experience from kindergarten through high school.

Science for All Americans, the first report of Project 2061, has little to say about what ails the educational system, points no finger of blame, prescribes no specific remedies. Rather, its basic purpose is to characterize scientific literacy. Thus, its recommendations are presented in the form of basic learning goals for all American children. A fundamental premise of Project 2061 is that the schools do not need to be asked to teach more and more, but to teach less so that it can be taught better. Accordingly, the recommendations given in *Science for All Americans* for a common core of learning are limited to the ideas and skills that have the greatest scientific and educational significance.

Phase II involves teams of educators and scientists transforming *Science for All Americans* into several alternative curriculum models for the use of school districts and states. During this phase, the project is also drawing up blueprints for reform related to the education of teachers, materials and technologies for teaching, testing, the organization of schooling, eduational policies, and educational research. While engaged in creating these new resources, Project 2061 is trying to significantly enlarge the nation's pool of experts in school science curriculum reform and is continuing its effort to publicize the need for nationwide scientific literacy.

Phase III will be a widespread collaborative effort, lasting a decade or longer, in which many groups active in educational reform will use the resources of Phases I and II to move the nation toward scientific literacy. Strategies for implementing the reform of education in science, mathematics, and technology in the nation's schools will be developed by those who have a stake in the effectiveness of the schools and who will take into account the history, economics, and politics of change.

RECOMMENDATIONS

The national council's recommendations address the basic dimensions of scientific literacy, which, in the most general terms, are:

• Being familiar with the natural world and recognizing both its diversity and its unity.

- Understanding key concepts and principles of science.

- Being aware of some of the important ways in which science, mathematics, and technology depend upon one another.

- Knowing that science, mathematics, and technology are human enterprises and knowing what that implies about their strengths and limitations.

- Having a capacity for scientific ways of thinking.

- Using scientific knowlege and ways of thinking for individual and social purposes.

The council's recommendations cover a broad array of topics. Many of these topics are already common in school curricula (for example, the structure of matter, the basic functions of cells, prevention of disease, communications technology, and different uses of numbers). However, the treatment of such topics tends to differ from the traditional in two ways.

One difference is that boundaries between traditional subject-matter categories are softened and connections are emphasized. Transformations of energy, for example, occur in physical, biological, and technological systems, and evolutionary change appears in stars, organisms, and societies.

A second difference is that the amount of detail that students are expected to retain is considerably less than in traditional science, mathematics, and technology courses. Ideas and thinking skills are emphasized at the expense of specialized vocabulary and memorized procedures. The sets of ideas that are chosen not only make some satisfying sense at a simple level but also provide a lasting foundation for learning more. Details are treated as a means of enhancing, not guaranteeing, students' understanding of a general idea. The council believes, for example, that basic scientific literacy implies knowing that the chief function of living cells is assembling protein molecules according to instructions coded in DNA molecules, but that it does not imply knowing such terms as "ribosome" or "deoxyribonucleic acid."

The national council's recommendations include some topics that are not common in school curricula. Among those topics are the nature of the scientific enterprise, and how science, mathematics, and technology relate to one another and to the social system in general. The council also calls for some knowledge of the most important episodes in the history of science and technology, and of the major conceptual themes that run through almost all scientific thinking.

The council's recommendations, which in the report are presented in 12 chapters, can be summarized in four general categories: The Scientific Endeavor, Scientific Views of the World, Perspectives on Science, and Scientific Habits of Mind.

THE SCIENTIFIC ENDEAVOR

All students should leave school with an awareness of what the scientific endeavor is and how it relates to their culture and their lives. This awareness should include understanding the following:

- The scientific endeavor stems from the union of science, mathematics, and technology. Technology provides science and mathematics with tools and techniques that are essential for inquiry and often suggests new lines of investigation. In the past, new technologies were based on accumulated practical knowledge, but today they are more often based on a scientific understanding of the principles that underlie how things behave. Mathematics is itself a science, but it also provides the chief language of the natural sciences and a powerful analytical tool widely used in both science and technology.
- Science, mathematics, and technology have roots going far back into history and into every part of the world. Just as all peoples have been inventive, shaping tools and developing techniques for modifying their environment, so too they have been curious about nature and how it works. Although modern science - which is truly international - is only a few centuries old, aspects of it (especially in mathematics and astronomy), can be traced back to the early Egyptian, Greek, Chinese, and Arabic cultures.
- Science, mathematics, and technology are expressions of both human ingenuity and human limitations with intellectual, practical, emotional, aesthetic, and ethical dimensions. Progress in these fields results from the cumulative efforts of human beings with diverse interests, talents, and personalities, although social barriers have led to the underrepresentation of women and minorities.
- The various natural and social sciences differ from each other somewhat in subject matter and technique, yet they share certain values, philosophical views about knowledge, and ways of learning about the world. All of the sciences presume that the things and events in the universe occur in consistent patterns that are comprehensible through careful and systematic study. Although they all aim at producing verifiable knowledge, none of them claims to produce knowledge that is absolutely true and beyond change.
- The subject matter investigated and techniques used within the various sciences change with time and the development of new instruments, and the boundaries of the scientific disciplines are constantly shifting. Even so, the general attributes of scientific inquiry persist. Descriptive, experimental, and historical approaches are used, depending on the phenomena being studied and the tools at hand. However, the approaches are all alike in their demand for evidence, their use of testable hypotheses and logical reasoning, their search for explanatory and predictive theories, and their efforts to identify and avoid bias.
- Mathematics is the science of abstract patterns and relationships. As a

theoretical discipline, it explores the possible relationships among abstractions without concern for whether they have counterparts in the real world. It often turns out, however, that discoveries in pure mathematics have surprising and altogether unanticipated practical value. As an applied science, mathematics deals with problems that originate in the natural and social sciences and in the everyday world of experience. In trying to solve such problems, it sometimes happens that fundamental mathematical discoveries are made.

- Whether theoretical or applied, mathematics is a creative process rather than one of using memorized rules to calculate answers. Mathematical processes include representing some aspects of things abstractly, manipulating the abstractions logically to find new relationships between them, and seeing whether the new relationships say something useful about the original things. The things studied in this way may be objects, collections, events, processes, ideas, numbers, or other mathematical abstractions.
- In the broadest sense, technology extends our abilities to change the world: to cut, shape, or put together materials; to move things from one place to another; to reach farther with our hands, voices, senses, and minds. Engineering is a process of designing and building technological systems to achieve such changes. Engineers must take into account physical, economic, political, social, ecological, aesthetic, and ethical considerations, and make trade-offs among them.
- Technological and social systems strongly interact with each other. Social and economic forces determine which technologies will be undertaken, paid attention to, invested in, and used; technology, in turn, has always had an enormous impact on the nature of human society. Some of the social effects of technological change—benefits, costs, risks—can be anticipated, and some cannot.

SCIENTIFIC VIEWS OF THE WORLD

Knowlege of science, mathematics, and technology is valuable for everyone because it makes the world more comprehensible and more interesting. *Science for All Americans* does not advocate, however, that all students need to gain detailed knowledge of the scientific disciplines as such. Instead, the report recommends that students develop a set of cogent views of the world as illuminated by the concepts and principles of science. Such views include the following:

- The structure and evolution of the universe, with emphasis on the similarity of materials and forces found everywhere in it, the universe's response to a few general principles (such as universal gravitation and the conservation of energy), and ways in which the universe is investigated.

- The general features of the planet earth, including its location, motion, origin, and resources; the dynamics by which its surface is shaped and reshaped; the effect of living organisms on its surface and atmosphere; and how its landforms, oceans and rivers, climate, and resources have influenced where and how people live and how human history has unfolded.
- The basic concepts related to matter, energy, force and motion, with emphasis on their use in models to explain a vast and diverse array of natural phenomena from the birth of stars to the behavior of cells.
- The living environment, emphasizing the rich diversity of the earth's organisms and the surprising similarity in the structure and functions of their cells; the dependence of species on each other and on the physical environment; and the flow of matter and energy through the cycles of life.
- Biological evolution as a concept based on extensive geological and molecular evidence, as an explanation for the diversity and similarity of life forms, and as a central organizing principle for all of biology.
- The human organism as a biological, social, and technological species— including its similarities to other organisms, its unique capacity for learning, and the strong biological similarity among all humans in contrast to the large cultural differences among groups of them.
- The human life cycle through all stages of development and maturation, emphasizing the factors that contribute to birth of a healthy child, to the fullest development of human potential, and to improved life expectancy.
- The basic structure and functioning of the human body, seen as a complex system of cells and organs that serve the fundamental functions of deriving energy from food, protection against injury, internal coordination, and reproduction.
- Physical and mental health as they involve the interaction of biological, physiological, psychological, social, economic, cultural, and environmental factors, including the effects of food, exercise, drugs, and air and water quality.
- Medical technologies, including mechanical, chemical, electronic, biological, and genetic materials and techniques; their use in enhancing the functioning of the human body; their role in the detection, diagnosis, monitoring, and treatment of disease; and the ethical and economic issues raised by their use.
- Features of human social dynamics, including the consequences of the cultural setting into which a person is born, the nature and effects of class distinctions, the variations among societies in what is considered appropriate behavior, the social effects of group affiliation, and the role of technology in shaping social behavior.
- Social change and conflict, with emphasis on factors that stimulate or retard change, the significance of social trade-offs, causes of conflict, mechanisms for resolving conflict among groups and individuals, the role of governments in directing and moderating change, and the effects of the growing interdependence of world social and economic systems.

- The human population, including its size, density, and distribution, the technological factors that have led to its rapid increase and dominance, its impact on other species and the environment, and its future in relation to resources and their use.
- The nature of technologies, including agriculture, with emphasis on both the agricultural revolution in ancient times and the effects on twentieth-century agricultural productivity of the use of biological and chemical technologies; the acquisition, processing, and use of materials and energy, with particular attention to both the Industrial Revolution and the current revolution in manufacturing based on the use of computers; and information processing and communications, with emphasis on the impact of computers and electronic communications on contemporary society.
- The mathematics of symbols and symbolic relationships, emphasizing the kinds, properties, and uses of numbers and shapes; graphic and algebraic ways of expressing relationships among things; and coordinate systems as a means of relating numbers to geometry and geography.
- Probability, including the kinds of uncertainty that limit knowledge, methods of estimating and expressing probabilities, and the use of such methods in predicting results when large numbers are involved.
- Data analysis, with an emphasis on numerical and graphic ways of summarizing data, the nature and limitations of correlations, and the problem of sampling in data collection.
- Reasoning, including the nature and limitations of deductive logic, the uses and dangers of generalizing from a limited number of experiences, and reasoning by analogy.

PERSPECTIVES ON SCIENCE

Scientific literacy also includes seeing the scientific endeavor in the light of cultural and intellectual history and being familiar with some powerful ideas that cut across the landscape of science, mathematics, and technology. To that end, the national council recommends that all students develop the following perspectives on science:

- An awareness that scientific views of the world result both from a combination of evolutionary changes, consisting of many small discoveries accumulating over long periods of time, and from revolutionary changes, consisting of the rapid reorganization of ways of thinking about the world.
- Familiarity with some of the episodes in the history of science and technology that are of surpassing significance for our cultural heritage. Such milestones in the development of Western thought and action include Galileo's role in

changing our perception of our place in the universe; Newton's demonstration that the same laws apply to motion in the heavens and on earth; Darwin's observations of the variety and relatedness of life forms that led to his postulating a mechanism for how they came about; Lyell's documentation in layers of rock of the great age of the earth; and Pasteur's identification of infectious disease with tiny organisms that could be seen only with a microscope.

- An understanding of a few thematic ideas that have proven to be especially useful in thinking about how things work. These include the idea of systems as a unified whole in which each part is understandable only in relation to the other parts; of models as physical devices, drawings, equations, computer programs, or mental images that suggest how things work or might work; of stability and change in systems; and of the effects of scale on the behavior of objects and systems.

SCIENTIFIC HABITS OF MIND

Throughout history people have concerned themselves with the transmission of habits of mind—shared values, attitudes, and ways of thinking—from one generation to the next. Given the great and increasing impact of science and technology on every facet of contemporary life, part of scientific literacy consists of possessing certain scientific values, attitudes, and patterns of thought. Accordingly, the national council recommends that elementary and secondary education be modified as necessary to ensure that all students emerge with the following:

- The internalization of some of the value inherent in the practice of science, mathematics, and technology, especially respect for the use of evidence and logical reasoning in making arguments; honesty, curiosity, and openness to new ideas; and skepticism in evaluating claims and arguments.
- Informed, balanced beliefs about the social benefits of the scientific endeavor—beliefs based on the ways in which people use knowledge and technologies and also on the continuing need to develop new knowledge and technologies.
- A positive attitude toward being able to understand science and mathematics, deal with quantitative matters, think critically, measure accurately, and use ordinary tools and instruments (including calculators and computers).
- Computational skills, including the ability to make certain mental calculations rapidly and accurately; to perform calculations using paper and pencil and electronic calculators; and to estimate approximate answers when appropriate and to check on the reasonableness of other computations.

- Manipulation and observation skills, with emphasis on the correct use of measuring instruments; the ability to use a computer for storing and retrieving information; and the use of ordinary hand tools.
- Communication skills, including the ability to express basic ideas, instructions, and information clearly both orally and in writing, to organize information in tables and simple graphs, and to draw rough diagrams. Communicating effectively also includes the ability to read and comprehend science and technology news as presented in the popular print and broadcast media, as well as general reading skills.
- Critical-response skills that prepare people to carefully judge the assertions— especially those that invoke the mantle of science—made by advertisers, public figures, organizations, and the entertainment and news media, and to subject their own claims to the same kind of scrutiny so as to become less bound by prejudice and rationalization.

BRIDGES TO THE FUTURE

Four major steps are required to make significant headway in realizing the goals expressed in *Science for All Americans*; (1) develop new curriculum models; (2) improve the teaching of science, mathematics, and technology; (3) develop a realistic understanding of what it will take to achieve significant and lasting reform nationally; and (4) initiate collaborative action on many fronts. These steps must reflect the following considerations:

- The school curricula - from kindergarten through twelfth grade - used in today's schools were not designed to achieve the broad goals outlined in this report. To ensure the scientific literacy of all students, the curricula must be changed to reduce the sheer amount of material covered; to weaken or eliminate rigid disciplinary boundaries; to pay more attention to the connections among science, mathematics, and technology; to present the scientific endeavor as a social enterprise that influences—and is influenced by—human thought and action; and to foster scientific ways of thinking.
- The effective teaching of science, mathematics, and technology (or any other body of knowledge and skills) must be based on learning principals that derive from systematic research and from well-tested craft experience. Moreover, teaching related to scientific literacy needs to be consistent with the spirit and character of scientific inquiry and with scientific values. This includes starting with questions about phenomena rather than with answers to be learned; engaging students actively in the use of hypotheses, the collection and use of evidence, and the design of investigations and processes; providing students with hands-on experience with mechancial, electronic, and optical tools; placing a premium on students' curiosity and creativity; and frequently using a

Science Education in the United States: Issues, Crises and Priorities.

student team approach to learning.

- Educational reform must be comprehensive, focusing on the learning needs of all children, covering all grades and subjects, and dealing with all components and aspects of the educational system. Patching up this or that part of the system will accomplish little. Reform on a national scale will necessarily take a long time, given the size and complexity of the U.S. educational system and the decentralization of authority and resources. It will also require that positive conditions for change be established and that public support for reform be sustained for a decade or longer.

- Finally, to have any hope of success, reform must be collaborative and involve administrators, university faculty members, and community, business, labor and political leaders, as well as teachers, parents, and students themselves. To that end, *Science for All Americans* concludes with an agenda for action that suggests steps that individuals, institutions, organizations, and government agencies can take to work together toward reform. For its part, Project 2061 will continue to do what it can to keep scientific literacy and educational reform on the agenda of educators, scientists, policymakers, and the public.

There are no valid reasons—intellectual, social, or economic—why the United States cannot transform its schools to make it possible for all students to achieve scientific literacy. It is a matter of national commitment, determination, and a willingness to work together toward common goals. *Science of All Americans* is intended to help in clarifying those goals.

REFERENCES

American Association for the Advancement of Science. 1989. *Science for All Americans.*

Appley, Mortimer and Mader, Winifred B. 1989. *Social and Behavioral Sciences: Report of the Project 2061 Phase I Social and Behavioral Sciences Panel.* American Association of the Advancement of Science.

Blackwell, D. and Henkin, L. 1989. *Mathematics: Report of the Project 2061 Phase I Mathematics Panel.* American Association for the Advancement of Science.

Bugliarello, G. 1989. *Physical and Information Sciences and Engineering: Report of the Project 2061 Phase I Physical and Information Sciences and Engineering Panel.* American Association for Advancement of Science.

Clark, M. 1989. *Biological and Health Sciences: Report of the Project 2061 Phase I Biological and Health Sciences Panel.* American Association for the Advancement of Science.

Johnson, J.R. 1989. *Technology: Report of the Project 2061 Phase I Technology Panel.* American Association for the Advancement of Science.

Science Education in the United States: Issues, Crises and Priorities. Edited by S. K. Majumdar, L. M. Rosenfeld, P. A. Rubba, E. W. Miller and R. F. Schmalz. ©1991, The Pennsylvania Academy of Science.

Chapter Thirty-Five

SCOPE, SEQUENCE, AND COORDINATION OF SECONDARY SCIENCE: A RATIONALE[1]

BILL G. ALDRIDGE

Executive Director
National Science Teachers Association
1742 Connecticut Avenue., N.W.
Washington, D.C. 20009

THE SCOPE, SEQUENCE AND COORDINATION PROJECT

The Project on Scope, Sequence, and Coordination of Secondary School Science (SS&C), initiated by the National Science Teachers Association, is a major reform effort to restructure science teaching at the secondary level. NSTA's plan calls for the elimination of the tracking of students, recommends that all students study science every year for six years, and advocates the study of science as carefully sequenced, well-coordinated instruction in physics, chemistry, biology, and earth/space science. As opposed to the traditional "layer cake" curriculum in which science is taught in year-long discrete and compressed disciplines, the NSTA project provides for "spacing" the study of each of the sciences spread out over several years. Research on the "spacing effect" indicates that students can learn and retain new material better if they study it in spaced intervals rather than all at once, and this way they revisit a concept or idea at successively higher levels of abstraction.

The American Association for the Advancement of Science (AAAS) has also embarked on a comprehensive program of reform known as Project 2061. During the first three phases of the AAAS project, leading scientists and educators

---

[1]This document was prepared pursuant to grants from the National Science Foundation and the U.S. Department of Education as part of the NSTA Scope, Sequence, and Coordination of Secondary Science Education Project.

identified knowledge and skills deemed essential for improved science education. These outcomes have been adopted by NSTA as goals for Scope, Sequence and Coordination.

While Project 2061 and SS&C are quite different in their approaches and their time frames for implementation, both reform efforts share the same goals. Both projects contend that less content taught more effectively over successive years will result in greater scientific literacy of the general public. The SS&C project goes beyond this claim with the considered judgement that less content taught effectively over several years will, from the same pool of young people, produce more and better scientists and engineers.

The fundamental goal of the Scope, Sequence, and Coordination Project is to make science understandable by essentially all students. As so well expressed in the *California Framework:*

> The structure of the natural world requires language to explain it. But there are many ways to describe natural phenomena. Customarily, we recognize a phenomenon and give it a name . . . Sometimes in science curricula, remembering the names and their definitions seems to become an end in itself. But a name should not become more important than the phenomenon being described, or than its empirical or logical relationships with other phenomena.

In order to concentrate on experience with phenomena, the number of topics included in the restructured scope and sequence and their accompanying baggage of facts and terminology is greatly reduced. This leads naturally to less continuity than when one must "cover" everything. However, fewer topics will result in more in-depth student learning and far greater student retention.

NSTA's SS&C reform effort emphasizes appropriate "sequencing" of instruction, taking into account how students learn. In science, profound understanding comes from having concrete experiences with a phenomenon before it is given a name of a symbol. These experiences must also be in several different contexts before the concept becomes part of one's mental repertoire. With such prior hands-on experience, students can understand science. The practical applications of science should begin in the seventh grade with issues and phenomena of concern to students at a personal level and then move toward more global applications in the upper grades. As students mature, they are able to generalize from concrete, direct experiences to form abstractions and theories. With a sequenced approach students will no longer be expected to mindlessly memorize facts and information. With personal and societal applications, science will make sense and have relevance to their lives.

Another component of NSTA's SS&C Project is "coordination." Earth/space science, biology, chemistry, and physics have topical areas and processes in common. Coordination among these four subjects leads to awareness of the interdependence of the sciences and how they fit together as part of a larger body of knowledge. Project 2061 also emphasizes the importance of coordinating themes. NSTA has adopted themes identified by 2061 as the unifying threads of science and related applications.

The following chart illustrates one possible configuration of four science subjects taught over a six-year period. Notice that at first students experience most intensively the descriptive and phenomenological aspects of the sciences. The abstract and theoretical emphasis occurs in later years. Empirical and semi-quantitative treatments are emphasized in the middle years. Computers, technology, and practical applications should be integrated directly into each of these subjects.

EXAMPLE OF A REVISED SCIENCE CURRICULUM FOR GRADES 7 THROUGH 12 IN THE UNITED STATES

	Grade Level						Total Time Spent
	7	8	9	10	11	12	
Hours Per Week By Subject							
Biology	1	2	2	3	1	1	360
Chemistry	1	1	2	2	3	2	396
Physics	2	2	1	1	2	3	396
Earth/Space Science	3	2	2	1	1	1	360
Total Hours Per Week	7	7	7	7	7	7	
Emphasis	descriptive; phenomeno-logical		empirical; semi-quantitative		theoretical abstract		

The next chart illustrates how the model has been adapted for use in the California Framework:

EXAMPLE OF A REVISED SCIENCE CURRICULUM FOR GRADES 7 THROUGH 12 IN THE CALIFORNIA FRAMEWORK

	Grade Level						Total Time Spent
	7	8	9	10	11	12	
Hours Per Week By Subject							
Biology	2	2	1	1	1	1	288
Chemistry	1	1	2	2	1	1	288
Physics	1	1	1	1	2	2	288
Earth/Space Science	1	1	1	1	1	1	216
Total Hours Per Week	5	5	5	5	5	5	
Emphasis	descriptive; phenomeno-logical		empirical; semi-quantitative		theoretical abstract		

Why Reform is Needed

The time is right to embark on a reform effort in science education. As numerous reports and studies indicate, schools in the United States are failing to educate students for a world that depends more and more on sophisticated and rapidly changing science and technology. Surveys show that the majority of students leave our schools without a basic understanding of science, mathematics, or technology. Neither the demand for scientists and engineers is being met, nor are schools preparing future citizens with an adequate background of knowledge necessary to make decisions affecting their lives.

Students perceive currently structured, textbook-driven science subjects as difficult, boring, and having no relevance to their lives. Many opt out as soon as the system allows—over half our students don't take another science course after 10th grade. Only 19% of high school students take a course in physics and only about 40% take a course in chemistry.

There is an incorrect assumption widely held that only certain children are capable of learning science and mathematics. Students, identified at an early age as "most able" or "most intelligent" are tracked into future course work in science and math. These are often the most advantaged. The remainder are systematically filtered out of science and math. The research-based underlying principles of the Scope, Sequence, and Coordination initiative are that science is needed by everyone and that everyone is capable of learning and enjoying science. Students stay in science until they have had several years of a good learning experience in each of the coordinated science subjects.

The emphasis of SS&C is on science for *all* students. Truly gifted students will not be overlooked. They are likely to be a different mix of gender and ethnicity, and they will be rewarded for *quality* of learning and thinking rather than for rushing through and completing the course work ahead of their peers. Such gifted students, as well as other more enthusiastic and interested students, will be provided enrichment activities to maintain their interest. They can also be given additional responsibilities in cooperative learning situations, e.g., students teaching students.

Conditions for Reform

To be successful, a major reform effort requires change and accommodation on the part of science teachers and school administrators. Challenges relating to scheduling, teacher training, parental support, and assessment must be addressed. The Scope, Sequence, and Coordination Project recommends at least five class periods a week of science in each of six years. Adding more time to science requires creative scheduling decisions. Whatever the administrative decision regarding scheduling, there will be difficult problems of balance.

But, educational reform cannot take place without making some decisions which, at first, may not be popular.

Inservice teacher training must occur concurrently with curricular change. For seventh-grade teachers, the SS&C plan will mean focusing on four science subjects rather than on traditional middle school life science or earth science courses. This may require additional knowledge in those subject areas in which teachers currently may have only limited preparation. Interaction with colleagues who teach other science subjects is important in maintaining a properly sequenced and coordinated science program. NSTA recommends that teachers at the middle/junior high level teach no more than two science subjects. Even with two fields, most teachers will need to be updated in at least one subject area. Summer inservice and monthly one-day weekend education in each of the four discipline areas should be provided to address teacher needs. Teachers may have more classes of students, but if they teach the same students for at least two years, they will know their students better and will be able to keep track of their progress over a longer period of time.

Since specific science subjects will be taught only a few periods a week rather than everyday, homework and out of school explorations will be an essential extension of the classroom. Homework exercises will help maintain continuity in student learning. To insure community support, school personnel should meet with parents of all students in project classes to explain goals of the reform effort and how it will benefit their children. Parents should be made aware of the different expectations for out-of-class work and their role in helping students meet these expectations.

The priorities for SS&C suggest a vital shift of focus. The focus of science must move away from the textbook, tests, and even the teacher. It must move to the student. The textbook, the test, and the teacher are there to serve the student, not to control, dominate or make the student feel inadequate in the face of a massive unknowable mountain of information. Rather, the teacher is there to provide the circumstances in which learning can take place.

If the focus of the learning process is the student, what does this imply in practice? It implies that students will learn most readily when they are given the opportunity to challenge and develop their own theories, to collect their own data, to present their own outcomes of their investigations; in other words, to take responsibility for their own learning.

Within the existing models of science teaching, many teachers rightly complain that there is far too much to cover in textbooks. Some teachers respond by not using tests; and while many teachers are able to work without such a supportive framework, many more feel inadequate to teach without one. There is a direct parallel between students taking responsibility for learning, and teachers taking responsibility for teaching. No longer is the teacher obliged to be a holder of knowledge (encyclopedias can do that) but rather a teacher should be a manager of the learning environment—someone whose role is to liberate the potential of the students. The teacher's role is to create a physical and intellectual environment in which students are enabled to learn for themselves.

Assessment of Student Achievement

The main purpose of this science education reform effort is to increase student achievement and involve more students in science. The success of change cannot be determined unless evaluation is consistent with what students are expected to learn. Schools participating in the NSTA Scope, Sequence and Coordination Project need to develop a plan for assessment to determine if student participation and achievement are, in fact, improving. Although a proposal for a national assessment center is currently under review at the National Science Foundation, local areas must take the initiative in developing their own assessment instruments to be used in tandem with national efforts.

Currently, most classroom testing is at the level of recall of factual knowledge and information requiring little need for students to reason, think or relate laws or principles. To assess student outcomes, major changes in classroom testing and evaluation procedures will be needed.

Outcomes or performance assessment will place a strain on current practices in evaluation and grading. Currently, standardized tests, and most teacher-made tests, are designed to distribute student scores to show variation in student achievement. This design forces half of the students to be below the median in accomplishment, no matter how much they have learned. A shift in emphasis to expectations that all students will learn, but with qualitative differences in their performances that are to be valued, is asking for a major shift in the function of the school. Teaching for understanding entails such a shift.

Performance assessment should describe successively higher levels of student attainment of outcomes and raise the question, "Are these levels adequate for these students, at their current age and experience. Outcome statements are made operational by the kinds of problems and tasks students are asked to do to demonstrate their level of understanding at different times throughout their school experience.

Development of Scope, Sequence, and Coordination

To assist school districts implementing Scope, Sequence, and Coordination program, NSTA appointed four curriculum committees to identify science topics to be taught in each discipline at each grade level. Committee members were experts in the discipline areas and consisted of university professors, classroom teachers, curriculum developers, and textbook authors. Prior to determining the scope of science topics, they examined a number of sources, including the goals identified by Project 2061, the themes presented in the California Framework, the major textbook series, science trade books, and a number of curriculum projects. After an initial list of topics was formulated for all secondary grade levels, the committes concentrated on the seventh grade, since the first trials of the reform are to be implemented at this level. In selecting the topics to be taught in a restructured seventh-grade science curriculum, committee members worked with the following precepts inherent to the Scope, Sequence and Coordination Project.

- The four basic subject areas, earth/space science, biology, chemistry, and physics, are addressed each year, and the connections among them are emphasized.
- The coordinating themes identified by Project 2061 are used as unifying threads among the science disciplines. These themes are Systems, Models, Constancy, Patterns of Change, Evolution, and Scale.
- Science is shown to be open to inquiry and skepticism, and free of dogmatism or unsupported assertion by those in authority. The curriculum promotes student understanding of *how* we come to know, *why* we believe, and how we test and revise our thinking.
- Science is presented in connection with its applications to technology and its personal and societal implications.
- Students have the opportunity to *construct* the important ideas of science which are then developed in depth through inquiry and investigation.
- Vocabulary is used to facilitate understanding rather than as an end in itself. Terms are introduced only after students experience the phenomena.
- Texts are not the source of the curriculum but serve as references. Everyday materials, laboratory equipment, video, software, and other printed materials such as reference books and outside reading provide a substantial part of the student learning expcrience.
- Lessons provide opportunities for skill building in data collection, graphing, record keeping, and the use of language in verbal and written assignments.
- Enhancement activities or "extras for experts" are provided for the more enthusiastic and interested students.
- Of particular importance is that instruction enhances skepticism, critical thinking, and logical reasoning skills. During instruction teachers need to ask, "How do you know? Why do you believe? and What is the evidence?" Thinking and reasoning skills to be fostered include students' ability to:
  - control variables to determine relationships.
  - understand the meaning and use of ratio, proportion, and scaling laws.
  - draw inferences from evidence and discriminate explicitly between evidence (or the observations) on one hand and the inference being drawn on the other.

*NSTA acknowledges the invaluable assistance of all members of the curriculum committees working on Scope, Sequence, and Coordination. In particular, NSTA recognizes the extra efforts of committee members James Robinson, Arnold Arons, and Floyd Mattheis.

- recognize gaps in available information or knowledge when such gaps exist.
- Define technical terms operationally *after* the idea behind the term has been developed.

In summary, the selection of topics and activities for seventh-grade science will be based on relevance of the instruction to the students themselves, their lives, their future, and their immediate environment. Applications will relate to the student's personal lives first. These will be broadened to include application at the level of family, community, and global concerns over the six years of study. There will be a minimum of symbolic or mathematical abstractions in seventh grade science. Students will be presented with experience first, then terms and then reinforcement with applications. The emphasis at first will be on concrete rather than abstract ideas and on lessons that involve hands-on activities and experience with natural phenomena.

Scope, Sequence, and Coordination: The Supporting Literature

The purpose of identifying the references listed below is to provide documentation for those interested or participating in the NSTA Project on Scope, Sequence, and Coordination. These references were selected to give teachers and support staff a limited number of relevant papers that reflect current research and thinking regarding science teaching and learning. The list is not intended to be comprehensive. References cited in these selections provide additional information for those who wish to read further. It is recommended that these materials first be read and discussed by teachers and support staff and then the ideas presented to parents. Discussions should deal with both the meaning of what is presented and how it would translate into classroom practice.

American Association for the Advancement of Science (1989) *Science for All Americans: Project 2061.* Washington, DC: American Association for the Advancement of Science.
    Major recommendations to change the substance and character of science education are presented on the nature of science, the nature of mathematics, the nature of technology, the physical setting, the living environment, the human organism, human society, the designed world, the mathematical world, historical perspectives, common themes, and habits of mind. The common themes that cut across all sections provide an important consideration for curriculum development and emphasis.
Anderson, Charles W. (1987) Strategic teaching in science. In *Strategic Teaching and Learning: Cognitive Instruction in the Content Areas.* Jones, Beau, Fly,

et al., ed. Alexandria, VA: Association for Supervision and Curriculum Development.

Although the science chapter is the most relevant, the first three chapters by the editors contain useful material on strategic thinking. Anderson's major concern is the teaching of science for understanding. He carefully discusses contrasting views of science teaching such as presenting facts, rules, and definitions vs. developing the science processes.

Blumberg, Fran, Epstein, M., MacDonald, W., and I. Mullis (1986) *A Pilot Study of Higher-Order Thinking Skills Assessment Techniques in Science and Mathematics. Final Report, Part 1.* Princeton, NJ: National Assessment of Education Progress.

This project developed a conceptual framework of higher order thinking skills in science and mathematics which was used to construct prototype exercises, including hands-on activities in which students were asked to solve problems, conduct investigations, or respond to questions using materials and equipment. Problems and results of the pilot tests are presented.

Blumberg, Fran, Epstein, M., MacDonald, W., and I. Mullis (1986) *A Pilot Study of Higher-Order Thinking Skills Assessment Techniques in Science and Mathematics. Final Report, Part II, Pilot-Tested Tasks.* Princeton, NJ: National Assessment of Educational Progress.

This publication presents the instruments and samples of student responses to problems used in the study by the National Assessment of Educational Progress to develop ways to assess higher order thinking skills at grades three, seven, and eleven in both science and mathematics. Pilot-tested tasks included group tasks, station activities, and individual investigations.

Carnegie Council on Adolescent Development (1989) *Turning Points: Preparing American Youth for the 21st Century:* A Report of the Task Force on Education for Young Adolescents. Washington, DC: Carnegie Council on Adolescent Development.

This is a well-presented summary of the condition of adolescents in the 1980s and the need for changing educational structures and practices. The central section dealing with "Transforming the Education of Young Adolescents," proposed the creation of communities for learning in middle level schools by teaching a common core of knowledge.

Dempster, F. (1988) "The spacing effect: A case study in the failure to apply the results of psychological research." *American Psychologist,* 43. 627-634.

This article presents research that shows for a given amount of study time, spaced presentations are substantially more effective in increasing learning than intense, singular sessions. The author argues for using this knowledge of "spaced learning" for instructional purposes in the development of curricular frameworks.

Linn, Marcia C. (1986). Science. *Cognition and Instruction.* Dillon, R.F. and R.J. Sternberg, eds. New York: Academic Press.

Linn discusses the need for reform in science education and the advances in psychological research that have implications for enhancing student learning. She addresses recent advances in understanding student processing capacities, problem-solving strategies, and intuitive conceptions.

National Assessment of Education Progress (March, 1989) *Science Objective, 1990 Assessment.* Booklet No. 21-S-10. The Nation's Report Card, Princeton, NJ: Educational Testing Service.

The booklet presents the process by which the assessment objectives and purposes of school science were identified for the NAEP 1990 science assessment. The report highlights the assessment framework, the nature of science, and thinking skills in science.

National Science Board Commission of Precollege Education in Mathematics, Science, and Technology (1983) *Educating Americans for the 21st Century: A Plan of Action for Improving Mathematics, Science and Technology Education for all American Elementary and Secondary Students so their Achievement is the best in the World by 1995.* Washington, DC: National Science Foundation.

Student outcomes are proposed for science, mathematics, and technology education and for elementary, middle, and senior high school levels of education. Detailed recommendations are made for continuing education of science and mathematics teachers and for changes in preservice teacher education.

Piaget, Jean (1973) *To Understand is to Invent: The Future of Education.* New York: Grossman Publishers.

This short book presents the argument that students have direct experience with phenomena and construct their own explanations to develop cognitively. This "constructivist" approach to concept development has received a great deal of support from research in the cognitive sciences.

Resnick, Lauren B. (1987) *Education and Learning to Think.* Washington, DC: National Academy Press.

This report resulted from deliberations of the Committee on Mathematics, Science and Technology Education of the National Research Council. It describes what is meant by higher order thinking skills and addresses the issue of how schools can enable students to learn them. The monograph also considers the nature of thinking and learning, general reasoning, and the improvement of intelligence.

Robinson, James T. (1979) A critical look at grading and evaluation practices. *What Research Says to the Science Teacher.* Vol. 2. Rowe, Mary Budd, ed. Washington, DC: National Science Teachers Association.

This publication suggests four purposes for classroom evaluation:

course improvement, accountability, student development, and determining student grades. Each of these areas is discussed, and issues in their appropriate use are identified.

Stiggens, Richard J. (November 1987) *Measuring Thinking Skills through Classroom Assessment.* Portland, OR: Northwest Regional Laboratory.

Classroom assessment procedures of 36 teachers in grades 2-12 were studied in depth to determine the extent they measure students' higher order thinking skills in mathematics, science, social studies, and language arts. Results showed science teachers focus nearly two-thirds of their assessments on simple recall of facts and information.

The Task Force on Women, Minorities, and the Handicapped in Science and Technology. (December 1989) *Changing America: The New Face of Science and Engineering.* Washington, DC: National Science Foundation.

The final report from the Task Force offers specific actions for the key groups of decision makers affecting educational policy. Actions focus on keeping underrepresented groups in the science and engineering pipeline. Recommended actions for educators include, "Make science hands-on. Ensure that all students do science as well as read about it."

Dellarosa, Denise; Bovine, Jr., Lyle E., *Surface Form and the Spacing Effect.* Memory and Cognition, Vol. 13, No. 6, pp. 529-537, August 1985.

Although the spacing effect is one of the best known and most researched memory phenomena, it has yet to be satisfactorily explained. Recent explanatory attempts have fallen into two classes. The first class includes those explanations that attribute the spacing effect to increasing independence of encoding events with increasing intervals between repetitions. The best known and most widely studied of these is the encoding - variability hypothesis, which attributes higher recall of spaced repetitions to a greater likelihood that a repetition will be encoded to a different subjective context at longer intervals than shorter ones. The greater the number of retrieval routes, the greater the probability that the item will be retrieved.

Reynolds, James H., Glasa, Robert, *Effects of Repetition and Spaced Review Upon Retention of a Complex Learning Task.* Journal of Educational Psychology, Vol. 55, No. 5, pp.297-308, February 1964.

Two experiments were conducted to evaluate the effects of (a) amount of space-review repetition, and (b) the spacing of periodic review sequences upon retention of academic materials taught to junior high school students by programmed instruction methods. Repetition was varied by constructing programmed sequences which contained three different levels of stimulus-response repetitions for each of a number of scientific terms being taught. Results indicated that variations and repetition had only transitory effects upon retention, but that spaced review produced a significant facilitation in retention of the reviewed material.

Ascher, Carol. *Cooperative Integrated Learning in the Urban Classroom.* Report No. 10. Center for Research on Elementary and Middle School Education. Adapted for ERIC/CUE Digest, Number 30. ERIC Clearinghouse on Urban Education, New York, NY 1986. 5 p.

   Cooperative learning methods capitalize on the heterogeneous student bodies of most urban schools. They appear to foster better student achievement than do individualistic methods, to increase cross-ethnic friendships, and to improve students' self-esteem and positive attitudes toward other students and the school.

Bernagozzi, Tom. *The New Cooperative Learning and One Teacher's Approach.* Learning, Vol. 16 No. 6, pp. 38-43, February 1988

   A teacher describes how he modified a cooperative learning approach featuring heterogeneous grouping for his elementary school class, covering such topics as setting up teams, teaching a lesson, managing the evaluation and scoring system, and the approach's pitfalls and benefits. A list of resources and suggestions is presented.

The Action Council on Minority Education. *Education That Works: An Action Plan for the Education of Minorities.* Massachusetts Institute of Technology, Cambridge, Massachusetts. January 1990.

   This extremely important report makes very clear the need to stop all tracking of students and the goals for the implementation of the NSTA Scope, Sequence and Coordination structure as a major way of addressing the needs of all students, especially minorities.

Science Education in the United States: Issues, Crises and Priorities. Edited by S. K. Majumdar, L. M. Rosenfeld, P. A. Rubba, E. W. Miller and R. F. Schmalz. ©1991, The Pennsylvania Academy of Science.

Chapter Thirty-Six

OUTDOOR/ENVIRONMENTAL EDUCATION CENTERS
"Not Just A Field Trip"

DEBORAH A. SIMMONS[1], OWEN D. WINTERS[2] and DICK TOUVELL[3]

[1]Assistant Professor of Outdoor Education
Director of Resident Programs
Northern Illinois University
Lorado Taft Field Campus
Oregon, IL 61061
and
[2]Administrative Assistant and Publications Editor
Natural Science for Youth Foundation
Roswell, GA 30075
and
[3]Executive Director, Chippewa Nature Center
Midland, MI 48640

INTRODUCTION

"As if life wasn't tough enough already, now they want me to integrate some 'Environmental Education' into my curriculum this year! A three day/two night trip to Pinewood Ridge EE Center is scheduled in the Spring to top off 12 hours of preparatory lessons. Our Middle School Science Coordinator, Dr. Blackwood, tells us 'Just integrate the concepts into your present curriculum!' What is this 'Environmental Education' stuff, anyway?. . ."

In the last 20 years, since Environmental Education (EE) first took wing, an untold number of teachers have been placed in similar positions. And, we suspect, in the next 20 years an even greater number of teachers will suddenly need to (or want to) discover what EE really is, how an EE program integrates into a classroom curriculum and where a trip (or trips) to an Outdoor/EE center fits into the total package.

This chapter overviews the origins, goals, facilities, curricula, status, barriers, and trends of Outdoor/EE Centers. It is hoped that it will assist you in better understanding this young curriculum and its place in education. We begin with a brief examination of the milestones in the development of environmental education.

HISTORY AND ORIGIN

Two separately evolving movements can be seen as precursors of what became the environmental education movement of the 1970s. Well-known conservationists and educators in American History (e.g., John Muir, Teddy Roosevelt, Aldo Leopold, Rachel Carson) and national events (e.g., the dust bowl, creation of the Tennessee Valley Authority and the Civilian Conservation Corps) focused public attention on the environment and natural resources. Concurrently, progressive educators argued that education should be experiential and prepare children to become responsible members of society.

In the early part of this century, use of the outdoors, even within an educational context, focused primarily on the recreational aspects of the experience. But it was through his early work with the Life Camp and later National Camp that L.B. Sharp provided the philosophical underpinnings of a new field called Outdoor Education. His now famous statement "That which can be learned in the classroom should be taught there, and that which can best be learned in the out-of-doors should there be taught", provided a rationale for action. In response, schools in the 1930s and 40s developed and sponsored camping programs to provide children with concrete opportunities to learn about their world and to acquire social skills such as cooperation and social planning. At the same time, scouting, 4-H and other youth programs also began to include a nature study component. Soon, the movement evolved from a school camping program into an integrated outdoor education program. During the 1950s, the outdoors became a laboratory for the entire curriculum as students moved outside their classrooms to observe the world of nature.

Interest in the out-of-doors was heightened by the environmental movement of the late 1960s. Concerned citizens warned that the environment was threatened by our failure to protect endangered species and our lack of concern for a range of pollution problems. In 1970, the passage of the National Environmental Policy Act and the celebration of the first Earth Day led to the development of an unprecedented amount of materials to teach about the environment. Monies from Federal and State governments, as well as private and industrial grants and other support, were made available for teacher training, curriculum development, and EE facilities.

As can be readily seen, the origins of Outdoor/EE are lodged in a diversity of movements and events; thus, a direct lineage and definition is somewhat ephemeral. In an effort to provide structure, Disinger[2] presented an interesting and definitive treatise on the background of environmental education and the struggle to give it a singular definition. Ultimately, however, a casual operating understanding of environmental education is what finally touches practitioners in the field.

In practice, environmental education incorporates concepts popularly found in what has been called ecology, nature education, and conservation education. Also included are portions of the traditional school curriculum: science, mathematics, social studies, economics, language arts, and history. In "Outdoor Education," published as part of Phi Delta Kappa's Exemplary Practice Series,[3] it is argued that "hiking in the woods, collecting leaves, mastering a ropes course, cooking a meal over a campfire, plotting a compass course, and studying about the effects of pollution are all activities that can (still) be found in a curriculum guide for outdoor/environmental education." Still, many believe the mix must have added to it experiences that coalesce the lessons and learning—internalize it such that it will guide future beliefs, perceptions, choices and actions on the environment.

The goals of this relatively young movement have been gaining definition since the beginning of this century and have been clarified during the movement's rapid maturation in the past two decades. An analysis of these goals is equally fascinating.

GOALS OF ENVIRONMENTAL EDUCATION

Throughout the development of environmental education, many have debated its goals (see 4-6). Early on, Stapp[7] proposed that environmental education should work to develop a citizenry that "is aware of and concerned about the total environment and its associated problems, and which has the knowledge, attitudes, motivations, commitment, and skills to work individually and collectively toward solutions of current problems and the prevention of new ones." In 1977, official delegations from approximately 70 countries participated in an intergovernmental conference in Tbilisi, USSR[8] and proposed a set of five objectives for EE: awareness, knowledge, attitudes, skills and participation. Working from the Tbilisi objectives, Hungerford, Peyton and Wilke[9] proposed the "superordinate goal" for environmental education to be, "to aid citizens in becoming environmentally knowledgeable and, above all, skilled and dedicated citizens who are willing to work, individually and collectively, toward achieving and/or maintaining a dynamic equilibrium between quality of life and quality of the environment."

The goals and objectives described above suggest that encouraging environmentally sound behavior is a desired outcome of environmental education; that is to create changes in the way people relate to the Earth and its resources, and to foster positive changes through actions that will enable the Earth to continue to support people and all other lifeforms, indefinitely.

OUTDOOR/EE CENTER GOALS AND ACTIVITIES

Individual classrooms, scout troops, 4-H programs, state and federal fish and game commissions, state and national parks, and garden clubs all deliver EE programs. However, one of the largest and most influential delivery vehicles for the Outdoor/EE movement is the network of Outdoor/EE centers across America. In cooperation with school personnel, scout troops, and other educational leaders, the specialized teachers and facilities of Outdoor/EE centers have become one of the primary means of delivering EE nationwide. By examining the goals of these museums, nature centers, and residential field centers, a clearer view of how Outdoor/EE is being put into practice can be obtained.

In 1989, 1225 Outdoor/EE centers, representing an estimated 70% of the existing facilities nationwide, were identified and surveyed as part of the development of the Directory of Natural Science Centers, published by the Natural Science for Youth Foundation (NSYF).[10] A questionnaire was used to collect information on each center's mission and goals, program, audience, management and facilities.

As might be expected, the centers saw themselves serving similar audiences. Educating school-aged children was overwhelmingly endorsed as a goal of these centers. Nearly all (94.2%) of the responding centers indicated that educating school children was either a major or a minor audience goal. Similarly, educating adults was considered either the major or minor audience for 86.3% of the centers. Educating EE personnel was an audience goal for 56.1% of the centers.

As can be seen from Table 1, nature study was the most frequently mentioned (76.1%) "Program Content" goal. The second most often mentioned goal was encouraging environmentally sound behavior (74.4%) followed by teaching local natural history (71.5%), and influencing people's attitudes (66%). The remaining four program content goal statements were each mentioned by fewer than half of the centers as either major or minor goals.

At first glance it is probably somewhat surprising that Outdoor/EE centers endorsed such a variety of goals. However, an attempt to define the "typical" center provides some insight into the root of this diversity. The American Association of Museums,[11] defines a nature center as ". . . an organized and permanent nonprofit institution which is essentially educational, scientific, and cultural in purpose with professional staff, open to the public on some regular schedule. The Nature Center manages and interprets its lands, native plants and animals and facilities to promote an understanding of nature and natural processes. It conducts frequent environmental education programs and activities for the public."

A description of the "typical" nature center is, however, considerably more elusive. Why? Because there is great diversity among the centers. In search of the typical nature center, questions like, "How big is it?", "How many staff are employed there?" and "How large is its budget?" become difficult to answer. Information from the Directory of Natural Science Centers[10] presents a profile to help us understanding the varied composition of outdoor/environmental education centers.

TABLE 1

Center Goals

	Major Goal	Minor Goal	Combined Total
Audience			
Education School Children	82.7%	11.4%	94.1%
Education Adults	58.0	28.3	86.3
Training EE/Outdoor Educators	22.5	33.6	56.1
Program Content:			
Nature Study	57.0	19.1	76.1
Encouraging Environmentally Sound Behaviors	50.3	24.1	74.4
Teaching Local Natural History	46.9	24.6	71.5
Influencing People's Attitudes	41.9	24.1	66.0
Disseminating Local Environmental Issues Information	13.6	29.9	43.5
Developing Environmental Problem-Solving Skills	17.5	23.1	40.6
Helping People Develop Self-esteem	14.9	16.7	31.6
Environmental Activism	6.7	22.1	28.8

Management and Ownership

Who owns and operates the Outdoor/EE centers across the US and Canada? It is probably not too surprising to find that most are owned by public institutions. Over half of the centers are operated by some level of government. Approximately 39% are operated as private not-for-profits, with only 9% administered by educational institutions (colleges/universities and school districts).

Less than 3% are owned or operated by private for-profit or industrially sponsored organizations.

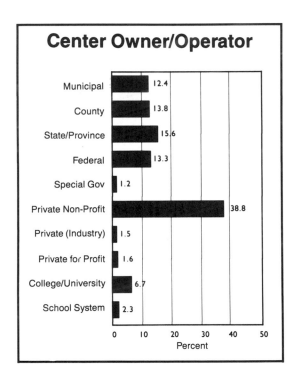

Center Owner/Operator

	Percent
Municipal	12.4
County	13.8
State/Province	15.6
Federal	13.3
Special Gov	1.2
Private Non-Profit	38.8
Private (Industry)	1.5
Private for Profit	1.6
College/University	6.7
School System	2.3

Setting

Where are Outdoor/EE centers typically located? Centers have been identified in every state and seven Canadian provinces. As might be expected, those states and provinces with the largest populations (California, New York, Florida and Ontario) report the largest number of centers. It also probably is not too surprising to note that 47% are found in rural areas, while 24.9% are in suburban settings, and only 18% are located in cities.

No matter where a center is located, however, chances are that it serves a broad range of populations. For example, 62.6% of the centers serve populations coming from rural, suburban, and urban settings on a regular basis. Few centers provide programs for a limited population; 4.2% serve only urban populations, 6.3% serve users from suburban areas only, and 4.3% only serve rural users.

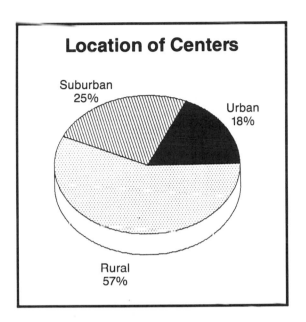

Facilities

Outdoor/environmental education centers tend to have exhibits (72.5%), hiking trails (65.9%), and self-guided nature trails (62.2%). Other types of facilities or program areas are far less commonly found: historic structures or sites (33.1%), captive wild animals (28.6%), animal rehabilitation facilities (12.89%), a greenhouse (10.7%) and farm animals (10.1%). Less than 10% of the centers reported having solar energy construction, a recycling center, a ropes or adventure course, a planetarium, or an observatory associated with their program.

Size and Use

Reflecting the diversity of types of natural science centers described in the Directory,[10] the size of the property associated with the center varied widely. While nearly 19% are located on five acres or less, 31.4% are over 500 acres in size.

The majority of centers provide programs for day users only, with approximately 20% providing a residential program. Of those providing programs for day users, two-thirds of the centers serve over 10,000 people annually, with 35.6% serving over 50,000 users. The numbers for residential users are equally impressive. Forty-three percent of those centers providing overnight programming

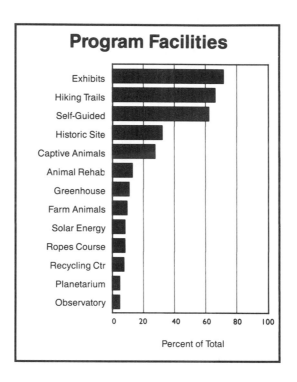

serve over 5,000 users annually, while approximately 20% serve between 3,000 and 5,000 users annually.

Accreditation

Every aspect of a center's operation, from its business practices to its program content, plays a significant role in its overall quality. User groups and fellow professionals sometimes judge a center by whether or not it has been accredited or recognized for the quality of its accomplishments or everyday operating standards. Until 1989, AAM's accreditation for nature centers was under a title and procedure used for evaluating museums. Now, a specific evaluation has been established just for nature centers, carrying with it a high level of prestige and organizational pride.

Only 15.2% of the centers report that they have gone through the formal process of recognition. Of those reporting that they have been recognized, 35.8% have been accredited by the American Camping Association, 46.1% under the previous American Association of Museums standards, and 12.4% have been recognized as a National Study Area. Finally, 3% report being accredited as an Institute for Earth Education Sharing Center.

In a way, the community of Outdoor/EE centers is much like the natural communities they teach about. They contain a rich diversity of resources which offers the teacher or group leader a valuable experience that is more than just a field trip. But, no matter whether a center could be considered "typical" or not, it is through its programs that its goals are truly manifested. The curricula available at the Outdoor/EE centers are equally diverse.

NATIONALLY AVAILABLE CURRICULA

Ideally, EE is highly interdisciplinary in nature. In practice, however, EE is largely perceived as belonging to the natural sciences and will, in all probability, continue to be dominated by them. Still, EE must reach out beyond the platform to garner support from the other disciplines. Selected EE curricula that integrate other subject areas can help to insure the interdisciplinary nature of EE.

Following is an annotated listing of major national curricula supporting EE. This review is not meant to describe the total universe of available curricula, but rather to provide some imagery of the programs that a teacher or center might select. EE curricula are usually published as either a package of individual activities that can be selected and used to enrich (or to take a "break" from) current classroom activities, or as a program of sequenced lessons that build toward specific goals in a systematic manner. In practice, most centers assemble experiences using portions of these curricula as well as their own specialized activities based on unique sites features and the expertise of their staff. Often, "custom made" experiences can be tailored to the interests and sophistication of the visiting class.

Audubon Adventures

The National Audubon Society has enjoyed a long tradition as a national leader in conservation and environmental education. Several million children learned about the natural world through their participation in the Junior

Audubon Clubs that were sponsored over the course of several decades.

Audubon Adventures is designed for elementary school classes and other groups of children in grades 3-6. A newspaper is issued (up to 32 copies/subscription) on a bimonthly basis, accompanied by a Leaders' Guide with background information and activity ideas. Also included are student membership cards, decals, and a classroom certificate of participation.

A wide range of topics is covered. The mix includes conceptually oriented as well as more specific stories and activities. The timely, relevant feature stories and activities are suitable for both short range and long term in-depth and cross-curricular studies. Currently, 300,000 children in 11,000 classrooms are using this resource. More than 50,000 minority children are enrolled through special funding programs in the major cities. For information, contact the Registrar, Audubon Adventures, National Audubon Society, 613 Riversville Road, Greenwich, CT 06831.

CLASS Project

Conservation Learning Activities for Science and Social Studies (CLASS) has been developed by the National Wildlife Federation (NWF) to enhance existing middle school science and social studies curricula. The investigations are centered around six content areas; energy use, environmental issues, forest watershed management, a hazardous substances, wetlands and wildlife habitat. The materials include a binder, illustrations, student worksheets, teacher transparency worksheets, and six color posters. CLASS can be ordered from School Programs, The National Wildlife Federation, 8925 Leesburg Pike, Vienna, VA 22184-0001.

In addition, the NWF publishes *Nature Scope,* a series of 15 volumes, each dealing with a different aspects of nature study. Designed for elementary students, each volume contains dozens of "ready to go" ideas and lessons, both indoor and outdoor, including pages of puzzles, mazes, and hands-on activities.

Supporting and augmenting classroom lessons, NWF publications include *National Wildlife* and *International Wildlife* magazines, *Ranger Rick* (for elementary students), *Your Big Backyard* (for preschoolers), and many other books, learning kits and videos. For more information contact The National Wildlife Federation, 1400 Sixteenth, NW, Washington, D.C. 20036-2266, (800) 432-6564.

Sunship Earth and Earthkeepers

Sunship Earth and Earthkeepers are programs that were developed by the Institute for Earth Education. Both are designed to be used as part of a complete environmental education program, not as supplements to existing school curriculum.

Sunship Earth, a five-day program for upper elementary students, focuses on key ecological concepts by setting up a "Sunship Study Station", in which the students are given passports to be stamped as they move through different stations. The book, *Sunship Earth,* describes the working of the program and the underlying philosophy used in the preparation of Earth Education materials. A complete Sunship Study Station Program Package, including duplication masters, teacher briefing sheet and cue cards are available.

Earthkeepers is a two and one-half to three day program for 10-12 years old students. The learners must earn special keys that open real locks on boxes that contain secret meaning about "E.M.," a mysterious, unseen character who shares the marvels and wonders of the earth. Two of the four keys must be earned back at school, and the ultimate goal of *Earthkeepers* is for students to gain an understanding of how life functions ecologically on Earth, and how we are both a part of environmental problems and their solution.

Both books, plus more information on new Earth Education programs and other related publications may be obtained by writing to The Institute for Earth Education, Box 288, Warrenville, IL 60555.

Essence

The American Geological Institute and the National Science Foundation sponsored the development of this resource which deals primarily with attitudes towards learning, classroom environment, teacher-student trust and student freedom in determining learning directions. The materials consist of brightly illustrated cards with suggested actions that leave much to the student's imagination and creativity. Many of the actions involve some type of expression on the part of the student, such as drawing or acting out their experience. The activities were designed to be "transdiciplinary," allowing the students to decide which direction to take the assignments, and "trans-grade level," serving students from elementary school through college.

Additional information may be gained by contacting ESSENTIA, Evergreen State College, Olympia, WA 98505, (206) 866-6784.

The Green Box

The Green Box contains an "open access curriculum" for environmental education, emphasizing humanistic principles of learning, with the goal of nurturing children to become "environmentally healthy adults". Students are encouraged to open the box themselves and cycle through the 218 activity cards that invite doing, thinking and showing experiences that employ school and classroom settings, environmental sites in the community and outdoor wilderness environments. The curriculum is designed for grades K-8 and is geared to the processes of learning as well as to the content. A 1989 revision added 40 new activities addressing new concerns dealing with global issues. For information contact The Humboldt County Office of Education, Environmental Education Program, 901 Myrtle Ave., Eureka, CA 95501, (707) 445-7082.

LIVING LIGHTLY
IN THE CITY
AN URBAN ENVIRONMENTAL EDUCATION CURRICULUM

Living Lightly

Living Lightly in the City contains a variety of activity ideas and resource materials to help urban children bridge the gap between the world of nature and their own urban-centered environment. Volume I concentrates on stimulating primary grade (K-3) children to explore their surroundings. Volume II focuses on grades 4-6 and covers a broad range of topics to help teachers infuse environmental studies into the existing curriculum.

Living Lightly on the Planet was designed to inform and motivate junior and senior high students. Both volumes contain activities that were created to be used in existing science and social studies classes. Volume I, for grades 7-9, concentrates on understanding of global issues while Volume II, for grades 10-12, emphasizes human actions and personal lifestyles. For information on these two curricula contact the Schlitz Audubon Center, 1111 E. Brown Deer Road, Milwaukee, WI 53217, (414) 352-2880.

OBIS

The University of California at Berkeley developed OBIS (Outdoor Biology Instructional Strategies) as a quick and easy supplement for a science program. The activities are geared for ages 10-15, although most can be modified for higher and lower levels. All can be accomplished in almost any outdoor setting using everyday items. The activities are printed on separate folders and each includes background information, materials and preparation, action, follow-up questions and suggestions for further study. Contact Delta Education, Inc., P.O. Box M, Nashua, NH 03061, (800) 258-1302, for information.

Project Learning Tree

Project Learning Tree is an interdisciplinary, supplementary program designed to provide students with opportunities to explore the existence of renewable and non-renewable resources, and human dependence on them. The activities can be used as a basis for a course of study or used as supplements to existing curriculum.

Each of the two volumes (K-6 & 7-12) contains over 80 activities. The ultimate goal of these is to make students aware of their own personal choices and actions towards environmental resources. The PLT books can be obtained only by attending a training workshop, usually through a state-sponsored program. More information on PLT can be obtained from The American Forest Institute, 1619 Massachusetts Avenue, NW, Washington, D.C. 20036, (202) 463-2468.

Project WILD

Project WILD is an interdisciplinary, supplementary environmental and conservation education program for educators of K-12 students. It is designed for easy integration into school subject and skill areas. Each activity may be used individually or in conjunction with others, and fits into a well-defined conceptual framework designed to take the learner from awareness and appreciation of wildlife to responsible human actions.

Each of the two original volumes (elementary and secondary) offers more than 80 activities. The newer *Aquatic WILD* containing 40 activities for K-12. There is no cost for the materials, but they must be obtained by attending a training session. Contact Project WILD, Salina Star Route, Boulder, CO 80302, (303) 444-2390 for more information.

REEP

REEP (Regional Environmental Education Program), administered by the Schuylkill Center for Environmental Education in Philadelphia, is a K-12 environmental education curriculum written by teachers, for teachers. It has been endorsed by the Pennsylvania Department of Education as meeting the State's mandate for environmental education.

Each grade level unit focuses on one ecological concept, with concepts first introduced in the elementary grades, and then repeated in more depth at higher levels for continuity across the curriculum. REEP units follow a definite scope and sequence and is not a collection of activities. Sequential learning within each unit involves experiences in awareness, concept building, lifestyle analysis and personal action. Although classroom-focused, one lesson in each unit involves an outdoor experience. The curriculum may be implemented anywhere. Please contact Nancy Christie, The Schuylkill Environmental Center, 8480 Hagy's Mill Road, Philadelphia, PA 19128, (215) 482-7300 for information.

THE FUTURE OF OUTDOOR/EE

In training our attention on the role of the Outdoor/EE center, it must not be forgotten that classroom teachers play an integral role in any environmental education program. Without the classroom teacher's enthusiastic participation, an on-going environmental education program is severely handicapped. According to a New Jersey survey,[12] teachers see environmental education playing a valuable role in the school program. Unfortunately, these positive attitudes toward environmental education describe only part of the picture. Even with the support of curriculum packages and Outdoor/EE center staff to help run programs, environmental education is not universally practiced within American schools.

It seems that although there may be a will, environmental education is, at times, stalled by a variety of barriers. A study conducted in Idaho by Ham and Sewing[13] typifies the situation across the country. Results of interviews with classroom teachers indicated that lack of time (both in the school day and for preparation) was the number one barrier to environmental education. Other important logistical barriers were lack of instructional materials and lack of funding. Conceptual barriers included a nearly exclusive focus on science and an emphasis on the cognitive aspects of EE. Another barrier stemmed from teachers' misgivings about their own competence to teach EE.

Overcoming these barriers will not be a simple task. It is essential that teachers are given proper training to develop a sense of competency and comfort in teaching this essential portion of the curriculum and support in the form of time, materials and resources to deliver quality EE.

Nonetheless, by focusing on the barriers to providing environmental education experience it may be too easy to underestimate the very real impact environmental education has had over the past 20 years. A synthesis of the research conducted on the effectiveness of various programs (e.g. classroom activities, field trips, trips to resident field centers) suggests that environmental education programs have been valuable in helping students gain ecological knowledge, form positive attitudes toward the environment, and develop problem-solving skills.[14]

But even putting the research evidence aside, twenty years after the first Earth Day in 1970, one only has to observe what is happening in society to see if positive changes are occurring. Another Earth Day in grand proportions was generated in 1990, the toughest national air quality act ever was legislated, and it is truly news when oil spills or when rainforests are being clearcut. In spite of the momentous problems humans have created, society is learning to change and is demanding action.

But where will EE be as we enter a new century? Utilizing a Delphi poll, The National Science for Youth Foundation asked 18 leaders of the EE movement in America to peer into a collective crystal ball to forecast future trends in the field of environmental education.[15] These leaders predicted the following events would occur by the year 2000:

•Students in K-12 schooling nationally will spend up to 5% of their instructional time engaged in EE.
•Congress will mandate some form of EE in our nation's schools.
•A majority of states will have established a state Office of Environmental Education.

Additionally, the telephone survey conducted as a part of the Delphi Poll in August 1989 indicated the following:

•61% forecast a term replacing "environmental education" by 2006.
•78% agreed that most nature centers will establish both daycare and latchkey programs, and have them by 1998.
•56% saw the possibility of one in five schools nationally devoting empty classroom space full-time to a mini-nature center run by the community nature center.

SUMMARY AND CONCLUSIONS

In the previous two decades, educational imperatives to produce an environmentally literate and actuated society have supported the development of specialized curricula and programs throughout the United States. New EE centers have been created and traditional camps and outdoor recreation centers have been retooled to focus on developing an environmentally literate citizenry. In the decades ahead, the involvement of Outdoor/EE centers will become more predominant as society demands the teaching of alternatives to environmentally self-destructive thinking and behavior. Most certainly, if the current environmental crisis worsens, as predicted in the Global 2000 report,[6] the next decade will see a mandate for an even greater focus on a learning that encourages environmentally responsible behavior.

Outdoor/EE centers will respond to this demand by delivering a wide variety of experiences and lessons that will better integrate with the overall needs and interests of user groups. And to meet these needs, a closer alliance between schools and centers will be fostered.

The lessons of the past and the vision of our future guarantee that Outdoor/EE is here to stay. EE is more than a field trip; it is essential learning! Until the time people no longer value a quality environment and care not at all for the future of unborn generations, educational experiences in the out-of-doors will continue to be a vital part of developing an environmentally moral individual.

REFERENCES

1. Sharp, L.B. May 1943. Outside the Classroom. *The Educational Forum.* 7(4):363-364.
2. Disinger, J.F. 1983. Environmental Education's Definitional Problem. *ERIC Information Bulletin. No. 2.*
3. Sartorius, S. and R. Briggs, Editors. 1987. Outdoor Education. Phi Delta Kappa's Exemplary Practice Series, Bloomington, IN.
4. Harvey, G.D. 1977. Environmental Education: A Delineation of Substantive Structure. *Dissertation Abstracts International.* 38:611A-612A.
5. Childress, R.B. 1978. Public School Environmental Education Curricula: A National Profile. *Journal of Environmental Education* 9(3):3-11.
6. Volk, G.L. 1983. A National Survey of Curriculum Needs as Perceived by Professional Environmental Educators. Unpublished Dissertation, Southern Illinois University, Carbondale, IL.
7. Stapp, W.B. et al. 1969. The concept of Environmental Education. *The Journal of Environmental Education.* 1(1):30-31.

8. Tbilisi Intergovernmental Conference on Environmental Education. 1978. Toward an Action Plan: A Report on the Tbilisi Conference on Environmental Education. A paper developed by the FICE Subcommittee on Environmental Education. U.S. Government Printing Office, Washington, D.C. Stock No. 017-080-01838-1.

9. Hungerford, H., R.B. Peyton and R. Wilke. 1980. Goals for Curriculum Development in Environmental Education. *Journal of Environmental Education.* 11(3):42-46.

10. *Directory of Natural Science Centers.* 1990. 6th ed. The Natural Science for Youth Foundation, 130 Azalea Drive, Roswell, GA 30075.

11. *Nature Center Definition.* 1986. American Association of Museums. 1225 Eye St., N.W., Washington, DC.

12. Simmons, D.A. 1988. The Teacher's Perspective of the Residential EE Experience. *The Journal of Environmental Education.* 19(2):35-42.

13. Ham, S.H. and D.R. Sewing. 1988. Barriers to Environmental Education. *The Journal of Environmental Education.* 19(2):17-23.

14. Iozzi, L. Editor. 1984. *Summary of Research in Environmental Education. 1971-1982, Monographs in Environmental Education and Environmental Studies,* Vol. II. North American Association for Environmental Education, Troy, OH.

15. Weilbacher, M. 1989. Environmental Education Leaders Predict Future Trends. NSYF's Blue Ribbon Report #3. *Natural Science Center News.* 4(3):Summer.

16. U.S. Council on Environmental Quality. 1980. *The Global 2000 Report to the President.* U.S. Government Printing Office, Washington, DC.

Science Education in the United States: Issues, Crises and Priorities. Edited by S. K. Majumdar, L. M. Rosenfeld, P. A. Rubba, E. W. Miller and R. F. Schmalz. ©1991, The Pennsylvania Academy of Science.

Chapter Thirty-Seven

SCIENCE EDUCATION OUTREACH: MUSEUMS, SCIENCE CENTERS AND ZOOS

GRETCHEN M. ALEXANDER[1], CAROL FIALKOWSKI[2], JUDEE HANSEN[3] and MAIJA SEDZIELARZ[4]

[1]Executive Director
West 40 Educational Service Center
North Lake, IL 60164
[2]Associate Director for Education and Exhibits
The Chicago Academy of Sciences
Chicago, IL 60614
[3]Education Specialist
Brookfield Zoo
Chicago Zoological Society
Brookfield, IL 60513
[4]Coordinator of Teacher Programs
Field Museum of Natural History
Chicago, IL 60605

There is growing recognition that science museums, technology centers, zoos, aquaria, and botanical gardens (all considered museums) can and should have a major impact on public understanding of science. The educational community traditionally has long looked to museums as a resource to enrich the school science curriculum, and more recently, to train their teachers in science teaching techniques. When private, state, and federal agencies support science museums, they do so precisely because the programs promise to enhance public awareness of issues that must be understood by an informed citizenry. And when citizens walk into a museum, zoo, or garden, they expect to learn.

MUSEUMS AS EDUCATIONAL INSTITUTIONS

Most museums were organized for curatorial purposes of collection and preservation of natural specimens and human-made objects, and to do research individual mood, interest, and state of mind.

By coming to a museum people have expressed some interest in the subject matter. They have committed time, effort, and money to the endeavor. They have indicated motivation to be engaged in learning if they are presented with the right vehicle, the appropriate entry point. Some visitors have backed up their commitment to learning and the institution by purchasing memberships.

The responsibility for education in museums at one time was solely that of the museum educator. Today, there is growing agreement that responsibility should be institution-wide, and shared among appropriate staff within and appropriate advisors outside the institution. As museums are sorting out who should select the messages to be communicated to the visitor, they are also concentrating on how those messages can be conveyed within the museum setting. Increasingly, museums are recognizing that, because they are *informal* educational resources, they require a new pedagogy—one that capitalizes on the sense of discovery their settings may inspire in learners. In school, if a learner does not learn, he or she fails. If a visitor to a museum does not learn, the institution has failed.

A museum maxim has it that people visit museums three times in their lives: as a school child, a parent, and as a grandparent. Such an outlook can be a real incentive for programming, but the truth is that people attend in a sometimes bewildering diversity of occasions and times. Unlike the school group audience that is usually preregistered, of a known age and (at least of stated) interest, the general public audience comes in all ages, infant to elderly, and varying group size and characteristics (family, individuals, out-of-town tourists and neighborhood children). Museums generally do not know the interests or educational background of this audience in advance.

This seeming curse of the museum world also has a unique potential: to offer visitor satisfaction and success where school science education may have failed, and to present topics of individual interest or relevancy where the school science curriculum necessarily must condense or eliminate. To be a non-threatening valuable resource available to adults as well as children, museums need to consider and answer many questions:

1. What motivates this diverse audience to come to a museum, science center, or zoo?
2. What can museums do to encourage the non-visitor to come?
3. How do museum audiences benefit from museum science education programs?

Today, exhibit planning teams use a growing body of visitor research to understand what visitors know, what they want to know, and how they want to learn about the subject matter of the exhibition. For example, preliminary to designing a major exhibit on Africa, the Field Museum of Natural History in Chicago orchestrated "town meetings" to gather information about the communities' beliefs, attitudes, and values regarding that continent. Increasingly, visitors are made partners in the exhibit development process; and planners are committed to testing and refining exhibits based on how visitors will receive them.

FORMAL AND INFORMAL LEARNING

Formal education typically refers to learning that occurs in a standard teacher/learner environment—the school classroom. Informal education describes learning that takes place just about everywhere else—on the playground, in social situations, within the family. Museum learning is classified as informal and is characterized by the following ten descriptors:

1. socially interactive
2. self-directed
3. self paced
4. voluntary
5. immediate gratification/ response
6. non-linear
7. object based
8. multi-sensory
9. enjoyable experience
10. intrinsically motivated

As we recall our experiences in school, we see that these descriptors do not necessarily apply to formal learning. In fact, except in a few cases, the conditions describing traditional classroom learning are the opposite:

1. individualistic
2. teacher directed, paced
3. involuntary
4. long term gratification/ response
5. linear
6. visually based
7. two dimensional
8. required, neutral, experience

The characteristics of formal and informal learning do not make one form better than, worse than, or more "right" than the other. Formal and informal learning are simply different. People can be engaged in both formal and informal learning simultaneously. However, after individuals leave the formal school setting most of their life-long learning experiences depend on informal sources such as television, books, other individuals, and cultural institutions. A holistic approach to education requires a combination of both formal and informal approaches to learning, and suggests cooperative ventures on the part of schools and museums. The use of museum collections as the "stuff" of science is one example.

Educators are aware of different learning styles. The effectiveness of instruction depends in part on the degree to which information is presented in ways that match the individual's learning style, e.g., visual, verbal, tactile. Linda Williams in her article, "Apples, Oranges and Sometimes Unicorns: Appreciating the Differences Between Individual Learners," cites the need for first hand experiences to develop meaningful understanding of concepts.

Who can provide the real objects from the past and present? Experiences that can't be matched by watching films, reading books, and "hands-on" classroom activities are available in the respositories of museums. Collections provide the essence of a museum's existence. These collections also provide the "stuff" of science learners of all ages need.

In a rapidly urbanizing world, children's observations of the natural world may be limited to the weeds of the corner vacant lot and the ants on the sidewalk. Night skies can be obliterated by densely packed buildings, light and air pollution. Museums in urban settings can help students make sense of the natural world of the city and convey the excitement of the wider world beyond.

Some teachers have discovered ways to integrate curriculum needs with the unique resources of local museums. There is synergy in the combined efforts— the prepared teacher and the museum. Moreover, in increasing numbers museums are offering a variety of options to assist teacher planning. Publications with lesson plans for museum and classroom use, in-service orientation sessions, formal workshops (some for graduate or salary schedule credit), and teacher advisory committees to address the needs and problems of both teachers and the museum are some of the resources available through this high-potential area of collaboration.

In 1925, the American Association of Museums drafted a resolution to work more closely with the National Education Association to encourage use of museums and teach how to use the life long learning opportunities museums provide. That resolution suggested teachers' colleges should teach about museum resources and how they can be utilized. Working with pre-service teachers has been a goal of museums since then. However, not all teacher preparation programs include the use of community resources: Some possible explanations for this deficit may be the perception of "informal education" as less than a legitimate partner in the serious business of education, increasing numbers and complexity of topics to be addressed in science education ("how will we fit this all in?"), and perhaps a lack of experience among teacher preparation faculty in utilizing informal resources.

Still, museums need to be more aggressive in approaching teacher training institutions—provide programs for education students, discuss possibilities with college instructors, outline benefits, and frankly discuss perceived problems with education college staff. Museums that have not investigated the pre-service teacher audience need to consider the tremendous potential of working with future field trip planners.

A serendipitous by-product of museum programs for teachers can be the simple encouragement and support of science teachers and the creation and/or facilitation of networking opportunities that enhance school science education. Such programs can provide science teachers at every level with colleagues of similar interest and concerns. The P.R.I.S.M. (Philadelphia Renaissance in Science and Math) project of the Franklin Institute lead to the formation of a Philadelphia Elementary Science Education Assocation that meets on a regular basis at the museum.

COMMUNICATIONS AND EXPERIENTIAL LEARNING

Labels are perhaps the most traditional mode of exhibit interpretation. Yet reading labels and looking at exhibits are not compatible activities for most visitors. Through casual and systematic studies of visitor use of signage, much has been learned about designing labels that:

1. attract attention,
2. hold attention, and
3. communicate the intended message.

The most effective labels engage the visitor and increase the meaning of the interaction with the exhibit. The often-read labels at the Monterey Bay Aquarium in California are examples of the state of this art.

Bringing true hands-on, experiential learning into the galleries is a difficult business when collections are hands-off, irreplaceable artifacts or living animals. Museums have met this challenge through discovery rooms and gallery programs that provide touchable castoffs from the collection, reproductions, and creative tie-in activities. Technology has presented solutions to interpretive problems. At the Children's Museum in Indianapolis, children can "try on" clothes from other countries by viewing a video monitor that places their faces over various costumes.

Zoos and aquariums occasionally provide opportunities to touch living collections in special petting areas and touch tanks; but for matters of practicality and philosophy, other ways to connect visitors with the living "objects" must be found. At the Brookfield Zoo, outside Chicago, a new interactive exhibit "Be A Bird" presents visitors with the chance to temporarily change places with a bird in hopes of enhancing public empathy for the interest in birds. Computer games, devices that illustrate principles of flight, and active demonstrations give a sense of what it is like to "eat like a bird," "fly like a bird," and more. The exhibit is surrounded by the collection so that visitors can observe the living representatives of the concepts they've discovered. Throughout the development of "Be A Bird," formative evaluation was employed to gauge visitors' needs and expectations and to judge the effectiveness of exhibit components. A summative study of the exhibit, compared with base-line data gathered before it was built, will assess to what degree the project is meeting its goals.

Perhaps the ultimate commitment of experiential learning is the Exploratorium in San Francisco, the vision of the late physicist, Frank Oppenheimer. No "do not touch" signs are found there for Dr. Oppenheimer believed that humans require a multiplicity of direct experiences with natural phenomena in order to truly perceive the commonalities that connect them.

The Exploratorium has been described as a smorgasbord of experience connected by a unifying theme - human perception. In this non-threatening "fun house" of over 500 exhibits visitors manipulate, create, sort out, and feel the

shoes, sinking into chairs, sighing, "Never Again," and, questioning the educational merit of the traditional day away from the classroom. Museums are helping teachers plan for field trips and integrate them into instructional units. The Bronx Zoo, for example, has created ZEST (Zoos for Effective Science Teaching), which trains middle and high school teachers to use the zoo as a resource for teaching biological and physical sciences. Later, trainees train colleagues. The program has reached ⅓ of a million students in three years.

Museum educators want to share their experiences and insight to help classroom teachers make the most of a museum visit, maximize the effectiveness of the experience in the face of scaled-back transportation dollars. Museum orientations stress that museums and zoos have the real thing (primary sources) and can be laboratories for application of concepts studied in the classroom. They can provide raw data and materials (e.g. wolf packs, rows and rows of birds or rocks and minerals, mirrors of every size and description) to allow students to make their own observations, see patterns, devise classification systems, and create original hypotheses to test. Museums can provide teachers access to collections and phenomena not readily available in the classroom.

Moreover, museums are recognizing that they can reach many more students and contribute more significantly to science education by educating the educators. There is growing concern that many elementary teachers are unprepared in science as a subject and in methods for teaching science. The self-directed approach to learning, to which museums so often subscribe, is proving an effective "de-mystifier" of science for teachers.

At the Pacific Science Center in Olympia, Washington, training programs provide teachers with enough background to answer students' questions and to facilitate interactive learning. The program has increased the amount of science teaching in the local school districts; in two years, students' science achievement test scores rose from the 77th to the 85th percentile.

The New York Hall of Science is training educators in the use of a portable planetarium that folds to fit into a teacher's car trunk, yet can expand student horizons by light years. The Franklin Institute in Philadelphia offers Teacher Overnighters, a professional "sleep-over" night of instruction and peer interaction. The Franklin Institute has also developed Museum to Go —kits of materials to enhance the school science curriculum, teacher training in use of the kits, and provisions for on-going support.

THE FUTURE

Science museums and technology centers are working to unveil and reveal the public's misconceptions about science and science concepts. Through audience research studies, museums are determining the entry level of their visitors in respect to science, including interests and questions. Knowing where potential "students" are in their thinking allows institutions and/or teachers to structure learning experiences to suit the user's needs. Exhibits, programs, and interpretive materials that recognize and acknowledge the character of the museum learning environment (voluntary, self-paced, enjoyable, socially interactive) are the most successful in achieving their educational goals. The real challenge of museum education is the integration of the particular characteristics of the institution and informal learning with the varied needs, levels, and audience types that frequent museums.

The science/technology/society approach (STS), based on learners "need to know," is a very effective tool or hook to involve visitors in active learning. A question about everyday technology that is an integral part of the learner's life engages the visitor in the discovery of how the technology works, the "science" of the product, and the implications for society. Science/technology/society is an area ripe for collaborations among museums, communities, and schools. The science museum has the "stuff," the school and the public have the need to know and the audiences. The partnerships bring about the development of a better product and clearer processes.

Museums are responding to the challenge of public science education by increased attention to the visitor in the form of random visitor interviews and unobtrusive observation of visit patterns (such as the simple process of gathering home zip codes at the admission desk). These observation efforts have helped museums plan more physical comforts (e.g., more frequent seating areas, infant care facilities, streamlined ticket procedure), target areas for outreach efforts by museum personnel, and/or devise more accessible and easily understood exhibits.

With increasing frequency museum personnel include evaluation as an integral part of every program; not only to measure cognitive learning (an often slippery measure) and affective response, but growth in skill levels as well. Formative evaulation, or testing of prototype exhibits, is becoming more than an exception. Surveys have been done of non-visitors to understand why museums are not attractive to some segments of the population. Increasing sophistication of museum research and more studies in a variety of museums, locations, and of various audience segments help put a face on the "unknown" visitor.

An encouraging trend in visitor research, befitting the diversity of the museum audience, is the bringing together of expertise from many fields. A recent issue of *The Journal of Museum Education* included articles by professors of education, developmental and environmental psychologists, and teachers.

It could be argued that the expansion of museum educational programming is self serving. However, museums are dedicated to attracting and keeping a more diverse audience. Admittedly, museums must look at the competition for the publics' leisure time and dollars from movies, theme parks, and sports events. Museum survival depends on appropriate market strategies. Museums have responded by creating diverse programming to attract diverse audiences. The question remains: "To what degree do cultural institutions such as museums, science centers and zoos, contribute to the scientific literacy of the public?" That answer will be forthcoming only if long-term studies are undertaken to measure the effects of programming to enhance science education. An easier question for now is, "What characterizes the programs that may be considered models for the future?" These model programs:

1. are an outgrowth of an institutional commitment to public understanding of science;
2. are based on objects the institution presents, sound scientific information, and sound learning theory;
3. assess the needs, expectations, beliefs, and attitudes of the targeted audiences;
4. have clearly defined goals designed to effect change in the audience;
5. have measurable objectives;
6. have programs for assessing attainment of objectives; and
7. have a commitment to refinement and maintenance based upon ongoing evaluation.

In short, the institutions most apt to positively affect scientific literacy are those that have accurately identified a gap in public science education, and are filling that gap by doing what cultural institutions do best—providing opportunities for visitors to directly sense their connectedness with the fundamental forces of nature.

REFERENCES

Bailey, Anne. 1988. Trends in Science Museums: Ending Science Illiteracy. *Museum News.* 66:50-53.

Berkovits, Annette. 1989. How to Add ZEST To Your Education Program. *AAZPA Annual Conference Proceedings.* pp.131-136.

Blackmon, Carolyn, Teresa LaMaster, Lisa Roberts, and Beverly Serrell. 1988. *Open Conversations — Strategies for Professional Development in Museums.* Field Museum of Natural History, Chicago, IL.

Falk, John H. 1984. Public Institutions for Personal Learning. *The Museologist.* 46:24-27.

Hansen, Judee. 1989. Be A Bird: Formative Evauation for Interactive Exhibit Development. *Journal of the International Association of Zoo Educators.* 21:56-63.

Hein, George. (Ed.). 1990. *The Journal of Museum Education.* vol. 15:#1

Heltne, Paul G. and L. Marquardt. 1988. *Science Learning in the Informal Setting.* The Chicago Academy of Sciences, Chicago, IL.

Hotchkiss, Nancy. 1989. Programs You Can Sink Your Teeth Into. *Journal of the International Association of Zoo Educators.* 21:112-116.

Huyler, Doug. 1989. Learning Through Doing: The Slimbridge Explorer. *Journal of the International Association of Zoo Educators.* 21:107-111.

King, Karen, et. al. 1990. Standards: A Hallmark in the Evolution of Museum Education. *Museum News.* 69:78-80.

Klein, Larry. 1986. *Exhibits: Planning and Design.* Madison Square Press, New York, pp. 208-211.

Lehman, Susan Nichols (Ed.). 1981. *Museum School Partnerships.* Center for Museum Education, Washington, D.C.

Serrell, Beverly and Yellis, Ken (Eds.). 1987. Evaluation. *The Journal of Museum Education.* 12.

Voris, Helen H., Maija Sedzielarz, and Carolyn Blackmon. 1986. *Teach the Mind, Touch the Spirit: A Guide to Focused Field Trips.* Field Museum of Natural History, Chicago, IL.

Science Education in the United States: Issues, Crises and Priorities. Edited by S. K. Majumdar, L. M. Rosenfeld, P. A. Rubba, E. W. Miller and R. F. Schmalz. ©1991, The Pennsylvania Academy of Science.

Chapter Thirty-Eight

A BUSINESS/EDUCATION PARTNERSHIP TO TRANSFORM MATH/SCIENCE EDUCATION IN BATTLE CREEK MICHIGAN

WILLIAM E. LaMOTHE[1] and MICHAEL J. BITAR[2]

[1]Chairman of the Board and CEO
Kellogg's
Battle Creek, Michigan 49016
and
[2]Superintendent of Schools
Battle Creek Public Schools
3 W. Van Buren Street
Battle Creek, Michigan 49017

Battle Creek area civic, business, professional and educational leaders are serious about changing math and science education for K-12 students in Calhoun County. As a first step, a joint project was undertaken which led to the opening of the Battle Creek Area Math/Science Center during the fall of 1990.

Driven by the declining math/science student test scores and the low rankings of United States students on international math and science exams, a deep commitment formed to make area students the best trained in the world in science, math and technology.

During the summer and fall of 1989, an effort initiated by the Battle Creek Community Foundation brought together a number of interested business, education and government leaders to informally discuss the community's educational program. Based upon expressions of concern and interest, the idea of a math/science center emerged. Generous financial contributions by Nippondenso Manufacturing, U.S.A., II Stanley and the W.K. Kellogg Foundation allowed for planning to continue at an accelerated rate unmatched by other educational reform movements. This fast-paced project allowed the Battle Creek Area Math/Science Center to move from conceptualization to reality in one year.

The Battle Creek Area Math/Science Center, being housed in 45,000 square feet of the former Springfield High School, provides elementary and secondary staff, as well as a technologically rich environment to:

1. Enhance programs in home schools for all K-12 grade students.
   a. Provide special motivational programs at home schools.
   b. Provide expertise and leadership in curriculum development.
   c. Work with business and governmental agencies to improve quality and quantity of math/science equipment in home schools.
   d. Develop a list of human and material resources available to area classroom teachers.
   e. Respond to requests for assistance from teachers.
2. Provide staff development activities to increase the teaching proficiency and efficiency of local math/science teachers.
   a. Make training available when new curriculum models are marketed.
   b. Provide immediate and long-term support when new texts and/or programs are adopted.
   c. Train teachers in current developments within their specialty area.
   d. Make the latest technology available for teacher use. Provide the educational programs teachers need to undersand and profit from advances in technology.
3. Provide enrichment activities during the school day, after school, weekends, and during the summer which will be available to all interested area students.
   a. Special presentations to students by world renowned scientists and mathematicians.
   b. Classes given by area scientists and mathematicians at their places of employment and at the Center.
   c. Topical symposia of current interest to students.
   d. Sponsorship of clubs and/or small group competitions.
   e. Opportunity for individual or small group research.
   f. Coordination of activities with other educational programs currently existing.
4. Conduct a pull-out program five half-days per week for interested and able math/science students.
   a. Phase in one grade per year beginning with 9th grade.
   b. Students selected from tri-county area on a competitive basis.
   c. Maximum student population of 300 in 9th-12th grades after four years of operation.
   d. Integrate math/science/computers.
   e. Accelerate and enrich the curriculum. Use community and business resource people to provide practical applications for student learning experiences.
5. Operate an elementary math and science materials distribution center.

Science Education in the United States: Issues, Crises and Priorities. Edited by S. K. Majumdar, L. M. Rosenfeld, P. A. Rubba, E. W. Miller and R. F. Schmalz. ©1991, The Pennsylvania Academy of Science.

Chapter Thirty-Nine

THE WILLIAM PENN FOUNDATION: MINORITIES IN HIGHER EDUCATION INITIATIVE

BERNARD C. WATSON

President and CEO
The William Penn Foundation
1630 Locust Street
Philadelphia, Pennsylvania 19103

At the regional level, the William Penn Foundation has been addressing the need to interest more African-American and Hispanic students in science and math-based careers for a long time. As early as 1976, the Foundation supported the work of PRIME (*P*hildelphia *R*egional *I*ntroduction of *M*inorities to *E*ngineering), which provides enriched experiences and encouragement for talented minority students to pursue science and math careers. In an effort to reach deeper into the pool of potential scientific and technical workers, the Foundation initiated its *Summer Enrichment in Math and Science,* a project in its sixth year which involves the School District of Philadelphia, eight college campuses, and *average* students in projects whose express purpose it to say to young people that science is interesting as well as difficult, understandable as well as exact, and engaging in ways that they rarely get a chance to experience at the high school level. The following schools were funded as part of the Summer Enrichment in Math and Science:

Bryn Mawr College
Cheyney University
Community College of Philadelphia
Foundation of the University
    of Medicine and Dentristry of New Jersey
Haverford College
Swarthmore College
Temple University
University of Pennsylvania

This project, because it focuses on young people who are not necessarily already interested in science, may be the least well understood of all of the Foundation's efforts in this area. Our premise has been that not only nobel laureates and rocket scientists must study science, but that programs must reach the kinds of young people who in the past may have become auto mechanics. They must be stimulated into studying more science, and better science, to become our future workforce.

As early as 1984, the William Penn Foundation also made a grant to one of Pennsylvania's historically black colleges—Cheyney University—to support a summer institute in computer science. In 1988, we continued the tradition of that support with a grant to Lincoln University, another historically black college, to create its Center for the Preparation of Minority Students for Higher Education in Science and Engineering. From the very beginning of our efforts in the area, we have recognized that historically black colleges and universities have an important role to play. Indeed, when one looks at the degrees granted in the sciences, it is clear that these schools graduate the vast majority of scientifically trained African American students. You cannot unclog the pipeline without strengthening these institutions.

Our history of science and math-related grantmaking would be incomplete without mentioning our efforts to improve science and math teaching in the School District of Philadelphia. Our board of directors recently approved a three-year grant to establish a Mathematics Center for Urban Schools at Temple University. Here, with a professor of mathematics education whom superintendent Constance Clayton calls "our own Jaime Escalante," we have supported a teacher-training initiative to improve the teaching of algebra and calculus among ordinary high school math teachers. We believe that our students can learn more; that teachers want to teach more. We must do a better job of harnessing student ability and teacher desire to the technology of *doing it.* We worry that these things—teacher training and reaching average students—are not "sexy" enough for many grantmakers, but we believe that this is where a huge impact can be made.

This is the backdrop that framed the development of our Minorities in Higher Education Initiative. The Initiative itself was almost two years in development and involved consultation with many of the best minds currently working on the problems of science and math education in this country. Most of the problems we identified are *national* problems: the near-collapse of the teaching function at the college level; the difficulty African-American students face in establishing close relationships with faculty who are capable of nurturing their talents and ambitions; the unavailability of resources for meaningful undergraduate research.

But we should also point out that the projects we chose to fund were run by institutions which themselves had long histories of working on the problem, and because of that, they were able to tell us what works. Pre-freshman summer programs *work*. Structured group study sessions *work*. Think about it: when was the last time you solved an important problem on your job *by yourself?* Collaboration, pursuit of academic excellence, connecting undergraduates to both each other and to the faculties of their institutions formed the core of our initiative.

The historically black colleges and universities funded under this initiative were also charged with increasing the number of students heading to graduate school. Too many minority students are spirited into the working world without even considering the option of graduate school and education at the doctoral level. We want to place that option squarely before them, and give them a fighting chance to choose. We hope that with gentle nudging, many of them will choose to help replenish the professoriate.

If it is true that every crisis also represents an opportunity, our work is cut out for us. Intervention is needed on so many levels that everyone can play a part. We hope that our initiative demonstrates that even regional foundations can carve out a piece of this national problem and get to work on it. Our pragmatist heritage should prod us to study, certainly, but study to act.

THE 1990 FOUNDATION INITIATIVE

- The Minorities in Higher Education Initiative was approved by the William Penn Foundation Board of Directors in November 1989.
- The initiative was designed to focus on increasing the number of African-American and Hispanic students who graduate with degrees in fields where minorities are critically underrepresented and are vital to the future competitiveness of the nation: math, the sciences, engineering, and teaching.
- The Request for Proposals was issued in January 1990 and specified that the Foundation anticipated making three-year grants with the possibility of an additional two years—for a five-year overall commitment of up to $10.1 million.
- The RFP indicated that support would be available for the following types of programs: pre-college summer enrichment programs; innovative tutoring programs, especially those premised on group study and enrichment rather than remediation; programs to promote faculty mentoring; undergraduate research and internships; efforts to encourage minority students to pursue graduate education.
- Requests for Proposals were sent to 18 colleges and universities, and the following schools were funded as part of this initiative:
    Carnegie Mellon University
    Columbia University Teachers College
    University of Delaware

Drexel University
Morehouse College
North Carolina Agricultural & Technical
   State University
University of Pennsylvania
Pennsylvania State University
University of Pittsburgh
Rutgers University
Spelman College
Temple University
Tuskegee University
Xavier University of Louisiana
Camden County College

Science Education in the United States: Issues, Crises and Priorities. Edited by S. K. Majumdar, L. M. Rosenfeld, P. A. Rubba, E. W. Miller and R. F. Schmalz. ©1991, The Pennsylvania Academy of Science.

Chapter Forty

TECHNOLOGICAL LITERACY FOR THE NEW MAJORITY

LEONARD J. WAKS

Professor of Science, Technology and Society
The Pennsylvania State University
University Park, PA 16802

Today, citizens are faced with personal and social value, life-style choices and public policy issues that are beyond the scope of traditional education. Today's responsible citizens must understand and be able to act upon innovations and discoveries, such as life-extension, the strategic defense in space, and release of genetically engineered organisms into the environment, that threaten the quality of our lives, our natural environment and future generations, and even our democratic institutions (Prewitt, 1983). At present these understandings and capabilities are not widely held by citizens.

Science, Technology and Society (STS) is an educational innovation designed to promote responsible citizenship in our technologically dominated era (Waks, 1987). STS lessons and units can provide awareness of these new choices and issues, opportunities for inquiry and decision-making about them, and contexts for responsible action to resolve them (Rubba and Wiesenmayer, 1988). Yet, the fabric of American society is in the midst of monumental demographic change that will make disadvantaged minority groups, groups our educational system presently fails, the "new majority" of the school population. In this chapter I want to put forward three ideas that link the education of disadvantaged minority groups and STS education.

First, it is critical for our society, not just on grounds of equity but for our national survival, to reverse the pattern of school-related difficulties and failures among members of disadvantaged minority groups;

Second, Science, Technology and Society (STS) as a curriculum emphasis has an important and unique contribution to make to the secondary science education of urban and minority students; and

Third, in order for STS to make this contribution, it must be shaped to empower the learning of disadvantaged minority youngsters in urban schools, and this involves new role-relations among teachers and school leaders, learners, and members of the urban minority communities.

Education and Minority Groups: The Changing Context

In the 1960s and 1970s, the problems of urban minority education were interpreted in the educational policy arena largely in terms of "equity." The educational system was perceived to be doing an acceptable job for the "white majority," but (despite the achievements of the civil rights era) still failing with too many members of disadvantaged minority groups. While considerations of fairness demanded that we do better, minority education was not perceived as central to the economic, political or cultural well-being of our society as a whole.

Policy-oriented social science studies during those years, such as the widely-publicized reports by James Coleman (1966) and Christopher Jencks (1972), considered such questions as whether school could account for variability in academic achievement and economic success in later life. Arthur Jensen and others proposed genetic explanations of minority school failure.

In the 1980s the discussion of minority education has shifted from equity issues to concerns about national survival. Partly as a result, policy oriented research has shifted from a relatively passive attempt to account for the pattern of failure to a more active stance—attempting to identify and replicate success. This is well illustrated in the "effective schools" literature. Ulric Neisser (1986) in a definitive review of recent research, has noted the "ecological" character of recent discussions, in which causal efficacy is assigned to the impact of the social caste system, to cultural factors that conflict with the culture of schooling, and in-school organization and pedagogical practices that weaken and disrupt student attempts to learn.

Behind this shift in research is the shifting demographic pattern in our nation and its schools, colleges, and work force.

Changing Demographics of American Education

When I began my study of STS for urban and minority learners in 1985, this changing demographic pattern of American education had not yet assumed the central place it now occupies in policy discussion. Minority education was still largely focused on equity concerns, its importance for the well-being and even survival of our nation was not yet widely appreciated. The demographic data and their implications for the educational system had just been effectively presented in a report authored by Harold Hodgkinson (1985) for the Institute for Educational Leadership, *All One System: Demographics of Education, Kindergarten through Graduate School.* The basic demographic facts are now better known, but their implications for curriculum policy are still being digested by educators. In what follows some data are presented and their implications for education brought into focus.

Births. While the white population is failing to reproduce itself, the non-white groups are either holding steady or increasing in absolute terms, and are thus gaining larger relative share of the population. In order to maintain a steady state, the average female must bear about 2.1 children. The birth rate of the white population is 1.7 per female; of Puerto Ricans it is 2.1. For African-Americans the birth rate is 2.4 and for Mexican Americans it is 2.9. So as these children enter the schools and the work force, the demographic patterns in schools and workplaces change.

Where the school age population is increasing, it is due to larger numbers of minority children. Where school population is decreasing, it is due to a declining number of white children.

Age. The minority non-white population is younger. The average age of the white population is 31, African-American 25, and Hispanic 22. This means that the average white female is nearing the end of her child-bearing years while the average Hispanic female is entering hers. So the demographic change noted above will accelerate.

Aging Population. The population as a whole is aging—the proportion of older Americans is increasing. In 1950 each retired person was supported by 17 workers. In 1992, this ratio will fall to one in three, and one of the three workers will be from a minority group.

Family Status. Normative patterns of family life and support systems for children are changing. In 1950 over 60% of all households with school age children were "typical" households with a working father, a non-working mother, and 2 or more children at home. In 1985 the percentage of households "typical" in this way had fallen to 7%, and was rapidly falling. Indeed 60% of all school age children will experience living with just one parent (generally a working or non-working mother) before the school leaving age of 18.

The implication is that expectations about the home background and support systems in place for school-age children must be reconsidered and policy changes made to accommodate to new norms.

The nation is also experiencing an epidemic of out-of-wedlock births, over half of which are to teen-aged parents. This is not a "minority" phenomenon: while the percentage of out of wedlock births is higher for non-whites, in absolute terms there are more white than minority out of wedlock teen births, and U.S. white females are more than twice as likely to have out of wedlock births in their teen years than their counterparts in any other industrial country.

The bottom line results of these changes are:
• Public schools everywhere will have larger percentages of children from disadvantaged minority groups, members of which have experienced and who continue to experience high proportions of school difficulties;
• Society as a whole, and the work force in particular, are changing to have an increasingly larger proportion of people from disadvantaged minority groups; and

• A higher percentage of school children in all groups will come from broken homes and will be born out of wedlock to teen age parents.

This is the challenge to the well-being and survival of our society. When these disadvantaged minority groups comprised a small proportion of school children, workers and citizens, it was possible for large percentages of individuals from these groups to fail without significant harm to the white population. The white majority was able to neglect the concerns of minorities, or even rig the system of educational and economic benefits against these groups, through segregation and job discrimination, without being forced to pay a high price.

But as these groups become a majority of the public school population and a very dominant proportion of the work force, they *eo ipso* become the key to the health of our society in economic, political, and cultural terms. They hold the key to our future. If they fail our American society as a whole fails.

Social and Educational Implications of the New Demographics

How, more specifically, will our society fail if we cannot reverse the school-related problems and failures of disadvantaged minority groups? Following Henry Levin (1986), we may understand the risks in terms of (1) creation of a dual society marked by social and political upheaval, (2) a conflict in higher education, (3) work force deterioration, and (4) a crisis of taxation and public costs.

1. *Dual Society.* Because of universal suffrage, disadvantaged groups can gain potential political power even if they fail to achieve economic equity. This raises the spectre of political conflict between "haves" and "have nots" organized around racial and ethnic themes. This would impose unbearable strains on our democratic institutions. Leaders such as Jesse Jackson, who have sought to build a rhetoric of coalition and harmony, may be replaced by minority leaders who, trading on the lack of knowledge and education of their constituents, appeal to their frustrations and baser instincts. Meanwhile, many in the white population may once again put forth "white power" advocates such as David Duke. In addition to the obvious inherent evils of intergroup conflict, such a politics of race and resentment also will certainly distract the nation from the pressing and complex issues now coming to the forefront in our technologically-dominated society.

2. *Conflict in Higher Education.* As the proportion of school aged students from disadvantaged groups grows, then even with high drop out rates such students will continue to increase as a proportion of the age cohort to be recruited for post-secondary education. Institutions of higher education will have to find ways of recruiting and retaining them, just to meet their minimum enrollment needs and "stay in business."

However, if school difficulties obstruct the learning of students from disadvantaged minority groups during their secondary school years, then either a large and increasing proportion of the college and university students will experience academic failure in college, or colleges and universities will be forced to shift increasingly to remedial functions and to lower academic standards, creating conflicts regarding the "culture" of these institutions. An early and bitter example of this was the battle over curriculum and standards at the City University of New York following the adoption of an "open admissions" policy.

The existence of academically under-prepared students recapitulates the "dual society" in higher education itself. "Conservative" forces rally to "preserve the character and standards" of the institutions, leading to intense conflict again organized largely along racial and ethnic lines about the appropriate mission, goals, standards, and curriculum of higher education. A recent battle in this war was the conflict over "Western Civilization" at Stanford. Conservatives sought to preserve the notion of Western Civilization as an organizing theme for general education, while many students and faculty members sought to develop a more "universal" conception of our cultural heritage, embracing the contributions of non-white, non-Western groups. Recent racial incidents at Michigan, Penn State, and other major Universities are also tied to the changing demographics and the new importance of non-white "non-traditional" students, with different learning histories, values and educational needs.

From the "conservative" side, victory in this battle means "preserving standards of excellence," maintaining "traditional" curriculum and conventions of academic achievement. Such a "victory" may be expected to have as a side effect intensified inter-group hostilities and intensified resentments among those in the work force disqualified from college completion, when our economic and political institutions can least bear these strains.

3. *Work Force Deterioration.* Until recently, poorly educated workers from disadvantaged minorities were frequently able to find jobs with low educational qualifications, especially in the industrial sector. However, this sector of the economy is declining, and the service sector is expanding. The service sector is very heterogeneous in terms of educational qualifications and skill requirements, but it appears to most observers that a growing proportion of the work force is not qualified for emerging jobs. This includes not merely professional and technical jobs, but even the new jobs at the low end of the skill spectrum.

The lack of fit between basic skills and workplace demands places U.S. business and industry at a competitive disadvantage with those of other developed and developing nations. In many international comparisons of academic achievement, especially in science and mathematics, the United States finishes near the bottom. While Japan and Korea can graduate over 90% of their cohort from secondary school and effectively convey basic skills and cognitive routines, we can graduate only 70%, many of whom remain functionally illiterate. These facts stunned the nation when publicized by the National Commission on Excellence in Education (1983) in "A Nation at Risk."

It is frequently argued that those entering the work force in the United States have lower levels of productive knowledge and skill than our competitors, and thus impose higher training costs and lower productivity on business and industry. This puts U.S. industry at a competitive disadvantage in the global technological economy, and that implies a lower tax base, an increasing balance of payments deficit, and a deflated value for U.S. currency. That erodes the value of individual wealth and income, implying a lower standard of living for Americans.

Concern about this is hardly confined to business leaders and conservative politicians. Liberal economists such as Lester Thurow of MIT emphasize the relationship between the "technological literacy" and "productive knowledge and skill" gaps and our trade and budget deficits. He points out that high-tech quality control workers need the basics of algebra and statistics to perform routine job functions. IBM must teach its workers this knowledge, while Japanese firms can take it for granted among their entry level workers, giving them lower costs of production and thus a competitive advantage. The conclusion of this sort of argument is that unless we can reverse our patterns of school failure for the rapidly growing segment of disadvantaged minorities in our schools, the comparative quality of the American work force will deteriorate further. Thus we will be forced to lower our standard of living either in the present, or, to pay for an ever-increasing foreign debt burden, even more dramatically in the future.

4. *Crisis of Public Costs.* Civil unrest, educational upheaval, economic decline—these all impose additional cost burdens for the public. If we cannot reverse school difficulties and failures for disadvantaged minority groups, then our society will have large numbers of alienated and unemployable or marginally employable teens and young adults. These individuals become the welfare mothers, the drug and alcohol abusers, the anti-social, mentally ill and criminal elements of society, and they impose large public costs.

Failure effectively to integrate members of disadvantaged minorities into the economic mainstream of society during the school years places increasing burdens on the taxpayers. The costs of the welfare and criminal justice systems are increasing at the very time when the economy, faced with foreign competition, escalating expenses and increasing debt, can least bear such costs.

The "Excellence" Solution: Raising Standards

A series of national education reports has focused national attention on the poor performance of the American educational system. The first proposed solution to the problem of low academic achievement was advanced under the banner of "excellence." Most of the reports, following the lead of "A Nation At Risk," used the language of "excellence." In practice, excellence meant raising standards, increasing school hours, emphasizing basics, establishing minimum competencies for high school graduation and minimum competency tests for teachers. National leaders directed leaders at the State and local levels to implement these reforms.

The impact on disadvantaged minority learners was predictable. While Governors put educational reform on the agenda, and state legislatures passed reform legislation, few states appropriated increased funds for improved performance. Standards were thus raised while many minority students were already failing to achieve the lower standards, and no additional assistance was provided. As a result, more minority students experienced frustration and failure and dropped out of school or failed to graduate. Harold Howe II (1983), Ernest Boyer (1984), and other leaders immediately predicted this unintended consequence of the "excellence" movement.

Minority educators were also quick to express their deep concerns. Faustine C. Jones-Wilson (1984) wrote that:

> The clear danger to blacks is that we might be ignored in the quest for excellence and quality since so many power holders seem to believe us incapable of attaining the highest standards of mental performance. . . (p. 98).

> Black students are failing competency tests in larger proportions than their white or Asian counterparts, and in some communities their Hispanic peers. As a result these youth will receive school attendance certificates, not high school diplomas, and will be unemployable unless remedial education and retesting are provided them.

> Black children, their parents, and their organized groups need to understand that the current mood is to test them out of the educational and employment pictures. It needs to be understood that the competition has increased for scarce places in education and employment. . . . The feeling is that people should merit education and employment opportunity, and merit is more than often determined by test scores (p. 109).

The solution to the problem of education for disadvantaged minorities does not lie merely in raising standards of achievement, although achievement will of course improve as the problem is addressed in more effective terms. As Boyer (1984) put it, "schools have less to do with 'standards' than with people."

Redefining the Problem

The problem of education for disadvantaged minorities needs to be redefined in terms of people, their perceptions and needs.

To begin this process we must recognize that thinking about the problem in terms of "ethnic minorities" is misleading. This term focuses on cultural difference (ethnicity) and low percentages of the over-all population (minorities). These factors are no longer central to the problem in the way they once were.

Before the civil rights era there was blatant social, political, economic and cultural discrimination in our society, North and South, supported by custom and law. This pattern of historic racism imposed a stigma upon non-white minorities, involving denial of the right to vote and participate in civic life, the imposition of a job ceiling and the assignment of low status employment, the denial of equal protection of their laws, prohibition of social relations, and invidious comparisons of cultural activities and products.

In the years after the civil rights movement and civil rights legislation, historic racism and its associated stigma has been, while not ended, reduced and altered. There are new opportunities for many members of once stigmatized non-white groups in education, in politics, in cultural life, and throughout all sectors of the economy including the professions, business and industry. This is one of the most significant facts of our times, but like most important "pluses" it has unintended "minuses."

One "minus" in the new situation is that those left behind face more difficult problems, with fewer supports. As individuals from the once stigmatized groups avail themselves of emerging opportunities, they move out of ghetto neighborhoods and the isolated "minority" community, and enter the "mainstream." To the extent that they do so on equal terms, with their cultural identities intact, they inter-act with members of the white population and continue to forge a more "universal" culture which partakes of the dynamic interaction of various ethnic groups in the mainstream.

Now we face a new problem of "residual" out-of-the-system individuals caught in the inner cities. This is not a problem of "ethnic minorities" but of people without essential social resources. As those more capable of making, and keen to make, an accommodation to the educational and economic mainstream leave the disadvantaged communities, those left behind in the inner cities remain in situations increasingly emptied of individuals striving to succeed in the normative pattern of hard work and delayed gratification leading from educational achievement to economic success. This means, in the residual communities there are fewer models of success through the "work ethic," people who are working, achieving, becoming economically, culturally, socially, and politically affiliated. The very serious problems of those who remain in isolated "ghetto" conditions are no longer simply problems of race or ethnicity.

The problem which remains is not one of aspirations but of means for achieving them. Studies have shown that the aspirations of ghetto youngsters are not any lower than middle class youngsters in suburbs. What ghetto children lack are clearly demarcated pathways to success, to the achievement of their aspirations. What they lack are not "middle class values," as much as middle class resources.

The demarcation of success pathways has never been entirely or even predominantly the business of the schools. Family and community goals, models, and support systems have always been primary. Different groups have used the schools in somewhat different ways to achieve entry into the mainstream. The schools provided paths for these groups, but the groups had to search for and then light these paths, so to speak, with cultural factors (goals, support systems, models, work ethic) they themselves provided. The schools were instruments which in different ways served the somewhat diverse purposes brought to them by various groups.

There is no "culture of poverty" if this phrase implies the disadvantaged groups lack aspirations to succeed. Instead there is a "poverty-dominated culture" among the residual disadvantaged population, in the sense that the support systems, models, and work ethic needed to use the schools for success have all eroded as the most school-adaptable members of these groups have entered the mainstream and left the ghetto.

Among inner city residents, school children and their parents and neighbors, schools are not widely perceived as pathways to success. On the contrary, as John Ogbu and James Comer have each documented, the school is frequently perceived by the ghetto community as a hostile force. Because of the performance of the educational system in the era of historic racism, when schools for disadvantaged minority groups were frequently rigged for failure, these perceptions have considerable historical justification and cannot be dismissed as mere paranoia.

But reactive perceptions add to the problem. Many have written about the ghetto residents feelings of being trapped. For many school children, school itself is a trap laid by the white population. Ghetto students speak of school effort and success as "the white man's way." This way of thinking blinds young people to the opportunities which may exist in their school situations. As John Ogbu puts it, the reasons for minority school learning difficulties are to be found both in the ways the wider society (including the schools) treats the minorities and the way the minorities themselves respond to the treatment (Ogbu, 1985, p. 864). His data suggest that Black children begin to internalize a distrustful attitude towards schools early in their school careers, and he notes that children with such an attitude have greater difficulty accepting and following school rules of behavior for achievement.

To summarize to this point: the school-aged children and youth and their parents have high aspirations for success but lack clearly demarcated pathways. They frequently see the schools as pathways to disappointment, frustration and failure, not success. Such perceptions can lead to reactive attitudes and behaviors which obstruct school commitment and academic achievement.

The problem of school commitment is compounded by the existence of a well-elaborated "oppositional" cultural alternative—which, as Ogbu notes, includes different styles of talking, thinking, feeling and acting. One important dimension of this alternative is the street culture and the "gang," with its associated behaviors of truancy, delinquency, petty crime, and violence.

The gang is certainly among the most misunderstood and devalued of social forms in the popular mind, but sociologists have frequently pointed to its positive as well as negative features, including loyalties, fraternity, socialization of group norms in an otherwise normless situation, etc. The gang and its associated pattern of delinquency is also a form of economic activity and a pathway to the criminal economy. In large cities the gang culture affects, in large ways or small, the lives of almost every inner city young person from early teen through late teen years (Kinsburg, 1975).

For many urban youth the question is "which way is right, street or school?" The children rely upon one another, and breaking with their peer culture is very threatening. To be marked as a "Tom," as going the "white man's way," is to face ostracism from the one system that provides meaningful support, clear norms and demarcated pathways of life. As Ogbu (1985) states the point:

> Inner-city black youngsters and similar children may define academic success as more appropriate for whites; therefore minority students who do well academically are regarded as "acting white" or "strange" and are subject to peer pressure to change. The dilemma is . . . that they often feel forced to choose between doing well in school and manufacturing their group membership in good standing (p. 866).

As Dr. Cora Turpin, a senior science teacher in Philadelphia, puts it, the kids have "nobody to fall back upon; life for them is a singular effort. The peer group is crucial to survival, and the kids must accommodate to their peers. They are tortured for setting high academic standards." And Maria Arguello, also a senior science teacher, notes, "the kids depend upon one another for support and even survival. So looking good in front of peers is crucial. If teachers threaten their image with peers, they are threatening their very survival."

Ogbu draws the conclusion that at the level of practice, teachers and schools needs to "develop programs to help these minority children learn how not to equate mastery of school culture with loss of group identity and security (Ogbu, 1985, p. 868, his emphasis).

The problem is further compounded by unequal provisions for educational finance, the growing shortage of superior teachers who can penetrate the inner world of the ghetto youngster, and deteriorating school buildings. These send a message to the students that the system doesn't care.

To draw the main implications of this section; the problem of "minority education" does not reside in either ethnic differences or in the small relative proportion of these groups in the population. The groups are growing in size and importance, and with the decline of historic racism many members of these groups are entering and making major contributions to the mainstream society. We are now faced with a new problem—a growing population of individuals left behind despite the inroads against historic racism.

This problem is not a "technical" problem of finding the "right educational methods" to teach disadvantaged youth, and especially not a problem about "low academic abilities." The problem is not really "how to teach these kids" at all, but the prior problem of "how to win their hearts and minds." This can only be done on terms which are understandable, believable, and appealing to them. Very little going on in comprehensive high schools now can meet this test. Mary Ann Raywid makes this point well:

There is obviously a serious mismatch between many of our high schools and the students they are supposed to be serving. We must change that and change it soon. If we are serious about wanting to keep prospective dropouts in school, then clearly what we must do is change the way they feel about school. They have to be convinced that education is of value, that it is worthwhile and can make a difference in their lives.

It is the simple fact that they hate school . . . so that if we want to keep "at risk" youngsters in school we are going to have to provide a different kind of school environment . . . a different kind of place, with a different kind of organizational structure and a different feel and flavor.

There is no way to simply "whitewash" this problem; rather we must change the relationship of school, learner and community in very fundamental ways. We have to make the relationship work, so that it can maintain learner commitment despite the large demands it places on the learners, along with frustrations and occasional setbacks. What might this look like?

Domination, Advocacy, and Empowerment

There are several conceptual frameworks in the literature for understanding the educational problems of disadvantaged minorities. Those that centered on the themes of domination, advocacy and empowerment (Cummins, 1986; Ogbu, 1985, 1986) would appear particularly relevant. In addition, there appears to be a need to augment these ideas to emphasize the active role of the learners in resisting (not merely being disabled by) inappropriate school routines (Welsh 1987).

To such researchers as John Ogbu and Jim Cummins, the school problems of disadvantaged minority members lie in the re-establishment in school of the very forms of social relationship in the larger society that disempower these youngsters, de-valuing them and weakening their energies for school learning. Whatever other changes may or may not be appropriate in the curriculum content, primary reform is "dependent on educators, collectively and individually, re-defining their roles with respect to minority students and communities (Cummins, 1986)." Role definitions and structures in the classroom, and the dynamics of role interactions, either empower or disable learners.

Cummins (1986) conducted analyses of cross-national empirical data on the academic achievement of individuals in low status groups. His research indicates that educational programs for such groups are successful to the extent that:

(1) minority student language and culture is incorporated in the school program;

(2) minority community participation is encouraged as an integral part of the learner's education;

(3) the methods of pedagogy promote intrinsic motivation on the part of students to use language actively to generate their own knowledge; and

(4) professional involved in assessment become advocates of learners rather than legitimating the location of the "problem" in the learner.

On this analysis the problem resides not so much in the content of the school curriculum, as in the relational context. Useful change will not result from new curriculum content alone, but will require a more "user friendly" context for schooling. The message to learners and community residents must be: the schools are here to assist us in improving our lives, expressing our unique personal and cultural identities, and solving our common problems. The schools must "feel right" to the learners and their communities.

Trubowitz notes that the feelings of young people are a "facet of all curriculum content," and that youngsters will not "accept or become involved with curriculum content that does not take their feelings into account." But when they "sense that their vital needs—as represented by their inner emotions—are understood and dealt with, then communication and cooperation are encouraged" (Trubowitz, 1968, p. 92).

A contextual change does not necessarily demand a wholesale revision of curriculum content. From a practical standpoint that is a good thing, because minority education leaders tend to reject wholesale curriculum content revision. They do not want minority children to be guinea pigs for educational experiments, but rather want the same "quality" education which in their view has worked for the white children who have experienced educational and economic success. When a user-friendly relational context is established and communicated convincingly, much of the old curriculum content can be reconfigured and infused with new meaning and relevance. This applies to many components of the discipline-based science curriculum. The area of STS (Science-Technology-Society) holds that potential as the American Chemical Society's CHEMCOM (Chemistry in the Community) course amply demonstrates.

The New Social Context of Schools and STS Education

What is the relationship between the new relational context prescribed by Cummins (1986) and STS as an emphasis and organizing theme in science education?

1. *Minority Language and Culture*. Language is the heart and soul of the learner. Language (reading, writing, thinking, speaking, listening and interpreting) is not merely a "skill," but rather it is voice, spirit, soul. Language implies deep cultural roots transformed through personal self-expression and construction of meaning.

As William Labov (1972) and others have demonstrated, Black American English is a highly subtle, nuanced, and expressive language. The issue faced is not whether Black or "Standard" English, or Spanish, etc., should be used as the language of instruction. Even if so-called standard English is used in instruction, the native language of the learner must still be understood and valued. As Charlotte Brooks has said, to deny the child's language is to deny the child.

Barry Kinsburg has emphasized that the gang and street culture is, as much as anything, a context for expressive language use. The leader of the gang is given the title of "runner," and Kinsburg notes that in Black ghetto argot, to "run it" means to talk or tell a story. He adds:

> The runner may be thought of as the "talker" or "story teller." Several gang researchers have stressed the high value placed on verbal ability by adolescents . . . . Members of gangs feel that verbal ability is a central characteristic of leaders . . . . Runners were the leaders of the daily routine. On the corners, we observed that leaders would often be in a center of a close semi-circle of gang members dominating the conversation of the group (Kinsburg, 1975, pp. 66-67).

However, the verbal ability of the "runners" and other inner city youth may not be detected by mainstream professionals or standardized tests. Kinsburg provides the telling example of one prominent Philadelphia gang "runner," "Deacon." Two psychologists who examined him had a dim view of his abilities. The first reported Deacon's verbal IQ to be 87, and his reading level at age 17.2 to be 1.7. His report called Deacon "slow-functioning" and stated that "in verbal efforts the boy indicated substantial limitations." The second psychologist ended his report by asserting that Deacon "appears without talent of any kind." (p. 19).

Yet Kinsburg's own observations of Deacon contradicted these reports:

> A debate with him required a great deal of skill, but even then one could never expect to have the last word. Other gang leaders, when asked why Deacon dominated, . . . stated "Deacon can rap for us: he knows what we feel." He displayed a quick and subtle mind, and was a great creator of phrases . . . . He was the elaborator, the teller, the one who would explain at length, and almost without argument, the thinking and feeling of the group. (Deacon was) the center of attention, telling jokes, engaging in verbal repartee... and performing an almost continuous monologue about the real and mythical activities of fellow gang members.

Such reports from the field should help us to understand why youngsters associate the culture of the school, and its aura of language constraint and correction, which they contrast with the expressive street culture, as a kind of personal and cultural homicide—they feel dead in school; as Raywid (1987) insists, they "hate and detest" the school experience.

Sidney Trubowitz (1968) notes with irony, that "children who can insult each other with a neat turn of phrase are not totally without language ability." He adds, "the use of highly figurative language, understood only by the initiated, may be the children's way of showing their power over society and authority" (p. 89).

STS issues, with their focus on values, on diverse cultural definitions of problems, and on diverse solutions based on the use of cultural resources rooted in diverse traditions, are a channel through which expressive language can flow. Young people can express their ideas, wishes, hopes, and plans, and discover that these are regarded as interesting, important, and "on target." Learners can give vent to their feelings of anger and frustration, and their need to have a greater control over their environment. Through this process, they can see that their own language permits an enjoyable play of imagination which is accepted and admired in school, that there is an adult audience for their experiences and hopes, and as Trubowitz (p. 89) puts it, "an acceptance of the idea of possible change."

2. *Active Community Participation.* Cummins notes that community participation must be encouraged as an integral part of the learner's education. This contrast sharply with the invidious comparison between the "civilized" culture of the school and the "barbarians" beyond the gates. It implies acknowledgment of the value and worth of viewpoints in the community. It means opening up the wall of separation between school and community, so that much learning is "community-based" and places youngsters in contact with adults in the community. And it means making the school a significant resource for the adult community, through continuing education, information and public awareness activities.

It does not mean "token" participation to satisfy some bureaucratic demand for participation, e.g. a rubber stamp parent curriculum committee. Empirical studies show that such forms of participation have no impact on school achievement, and we should hardly expect that they would.

Encouraging participation depends upon (a) recognizing that life in the disadvantaged minority communities has some positive elements that are educative in the best sense, and (b) making the problems of living in the community a relevant and important focus of learning.

STS education provides several means for enhancing relations between the school and the minority community. Issues such as solid waste management, chemical pollution of the air and water, urban health problems, changing workplace quality and declining industrial jobs, high-rise housing, noise pollution and crowding, heroic measures to save low weight babies, AIDS, and illicit drugs all bear on urban youth. STS units on such issues can draw upon the learners' experience of their urban environment and enrich it.

As Trubowitz (1968) notes:
Children are in daily contact with other city agencies. They include the Park Department, the Department of Sanitation, the Rapid Transit System, the Department of Gas and Electricity, and the Department of

Water Supply. By focusing on the work of city agencies, and by exploring in depth the effect of agency work, the school can help the children develop a clear understanding of this aspect of the world around them.

Leaders within the community, representing the professions, city administration, and community groups can be called into the classroom to address urban issues. Or students can learn about these problems at first hand through community-based learning activities.

The school can sponsor STS "short courses" and "briefings" so that parents and community residents can enlarge their awareness about the problems facing their communities. Community leaders can participate in such efforts, building support in the community to address problems by political action. By such participation, members of the community can strengthen their ties of loyalty and support for their schools. Through STS, students can also learn actively to seek solutions to problems they have passively endured, and their own learning activities can be part of the solution, e.g., in an environmental campaign.

3. *Intrinsic Motivation to Use Language Actively to Generate Knowledge.* This component implies that youngsters learn through defining situations in their own terms, defining problems and proposed solutions in terms of their own cultural values. It takes "active language use" to the level of "knowledge." This means breaking the cultural stereotype of knowledge as a static entity expressed in a foreign language in a textbook. It stands against the shaping and constraining of language implicit in passive textbook learning and the one culturally defined "right answer."

STS units contribute directly to breaking the cultural stereotype of knowledge, because in them the focus of "knowledge" shifts from the "textbook" problem and the one right answer to real problems encountered in the community. Learners' direct experience and that of community residents, their interpretations, analyses, and different solutions are relevant "knowledge" in STS.

4. *Assessment for Advocacy.* This kind of assessment means that the "evaluator" is open and actively seeks out the power, elegance, simplicity, intelligence, courage and other virtues in the responses learners actually make to situations which concern them. This compares with the gate keeper notion of evaluation, based on behavioral objectives, standardized tests and minimum competencies, the red pencil and the failing grade.

Assessment for advocacy means defining the educational situation holistically, in terms that learners find convincing, and with goals which are personally meaningful, however demanding. It goes beyond "judgment" and "standards" to actively discover worth and merit in the learner's mode of approaching the tasks at hand, building on what is positive and correcting what is negative in clear but non-judgmental terms. STS education lends itself to this mode of assessment, for its goal is not inert knowledge but the active participation and empowerment of learners.

STS in Secondary Science Education

STS education is an already established curriculum emphasis in science instruction. It is not a new invention geared for "low ability" students, but rather an innovation for all students which has won the enthusiastic endorsement of the leading professional associations and government agencies such as the National Science Teachers Association, The American Association for the Advancement of Science, and the Science Education Directorate of the National Science Foundation.

As defined in the literature of science education, and implemented as a curriculum emphasis, STS lends itself well to the changed role-definitions summarized above. Among currently authorized curriculum reforms in science education, it is the only one which does so. For this reason STS is the best currently available window for reform in the established science curriculum. This is its unique contribution.

In urban schools, however, STS must be implemented so as to emphasize these new role-definitions. Unfortunately, many ways of implementing STS fail to achieve this potential. An STS unit on "acid rain," that draws heavily on knowledge and concepts unfamiliar to urban learners, that approaches a problem far from their experience or concerns, and that is presented as authoritative knowledge by the teacher to passive learners who are expected to learn the right answers, will not make any difference in inner-city schools.

REFERENCES

Boyer, E. 1984.Reflections on the Great Debate of 1983. *Phi Delta Kappan.* 65(8):525-530.

Coleman, J. 1966. *Equality of Educational Opportunity (Summary).* Dept. of Health, Education and Welfare, Office of Education, Washington, DC.

Cummins, J. 1986. Empowering Minority Students: A Framework for Intervention. *Harvard Educational Review.* 56(1):18-36.

Hodgkinson, H. 1985. *All One System: Demographics of Education, Kindergarten through Graduate School.* Institute for Educational Leadership, Washington, DC.

Howe, H., II. 1983. Education Moves to Center Stage: An Overview of Recent Studies. *Phi Delta Kappan.* 65:167-172.

Jencks, C. 1972. *Inequality.* Basic Books, New York, NY.

Jones-Wilson, F.C. 1984. The State of Urban Education. *In: The State of Black America 1984.* National Urban League.

Kinsburg, B. 1975. *The Gang and the Community.* R & E Research Associates, San Francisco, CA.

Labov, W. 1972. The Logic of Non-Standard English. *In:* P.P. Giglioli, (Ed.). *Language and Social Context.* Penguin, pp. 179-216.

Levin, H. 1986. *Educational Reform for Disadvantaged Students: An Emerging Crisis.* National Education Association Professional Library, Westhaven, CT.

National Commission on Excellence in Education. *A Nation at Risk: The Imperative for School Reform.* U.S. Department of Education, Washington, DC.

Neisser, U. 1986. *The School Achievement of Minority Children: New Perspectives.* Erlbaum, Hillsdale, NJ.

Ogbu, J.U. 1985. Research Currents: Cultural-Ecological Influences on Minority School Learning. *Language Arts.* 62(8):860-869.

_____ . 1986. The Consequences of the American Caste System. Chapter 2. *In:* Neisser. *The School Achievement of Minority Children.*

Prewitt, K. 1983. Scientific Literacy and Democratic Theory. *Daedalus.* 112(2):49-64.

Raywid, M.A. 1987. Making School Work for the New Majority. *Journal of Negro Education.* 56(2):221-227.

Rubba, P. and R. Wiesenmayer. 1988. Goals and Competencies for Pre-College STS Education: Recommendations Based Upon Recent Literature in Environmental Education. *Journal of Environmental Education.* 19(4):38-44.

Trubowitz, S. 1968. *A Handbook for Teaching in the Ghetto School.* Quadrangle Books, Chicago, IL.

Waks, L.J. 1987. A Technological Literacy Credo. *Bulletin of Science, Technology and Society.* 7(½):357-366.

Welsh, C.E. 1987. Schooling and the Civic Exclusion of Latinos: Toward a Discourse of Dissonance. *Journal of Education.* 169(2):115-131.

Science Education in the United States: Issues, Crises and Priorities. Edited by S. K. Majumdar, L. M. Rosenfeld, P. A. Rubba, E. W. Miller and R. F. Schmalz. ©1991, The Pennsylvania Academy of Science.

Chapter Forty-One

THE UNDERUTILIZED MAJORITY: THE PARTICIPATION OF WOMEN IN SCIENCE

JANE BUTLER KAHLE[1] and LYNNETTE DANZL-TAUER[2]

[1]Condit Professor of
Science Education
McGuffey Hall
Miami University
Oxford, OH 45056
and
[2]Doctoral Candidate
Department of Biology
Purdue University
Lafayette, IN 47906

In a world that is becoming increasingly technologically oriented, it is crucial that all people understand and be able to use science in their everyday lives. Science and technology careers will make up an increasingly larger proportion of the future job market. For example, an overall increase of 15% is expected in scientific and engineering fields in the 1990's (Pool, 1990). Individuals without science and math skills will be at a severe disadvantage in terms of employment opportunities. In addition, scientific knowledge is necessary to make informed decisions on science and technology-related social issues such as acid rain, tropical deforestation, AIDS, and nuclear weapons.

Current practices in science education seem to result in differential attitudes, achievement, and participation patterns for male and female students. In nearly all cases, male students have more positive attitudes, higher levels of achievement, and greater participation rates in science courses and careers. This chapter will review attitudinal, experiential, and achievement differences, and will focus on participation problems and explore solutions.

THE PIPELINE TO SCIENTIFIC AND TECHNOLOGICAL CAREERS

The term "pipeline" has been widely applied to the educational processes through which students gain the skills, knowledge, and expertise to become part of the science and engineering infrastructure. It extends from elementary school through graduate school and post-doctoral studies. Students opt into the pipeline to scientific and technological careers by electing to take academic math courses in junior high school and advanced math and science courses in high school. In her analysis of the differential entrance and retention of majority and minority women and men, Berryman (1983) found that although girls and boys enroll equally in biology and algebra, fewer girls than boys continue to take chemistry, physics, trigonometry, and calculus. However, once majority or minority girls enter the pipeline, they tend to remain in it until the graduate and post-graduate years. On the other hand, male, non-Asian, minorities tend to leave the pipeline between high school and college. Recently, the final report of the Task Force on Women, Minorities, and the Handicapped in Science and Technology (1989) reinforced Berryman's findings. The problem as it concerns majority and minority women, therefore, is twofold: first, how to attract more girls into the pipeline during elementary and early secondary school; and, second, how to keep more doctoral, post-doctoral, and career women in it.

The Nature of the Pipeline

For our considerations, the pipeline to scientific and technological careers begins in grade 10, when millions of students decide whether to, or not to, enroll in elective science and math courses. Therefore, we will consider the nature of the pipeline from tenth grade through graduate school. However, the conditions and options that women face after leaving the pipeline also may affect girls' decisions to enter it.

According to the most recent report of the National Science Foundation on women and minorities in science and engineering, different course patterns continue to be prevalent for college bound high school girls and boys (NSF, 1990). In 1988, for example, girls on the average completed 3.6 years of high school math, compared to the 3.8 years taken by boys. Likewise, on the average, girls studied 3.1 years of natural science in contrast to the 3.3 years studied by boys. When those data are broken down by actual courses taken, 90% of college bound students take geometry, but smaller percentages of girls report taking trigonometry (52% girls vs. 59% boys) or calculus (15% girls vs. 21% boys). Furthermore, fewer girls (21%) than boys (24%) report enrolling in honor math courses. Whereas virtually all high school girls take biology and increasing numbers of them enroll in chemistry, only 35% of college bound women take physics, compared to 51% of the males. In the natural sciences, also, fewer girls (19%) than boys (21%) take honors courses. As stated in the recent report of the American Institute of Physics, the attrition rate is higher for women compared with men between high school chemistry and physics; for less than one third of the girls studying chemistry enroll in physics. On the other hand, over half of the boys continue on from chemistry to physics (Czujko & Bernstein, 1989).

At the undergraduate and graduate levels, the numbers of women in science and engineering majors has increased dramatically. In fact, NSF (1990) reports that between 1976 and 1986, the overall number of science and engineering baccalaureates earned by women has risen by 29%, which compares favorably to the 2% increase for men. However, Vetter (1987) demonstrates that those numbers are misleading in that they are an artifact of the increasing number of women who are enrolled in higher education. Furthermore, they are skewed by field. More than two-thirds of the women counted in the NSF report have received bachelor degrees in social science, psychology, or life science. However, in engineering 33% of all bachelor degrees are earned by men, compared to 6% received by women. Vetter warns that the number of women enrolled in college science, math, and engineering majors peaked in 1984 and, since then, has slowly declined. In the mid-1980s, freshman women indicated less interest in pursuing careers in engineering, computer science, physical science, and biology than did the women just a few years earlier. If freshman-year plans are a good indication of later career choices, the number of women in science and engineering can be expected to decrease in the near future!

The women who continue on to study science and engineering in colleges and universities are also more likely to leave the pipeline than are their male peers. A pioneering study by Graham (1978) found that undergraduate women had higher attrition rates from science majors regardless of ability level or success in science courses than did a group of comparable men. In addition, she found significant differences between undergraduate male and female science majors in number of extracurricular science experiences in high school and college and in level of career aspirations.

The type of undergraduate institution many women attend may affect their retention in the pipeline. For example, more women than men attend public two- and four-year colleges near their homes. The range of science offerings at such institutions may be restricted and engineering courses may be non-existent, making women less competitive than men in admissions to graduate programs in science and engineering in research universities. Even for women who matriculate at major research universities, barriers exist. Widnall (1988) analyzed the results of two student surveys, one at Stanford University and one at MIT, in drawing her conclusion that "women's perceived lower expectations lead directly to a loss of self-esteem and, over time, to lower performance - a self-fulfilling prophecy" (no page). The MIT women science majors responded that gender was a significant barrier to access to academic resources, and 20% of the Stanford women, compared to 7% of the men, reported experienci some form of discrimination during their undergraduate years. The undergraduate years, therefore, are a critical period for retaining women in the scientific pipeline.

Likewise, more women than men leave the scientific pipeline during their graduate studies, and those who stay in it take longer to reach the Ph.D. Documentation of, and causes for, the higher attrition rate of graduate women, compared with men, were the foci of Hite's (1983) study which analyzed attrition rates by field and by degree level at a major research university. Overall, 13% more women left doctoral programs than could be predicted by the general attrition rate for all doctoral students. When the conditions of the starting year, residence, age, and major/school were held constant, 10% of the women's additional attrition could not be explained. Hite concluded that women's higher attrition rate was due to gender.

Gender-related reasons for the attrition rate of graduate women in science have also been identified. For example, several studies have reported that although women and men hold equal numbers of university fellowships, women are less likely than men are to receive research assistantships (Vetter, 1987; NSF, 1990). Furthermore, Hornig (1988) and NSF (1990) point out that women, compared with men, pay more from their own funds for both their undergraduate and graduate studies.

Once women complete their education and successfully exit the pipeline, other barriers may prevent their full participation in scientific and technological careers. The lack of full participation by women may affect the decision of other young women to enter, and to stay in, the pipeline. For example, studies have found that women in science and math careers advance more slowly, have higher unemployment rates, and make less money than do men who have similar credentials (Vetter, 1987; Douglass, 1985). Talented undergraduate and graduate women preparing to be scientists may withdraw from the pipeline because they do not see increased numbers of successful women scientists.

Filters to the Pipeline

The pipeline issue has two perspectives: first, the entrance of interested and talented students and, second, the retention of them. What factors serve as filters to prevent either the entrance or retention of students in the scientific and technological pipeline?

Experiences and Skills

A critical filter is the number and types of science experiences and skills students gain in school. Girls and boys enter elementary school with equal interest in science but with unequal experiences in science (Kelly, 1985; Kahle & ̶akes, 1983; Iliams, 1985). During elementary school that difference, which favors boys, is exacerbated by differential experiences of girls and boys during science lessons. For example, in comparison with girls, boys more often use science instruments and materials (Kahle & Lakes, 1983); boys read more science-related books (Kahle & Lakes, 1983); boys interact with teachers more frequently (Sadker & Sadker, 1985; Hildebrand, 1987; Hyde, 1986); boys are assigned higher grades on science papers (Spear, 1984); and boys are asked more higher order cognitive questions in science lessons (Tobin & Garnett, 1987; Tobin & Gallagher, 1987).

Furthermore, international studies of student/teacher interaction patterns reveal sex-related differences. Morse and Handley's (1985) longitudinal study of junior high science students in Mississippi found that males had significantly more interaction and instructional questions than females did. Furthermore, the number of interactions initiated by boys increased over time, while girls appeared to become less assertive. In addition, boys received more feedback from teachers than did the girls and the differences increased as the students got older. Boys dominated classroom discussions, while girls waited to give solicited responses based on their prepared work. In addition, other researchers have noted that although teachers give boys specific instructions for completing a problem, they may show girls how to finish a scientific or technical task or they may do it for the girls (Whyte, 1986). Furthermore, differential teacher expectations for girls and boys in science affect the evaluation process and often result generally in lower achievement levels for girls, compared to boys, in science classes (Kahle, 1987).

Using both qualitative and quantitative methods, Tobin and his colleagues (Tobin & Gallagher, 1987; Tobin & Garnett, 1987) investigated gender differences in the classrooms of fifteen Australian high school science teachers. They found that classes were dominated by three to seven target students who tended to be male. Gender differences were most evident in whole class interactions, the activity structure used most often by the teachers. Boys were more involved in high level cognitive interactions, they raised their hands more often to answer teacher-initiated questions and were called on more often by teachers, and male students asked more questions. Teachers elaborated more on boys' answers than they did for answers provided by girls. Furthermore, they found that in small group work, especially laboratory activities, boys tended to dominate mixed sex groups. However, the authors did not identify gender-related differences in student-teacher interaction patterns when individual seatwork activities such as workbook assignments and copying notes were used. They reported that girls out-performed males in those activities.

Another filter that effectively screens girls from the science pipeline is the way in which science knowledge is assessed. Two recent analyses of the International Educational Achievement in Science results indicate that girls' achievement levels may be influenced by how science is taught and tested (Humrich, 1987; Kass, in press). Both the U.S. and Canadian surveys show that girls, compared with boys, achieve less well on the content questions in all areas of science at all ages tested (9-, 13-, and 17-year-olds). However, there are no significant differences in girls' and boys' performances on the sub-tests concerning process skills in the U.S. Humrich (1987) notes that standardized science achievement tests may not assess the scientific aptitude of girls but rather their ability to recall facts which they may consider boring and irrelevant to future education and career goals. As she explains:

> Manipulative process testing allows girls to participate in the fun part of science. They are tested individually so no boy can take over the experiment or tell them they are doing something wrong. Girls may feel freer to indulge in risk-taking when no one else is ready to pounce on their mistakes. Teaching science via process tasks may also be the way to encourage girls to study science. This does not mean continuance of present-day laboratories, but rather individual exercises in observation, recording data, stating of hypotheses, etc., with no right or wrong answers to be looked up in the back of the book (p. 44).

Kass' (in press) analysis of the Canadian results by gender yields comparable insights. She points out that "female levels of performance substantially overlap those of males and that observed differences, while consistent, are the 3% to 6% range when average proportions of respondents answering the test items correctly are compared" (p. 9). Kass uses the results of several attitude scales which accompanied the IEA science surveys to explain the achievement scores of girls and boys. She finds that girls' positive attitudes to schooling in general do not carry over to positive attitudes about science classes. She states:

> The relatively poorer attitude over the years of schooling of females compared with the attitude of males is paralleled by their relatively poorer science achievement as a group. This pattern seems to hold both over the years of schooling and over the science subjects (p. 14).

In comparison with boys, Kass notes a closer relationship between girls' science achievement levels and their expectations of success or failure in science as well as their opinions about the social implications of science. She suggests that, "current reconceptualizations of school science (i.e. curriculum which includes the richness of the human and personal experience of doing science in a pedagogically sound way) may hold promise for closing the gender gap in science achievement" (p. 14).

According to Linn, et al. (1987), girls' lower achievement levels in science result in both a lack of confidence and lower self-concepts concerning their ability to do science. Those attitudes have been cited by Kelly (1981) and Kass (in press) as reasons for the lower entrance and retention rates of girls, compared with boys, in the scientific/technological pipeline. Recent research, therefore, indicates that girls and boys have different experiences in school science, which lead to different skill levels as well as to differences in attitudes about and achievement levels in science. Since both attitudes and achievement levels in science affect entrance and retention in the pipeline (Kelly, 1985; Kahle, 1985; Widnall, 1988), both experiences and skills are effective filters to the scientific pipeline.

Masculine Image of Science

The stereotyped, masculine image of science and scientists is another filter to the pipeline. The results of studies in a number of countries (United States, Australia, New Zealand, and Norway) in which students were asked to "Draw-a-Scientist" indicate that internationally, a scientist is visualized as a white, near-sighted male who has little regard for his personal appearance (Chambers, 1983; Mason, Kahle, & Gardner, in press; Rennie, 1986; and Schibeci, 1986). These studies confirm the masculine image of scientists originally investigated by Mead and Metreaux (1957).

Within the sciences we find a shading of images; for example, students rate physics as more masculine than chemistry, which is followed by biology. Garratt (1986) explains the difference in the following way:

> Biology is perhaps perceived as being relevant to girls of all abilities, but only appropriate for boys of average ability. Conversely, physics may be seen as suitable for a broad ability band of boys, but only for girls of higher ability (p. 68).

The perception that physical science is tough, hard, and analytical leads to its more masculine image, an image that becomes an effective filter against girls' retention in the pipeline.

Sex Role Stereotyping of Gender Roles

Sex-role stereotyping of men's and women's roles in society is an acknowledged filter to the pipeline. Prior to the Girls in Science and Technology (GIST) project, it was assumed that stereotypic views of feminine appropriate roles influenced girls to avoid the physical sciences (Kelly, et al. 1984). The four-year British project, however, documented consistently that boys, from ages 11-14, held much stronger and more stereotypic views of appropriate careers for women than girls did. Indeed, casual expressions of disbelief or of disapproval from their male peers may be the most consistent and effective message girls receive concerning appropriateness of their behavior and interest in science. Boys' responses to an occupational stereotype inventory of 30 jobs indicated strong disapproval of 'masculine' activities for girls. Furthermore, when boys and girls were asked to rate themselves on masculine/feminine scales, the boys consistently described themselves as higher on the masculine scale and lower on the feminine one. Girls' self-ratings, on the other hand, were more moderate (Whyte, 1986). Kelly, et al. (1984) concluded that adolescent girls who perceived science as masculine performed less well in science. Often sex-role stereotyping of appropriate occupations, especially by boys, is an effective pipeline filter for adolescent girl

INCREASING THE NUMBER OF GIRLS AND
WOMEN IN THE PIPELINE

Both sociocultural and educational filters to the pipeline have been identified. In order to increase the participation levels of girls and women, strategies to remove those filters are needed. However, effective strategies must address the bases for the filters. A variety of factors have been suggested, and they fall into two large categories: biological factors or environmental ones.

Biological and Environmental Bases for Pipeline Filters

Over the years, a number of biological or physiological explanations have been proposed to account for differences in science achievement and attitudes between males and females. Biological theories have proposed that sex differences in scientific aptitude or ability were the result of characteristics such as genetic makeup, hormonal levels, and developmental patterns. The results of several types of studies have been used to support biological theories for cognitive and psychosocial sex differences. For example, differences exist between the sexes in nonhuman animals. Male rats are known to run complex mazes faster than female rats do, and males of various species are more aggressive than females are. Results from studies such as the IEA science survey, in which sex differences in science achievement are consistently found across all countries studies, also support the idea that biological differences between males and females may be responsible (IEA, 1988).

In 1981, Gray attributed sex differences in spatial ability to differences in genetic make-up, fueling the nature versus nurture controversy about causes for the under-representation of women in scientific fields. Spatial ability, defined by Gray as the ability to perceive and manipulate spatial relationships, is thought to predict success in geometry, quantitative thinking, and mechanical tasks and may be a factor in determining scientific ability. Males are generally thought to have greater spatial ability than females have (Linn & Peterson, 1983; Ben-Chiam, et. al., 1988), although differences within each sex are greater than those between sexes (Ben-Chiam, et al., 1988). Research on the magnitude of sex differences in spatial ability has been somewhat inconsistent. One problem has been the lack of a consistent operational definition for spatial ability. For example, Linn and Peterson (1983) looked at three aspects of spatial ability in their meta-analysis: spatial orientation, horizontality/verticality, and visualization. They found the largest and most consistent differences between males and females were in mental rotations and horizontality/verticality. Non-significant differences were found in visualization. In every case, differences were in favor of the males.

To account for the gender differences in spatial ability, Gray (1981) described biological mechanism involving a recessive spatial superiority gene found on the X chromosome.[2] A number of studies done in the 1960s and 1970s provided limited support for the X-linked gene hypothesis. For example, the distributions of scores on several spatial ability tests were found to be bimodal for men with the anti-mode of the distribution located at approximately the fiftieth percentile. Gray explained that each distribution represented one allele for spatial ability and that the frequency of each allele in the population was approximately 0.5. In this scenario, approximately half of the X chromosomes would contain the spatial superiority gene and half would contain the spatial inferiority gene. As would be predicted by the theory under these conditions, in several studies approximately 25% of the women scored above the fiftieth percentile of the male distribution. Using this model, one might expect that the distribution for females with Turner's syndrome would be similar to that of males, since the recessive genes are not masked.[3] However, such females are known to have poor spatial ability. To account for this finding, Gray stated that the effect of testosterone also might be important.

Curiously, Gray reported that the distribution of the females' scores on spatial ability tests was unimodal. Using the recessive spatial superiority gene hypothesis, one would expect that the distribution of female scores would be bimodal, with 75% of the individuals (those who are homozygous dominant and heterozygous) scoring poorly and 25% (those who are homozygous recessive) scoring well[4]. Although the unimodal distribution is never explained by Gray, his studies have been used to support a biological basis for a skill (spatial ability) that filters more girls than boys from the scientific pipeline.

A number of studies have failed to produce the patterns of correlations between spatial abilities of parents and their children predicted by the recessive spatial superiority gene hypothesis. Using Mendelian genetics, one would predict that there should be no correlation between fathers and sons, a small relationship between mothers and daughters, and a larger relationship between children and their opposite sex parent. Gray cited 27 different tests using a variety of spatial ability measures and found only three cases in which the father-son correlation was approximately zero. Correlations between mothers and daughters were higher than those between mothers and sons in about half of the studies. To account for this conflicting data, Gray concluded that other factors, such as environment and general intelligence, were likely to influence spatial ability. For example, modeling by children of their same sex parent might increase correlations in spatial ability. In addition, males and females often have different extracurricular interest, and it has been postulated that different activities, especially athletics, may account for some of the gender differences in spatial skills. Linn & Hyde (1989) have found that gender differences in spatial ability are declining and feel that they are likely to decrease even farther with increased female participation in extracurricular activities such as athletics.

Ben-Chaim, et al (1988) provide further support for the effect of environmental factors on spatial ability. The authors examined the relationships between spatial visualization and sex, grade, and socioeconomic status of approximately 1000 fifth through eighth grade students in a large midwestern city. They defined spatial visualization as "the ability to mentally manipulate, rotate, twist, or invert a pictorially presented stimulus object" (p. 51). After initial testing, students participated in an instructional unit which involved building and drawing three dimensional structures. Prior to instruction, there were significant differences in spatial visual ability between boys and girls, older and younger students, and students from different socioeconomic backgrounds. After three weeks, students were given a paper and pencil test with items that were, for the most part, different from the activities practiced. The results showed that all students had considerable improvements in spatial visual ability after instruction (as measured by the test), and the authors found that retention persisted a year after the instruction. Other studies also have found that training is effective in improving spatial skills and in reducing differences between males and females (Newcombe, 1982; Kelly, et al, 1984).

In reviewing genetic hypotheses to explain ability differences between males and females, Linn and Peterson (1985) conclude that genetic differences are unlikely to be of primary importance as very few characteristics are sex limited and "no known behavioral characteristics are sex linked, and the few biological characteristics [that are sex linked] are maladaptive" (p. 57). In addition, Hartl (1988) states that "in most cases, genetic variation that affects phenotypes within the normal range is not traceable to the effects of alleles or individual genes" (p. 18). In that case, we might expect that genetic differences due to sex should play a very small part in determining cognitive ability. Since training has been effective in improving spatial visual skills, it seems likely that environmental factors play a role in determining one's spatial visual acuity and that that filter can be effectively eliminated with experience and/or training.

A second biologically-based explanation for differences between males and females in cognitive abilities and psychosocial traits is hormonal differences at puberty. Increasing levels of testosterone in males and estrogen in females at puberty result in sex differences between males and females in body size, secondary hair growth, and fat deposition (Marieb, 1989). Hormone levels are known to influence some behavioral traits; for example, aggression and sexual and maternal behavior in many animals and, to some extent, in humans (Carlson, 1986). However, some differences, such as sex differences in spatial ability are known to be present before puberty and cannot be accounted for using the hormone level hypothesis (Linn & Peterson, 1985). It is also known that the endocrine system itself can be altered by behavioral and social interactions. Linn and Peterson (1985) conclude that "hormonal influence reflects both individual and contextual factors, and therefore cannot be given direct explanatory power" (p. 58).

Males and females also differ developmentally before puberty. Sherman (1967, 1978) has proposed that because girls mature earlier they have an advantage over boys in developing brain pathways that facilitate verbal proficiency. Since girls are reinforced for their verbal skills, Sherman feels it is less likely that girls will develop brain structures for alternate skills (such as spatial skills) for perception and communication. Early studies (e.g. Maccoby & Jacklin, 1974) concluded that females have higher verbal ability than males do and that this ability difference becomes manifest at approximately eleven years of age. However, later meta-analyses by Hyde (1981) and Plomin and Foch (1982) found that sex accounted for very little of the variance (1%) in verbal ability and that differences between males and females were stable across age. Through meta-analysis, Linn and Hyde (1989) demonstrated that gender differences in verbal ability "have declined essentially to zero" (p. 18).

After their review of a decade of research on possible biological factors, Linn and Peterson (1985) conclude that there is "no noncontroversial evidence for genetic explanations of gender differences in cognitive and psychosocial factors" (p. 53). They also state that "the evidence from the past decade of dramatic changes in educational and career choices of males and females clearly demonstrates the powerful impact of environmental factors on behavioral differences between males and females" (p. 53). As Sjoberg and Imsen (1988) conclude, biological explanations should not be used to defend that status quo or for ideological purposes; biology should be seen as a potentiality rather than as a restriction.

Further evidence for the influence of environmental factors on cognitive abilities and psychosocial traits is available. In a study of second grade classrooms, Leinhardt et al. (1979) found that teachers spent more instruction time and had more academic contact with girls during reading lessons and with boys during math lessons. Although the differences per interaction were small; for example, teacher interactions with boys in math were only nine seconds longer than those with girls, the authors estimated that over the course of a school year the boys received an additional six hours of math instruction. In this study, little achievement difference was found between boys and girls in either reading or math. Nevertheless, persistent differences could lead both to achievement differences as well as to perceptual differences about the appropriateness of science for girls and boys. For example, Parker and Offer (1986) found that when males and females had equal school science experience in terms of number and type of courses, sex differences in science achievement were negligible. Pallas and Alexander (1983) found that gender differences on the SAT-M decreased when differences in quantitative coursework were controlled.

In drawing conclusions on biological, cognitive, and psychosocial gender differences, Linn and Hyde (1989) state:

> Researchers have clarified the processes and situations that evoke gender differences revealing that differences are not general but specific and that many are responsive to training . . . [B]ecause the magnitude of gender differences is so clearly a function of the context or situation, we need

to focus on situations that minimize gender differences if the goal is to increase persistence and participation in mathematics and science courses" (pp.24 & 25).

Therefore, the underparticipation of girls in science courses and women in science careers is related to filters that are largely based on environmental conditions that may be affected by education.

Several theories have been developed to explain how environmental factors influence psychosocial traits and career decisions. Eagly (1987) uses social role theory to account for differences in male and female psychosocial traits. She feels that "sex differences are the product of the social role that regulate behavior in adult life" (p. 7), and that sex differences in adults are not necessarily accounted for by developmental factors. In this theory, gender role expectations encourage conformity to specific roles or characteristics. Experience in stereotyped roles leads to the development of skills and beliefs that in turn influence sex differences in social behavior. Unfortunately for women, the specific roles occupied by men tend to be of greater status and authority than those occupied by women.

Eccles (1989) identifies several environmental factors in developing a model to explain how girls make career choices. They include, "Causal attribution patterns for success and failure, the input of socializers (primarily parents and teachers), gender-role stereotypes, and one's perceptions of various possible tasks" (p. 4). Eccles explains that each of the factors contributes both to the perceived value of the activity and to one's expectation for future success. According to the model, expectations for success and perceived value influence the amount of effort one expends on an activity, the performance level that can be achieved, and the decision to participate or continue in the activity.

To test her theory, Eccles and her colleagues assessed the effect of expectations and values on students' decisions to pursue math and English courses. They found that sex differences became apparent in junior high and increased throughout high school. During that time, girls became less confident about their ability to do math and perceived math as less valuable. Similar changes were not apparent in male students, who did not seem to favor one subject over the other. They concluded that gender role expectations affected both boys' and girls' expectations for success in, as well as value of, a subject. In high school, girls were more likely than boys to choose advanced courses in subjects they were more confident about and which they valued; for example, English, foreign languages, and social sciences rather than math or physcial science courses. Eccles states, "We would predict exactly this result *unless* someone intervened and gave the females new information that would lead them to re-evaluate either their assessment of the importance of mathematics or their estimates of their mathematical aptitude, or both" (p. 9). Kahle's (1985) national study of high school biology teachers who were successful in encouraging girls as well as boys to continue in advanced science courses indicated that teachers could be effective intervention agents. In fact, students rated their teachers as second (to parents) in providing career information.⁵ As discussed in the next section, teachers and the educational environment can and do affect the retention of girls and women in science.

The Educational Environment

Flood (1988) notes that "schools are the first formal social settings whose charge it is to treat all the participants equitably . . . to maximize the range of choices and possibilities for all students" (p. 117). The behaviors and attitudes of teachers are especially important, and as Flood states, it is ". . . imperative that [the teachers] become models of equity and not just reflections of an inequitable system" (p. 117). The accessibility of teachers and their influence on students' course selections and career planning make schools a logical place to begin to change students' conceptions of what is appropriate and possible for them to achieve.

Since 1972, schools have had a legal responsibility to provide male and female students with equal encouragement, experiences, and opportunities in science and mathematics. Although often associated with athletic programs, Title IX mandates that:

No person in the United States shall, on the basis of sex, be excluded from participation in, be denied the benefits of, or be subjected to discrimination under any education program or activity receiving financial assistance (Educational Amendments of 1972: Title IX, in Hollander, 1978, p. 232).

Based on the recommendations of the President's Task Force on Women's Rights and Responsibilities, Title IX prohibits sex discrimination (with specific exceptions) in "recruiting and admissions, financial aid, textbooks and curriculum, housing facilities, counseling for careers, insurance and health care, single-sex groups and programs, extracurricular activities, and employment" (Hollander, 1978, p. 232). However, Title IX legislation has not had the impact on women's participation rates that was anticipated. As Dellinger (1979) writes:

For those who hoped Title IX would be a useful tool for seeking equity in American society, the [results] may be discouraging . . . Thus far no one could prove that Title IX has had the slightest impact on the familiar litany of statistics: one third of adult women lack a marketable job skill; one third of women employed are working in five low-paying occupations . . . (p. 31).

Although Title IX provides the legal basis for efforts to recruit and retain women in the scientific/technological pipeline, elimination of social/cultural barriers as well as modification of classroom activities are necessary for effective change.

A number of research studies have examined characteristics of teachers and instruction that create more equitable science classrooms, ones in which boys and girls participate and achieve equally. An examination of those characteristics will help educators change both their behaviors and the environments of their classrooms so that all students are involved in science. For example, Galton (1981) found that female high school students seem to prefer teaching styles which emphasized pupil initiated and maintained behavior directed towards experimental design and hypothesis testing. He hypothesized that the popularity of this teaching style was due to reduced chance of direct intervention by the teacher and "the fact of being encouraged to pursue an investigation independently is, in itself, an acknowledgement by an adult of the pupil's capacity" (p. 190). Furthermore, Melnick, Wheeler, and Gunning's (1986) qualitative study found that teachers who were successful in retaining girls in science courses used an individualized approach and incorporated behavioral objectives into their instruction. Some researchers have found that providing students with women scientists as role models and including non-sexist career information help to address several filters to the pipeline such as the masculine image of science and sex-role stereotyped social roles. For example, Smith and Erb (1986) examined the effect on attitudes of visits by women career models, information about women who made important contributions to science, and reading about young women who did science. Their results indicated that the use of role-models and careers information positively influenced the attitudes of both middle school boys and girls toward scientists, especially toward female scientists. Kahle (1985) and Melnick et al. (1986) also found that teachers who were successful in encouraging females to enroll in advanced science courses provided their students with non-sexist career information and women scientists as guest speakers. However, in Kahle's study (1985) the teachers also used non-sexist, teacher developed instructional materials as well as several distinct teaching techniques, including more frequent use of discussions, quizzes or tests, and laboratories.

However, instructional changes, designed around career information and topics in which girls had expressed interest, have not produced significant attitudinal or achievement changes (Kahle, 1987 and Kelly, et al., 1984). In addition, instructional changes that do not address differences in girls' and boys' skill and experience levels might actually be counter productive. Career information may foster girls' interest in science without helping them acquire the skills to actually do science. More effective retention of girls in science might be based on direct, skill-related activities that address the filters of experience, skills, and achievement levels. During the past decade, many instructional changes were tested as part of intervention projects which provided teachers with career information, lists of local women scientists, new curricula, and sample spatial activities, as well as with information about laboratory-based science, differential classroom participation patterns, and cooperative group learning. In a review of intervention programs that were designed to increase the participation of women in science and math, Stage et al. (1985) have found that nearly all successful programs placed a strong emphasis on academic skills. Research also indicates that activities such as hands-on, experiential laboratories and discussions on the applications of science and technology are important instructional strategies for involving girls as well as boys in science classes (Smail, 1987; Kahle, 1985). Finally, extracurricular science-related activities seem to promote the retention of females in the scientific/technological pipeline (Kahle, 1985, 1987; Kelly, et al. 1984).

SUMMARY

The concern about women's lack of full participation in scientific and technological fields is more than simply a concern about our gross national product or our ability to compete in international markets; it is a concern about the underutilization of talent, the underdevelopment of skill, and the lack of opportunity for over half of our population. The causes for women's under-participation are many and varied. As discussed, both biological and environmental factors have been proposed as the basis for women's lower achievement and/or interest levels in scientific areas. Emphasis on, and inadequate research about, biological factors has been particularly damaging to the entrance and success of girls and women in the scientific pipeline; for to aspire to a scientific career has meant that a girl was different and to succeed in one has meant that a woman was unique.

Today, it accepted that most differences in the achievement and retention patterns of young women and men in the scientific pipeline are related to environmental factors. Even within schools, those factors have proven to be effective filters. For example, fewer courses in math and science as well as fewer opportunities to use scientific equipment in science classes result in lower skill and achievement levels for girls, compared to boys. Since both skills and experiences are effective filters to the scientific and technological pipeline, changes within the education system can increase the participation levels of girls and women. Furthermore, as more girls enter the pipeline and as more women successfully exit it, both sex-role stereotyping of appropriate careers and of the image of science and scientists will change. Schools can be agents that transform rather than transmit societal roles and biases.

Both intervention projects and research studies have identified instructional activities, teaching behaviors, and institutional practices which work to deter girls and women in science. Other strategies, behaviors, and practices have been identified which motivate more diverse students to continue in the pipeline to scientific and technical careers. Our work, and that of others discussed in this chapter, indicates two things: first, that teachers can learn and use more equitable teaching strategies and, second, that the use of those strategies affects both the success and enrollment patterns of girls. The filters to the scientific pipeline are not impenetrable nor permanent. Increased entrance and enhance participation of girls and women in science courses and careers can be influenced by practices in schools and by the attitudes and behaviors of teachers.

NOTES

1. Vetter (1987) notes that in the U.S. bachelor degrees awarded to women have leveled off and that recent surveys of incoming freshman women indicate a continuation of that trend. She projects that in computer science and mathematics alone women graduates could drop from a high of 22,400 in 1986 to about 9,600 in 1989.
2. Chromosomes have been categorized as autosomes (number 1-22) and sex chromosomes (X and Y). In humans, the Y chromosome is responsible for determining the sex of the individual; individuals containing Y chromosomes are male. Normal males have one X and one Y chromosome, while normal females contain two X chromosomes. The X and Y chromosome function as a homologous pair, but the X chromosome is larger and contains more genes than does the Y. Males only have one copy of genes that are located only on the X chromosome, consequently recessive genes are not masked. In females, both X chromosomes must contain a recessive gene in order for it to be expressed.
3. Turner's syndrome is a genetic disorder in which individuals have only one sex chromosome, an X. Because there is only one sex chromosome, recessive genes would not be masked, and according to Gray's hypothesis, one would expect to find a greater number of individuals with superior spatial skills.
4. One biological explanation that was not tested by Gray is the possibility of incomplete dominance of the spatial inferiority gene.
5. High school students ranked teachers second, only to parents, among the adults with whom they had discussed careers. Counselors, professional in the field, and other family members were all ranked lower than teachers.

REFERENCES

Ben-Chiam, D., Lappan, G. & Hourang, R.T. (1988). The effect of instruction on spatial visualization skills of middle school girls and boys. *American Education Research Journal, 25,* 51-71.

Berryman, S.E. (1983). *Who will do science?* Washington, DC: The Rockefeller Foundation.

Carlson, N.R. (1986). *Physiology of Behavior* (3rd edition). Boston: Allyn and Bacon, Inc.

Chambers, D.W. (1983). Stereotypic images of the scientist: The draw-a-scientist test. *Science Education 67,* 255-265.

Czujko, R. & Bernstein, D. (1989, December). *Who takes science?* A report on student coursework in high school science and mathematics. (AIP Publication Number: R-345). New York: American Institute of Physics.

Dellinger, A.M. (1979). Title IX: The first six years. In M.A. McGhehey (Ed.), *Contemporary legal issues in education* (pp. 19-31). Topeka, KS: National Organization on Legal Problems in Education.

Douglass, C.B. (1985). Discrepancies between men and women in science: Results of a national survey of science educators. In J.B. Kahle (Ed.). *Women in Science* (pp. 148-168). Philadelphia: The Falmer Press.

Eagly, A. H. (1987). *Sex differences in social behavior: A social-role interpretation.* Hillsdale, NJ: Lawrence Erlbaum Associates, Publishers.

Eccles, J.S. (January, 1989). *Bringing young women to math and science.* Paper presented at the 155th Annual Meeting of the American Association for the Advancement of Science, San Franscisco.

Flood, C. (1988). Stereotyping and classroom interactions. In A.O. Carelli (Ed.), *Sex Equity in Education* (pp. 109-124). Springfield, IL: Charles C. Thomas Publisher.

Galton, M. (1981). Differential treatment of boy and girl pupils during science lessons. In A. Kelly (Ed.), *The Missing Half: Girls and Science Education* (pp. 180-191). Manchester: Manchester University Press.

Garratt, L. (1986). Gender differences in relation to science choice at A-level. *Educational Review, 38*(1), 67-76.

Graham, M.F. (1978). *Sex differences in science attrition.* Unpublished doctoral dissertation, State University of New York at Stony Brook, New York.

Gray, J.A. (1981). A biological basis for the sex differences in achievement in science? In A. Kelly (Ed.), *The Missing Half: Girls and Science Education* pp.42-58. Manchester: Manchester University Press.

Hartl, D.L. (1988) *A primer of population genetics* (Second edition). Sunderland, MA: Sinauer Associates, Inc.

Hildebrand, G. (1987). *Girls and the career relevance of science: A case study.* Unpublished master's thesis, Monash University, Melbourne, Victoria.

Hite, L.M. (1983). *A study of doctoral students compared by gender and type of field of study on factors of role congruence, perceived support from faculty, and perceived support from peers.* Unpublished doctoral thesis, Purdue University, West Lafayette, IN.

Hollander, P.A. (1978). *Legal handbook for educators.* Boulder, CO: Westview Press.

Hornig, L.S. (1988). Gender & science. In J.B. Kahle, J.Z. Daniels & J. Harding (Eds.), *Proceedings of 4th International Conference of Girls & Science & Technology.* West Lafayette, IN: Purdue University Press.

Humrich, E. (1987). Sex and science achievement. In *The second IEA science study - US,* revised edition. New York: Teachers College, Columbia University.

Hyde, J.S. (1986). Girls and science-strategies for change. *Pivot 3,* 28-30.

Hyde, J.S. (1981). How large are cognitive gender differences? *American Psychology, 36,* 892-901.

Iliams, C. (1985). Early school experiences may limit participation of women in science. In *Contributions to the third GASAT conference.* London: Chelsea College, University of London.

Kahle, J.B. (1987). SCORES: A project for change? *International Journal of Science Education, 9,* 325-333.

Kahle, J.B. (1985). Retention of girls in science: Case studies of secondary teachers. In J.B. Kahle (Ed.), *Women in Science: A report from the field* (pp. 49-76 & 193-229). London: The Falmer Press.

Kahle, J.B. & Lakes, M.K. (1983). The myth of equality in science classrooms. *Journal of Research in Science Teaching, 20,* 131-140.

Kass, H. (in press). The gender issue. *Science Education in Canada.*

Kelly, A. (1985). The construction of masculine science. *British Journal of Sociology of Education, 6,* 133-154.

Kelly, A. (Ed.) (1981). *The Missing Half: Girls and Science Education.* Manchester: Manchester University Press.

Kelly, A., Whyte, J., & Smail, B. (1984). *Girls into Science and Technology: Final report.* University of Manchester: Department of Sociology.

Leinhardt, G., Seewald, A.M. & Engel, M. (1979). Learning what's taught: Sex differences in instruction. *Journal of Educational Psychology, 71,* 432-439.

Linn, M.C. & Hyde, J.S. (1989). Gender, mathematics, and science. *Educational Researcher, 18*(8), 17-19, 22-27.

Linn, M.C.; DeBenedictis, T.; Delucchi, K.; Harris, A.; & Stage, E. (1987). Gender differences in national assessment of education progress science items: What does "I don't know" really mean? *Journal of Research in Science Teaching, 14*(3), 267-278.

Linn, M.C., & Peterson, A.C. (1985). Facts and assumptions about the nature of sex differences. In S.S. Klein (Ed.), *Handbook for achieving sex equity through education* (pp. 53-77). Baltimore, MD: Johns Hopkins University Press.

Linn, M.C., & Peterson, A.C. (1983). *Emergence and characterization of gender differences in spatial ability: A meta-analysis.* Berkeley: University of California, Adolescent Reasoning Project. In Linn, M.C., & Peterson, A.C. (1985). Facts and assumptions about the nature of sex differences. In S.S. Klein (Ed.), *Handbook for achieving sex equity through education* (pp. 53-77). Baltimore, MD: Johns Hopkins University Press.

Maccoby, M.E., & Jacklin, C.N. (1974). *The psychology of sex differences.* Palo Alto, CA: Stanford University Press.

Marieb, E.N. (1989). *Human anatomy and physiology* (pp. 928-929, 940-942). Redwood City, CA: The Benjamin/Cummings Publishing Company, Inc.

Mason, C.; Kahle, J.B.; & Gardner, A. (In Press). Draw-a-scientist test: Future implications. *School Science and Mathematics.*

Mead, M. & Metreaux, R. (1957). The image of the scientist among high school children. *Science, 126,* 384-389.

Melnick, S.L., Wheeler, C.W. & Gunnings, B.B. (1986). Can science teachers promote gender equity in their classrooms? How two teachers do it. *Journal of Educational Equity and Leadership,* 6, 5-25.

Morse, L.W., & Handley, H.M. (1985). Listening to adolescents: Gender differences in science classrooms. In L.C. Wilkenson and C.B. Mannett (Eds.), *Gender Influences in Classroom Interaction* pp. 37-56. Madison, WI: Academic Press.

National Science Foundation (NSF). (1990, January). *Women and minorities in science and engineering.* (NSF #90-301). Washington, DC: National Science Foundation.

Newcomb, N. (1982). Sex-related differences in spatial ability: Problems and current approaches. In M. Potegal (Ed.), *Spatial orientation: Developmental and biological bases* (pp. 223-250). New York: Academic Press.

Pallas, A.M., & Alexander, K.L. (1983). Sex differences in quantitative SAT performance: New evidence on the differential coursework hypothesis. *American Educational Research Journal, 20,* 165-182.

Parker, L.H., & Offer, J. (1986). Achievement certificate science: Sex differences in grades 1972-85. Paper presented at the Annual Conference of the Western Australia Science Education Association. Perth, In Parker, L.H. (1987). The choice point: A critical event in the science education of girls and boys. In B.J. Fraser and G.J. Giddings (Eds.), *Gender issues in science education.* Perth, Australia: Curtin University of Technology.

Plomin, R., & Foch, T.T. (1982). Sex differences and individual differences. *Child Development, 52,* 383-385.

Pool, R. (1990). Who will do science in the 1990s? *Science,* 248, 433-435.

Rennie, L.J. (1986). The image of a scientist: Perceptions of preservice teachers. Unpublished, paper University of Western Australia, Perth, WA.

Sadker, D. & Sadker, M. (1985). Is the ok classroom, ok? *Phi Delta Kappan, 55,* 358-361.

Schibeci, R.A. (1986). Images of science and scientists and science education. *Science Education, 70.* 139-149.

Sherman, J. (1978). *Sex-related cognitive differences: An essay on theory and evidence.* Springfield, IL: Charles C. Thomas. In Linn, M.C., & Peterson, A.C. (1985). Facts and assumptions about the nature of sex differences. In S.S. Klein (Ed.), *Handbook for achieving sex equity through education* (pp. 53-77). Baltimore, MD: Johns Hopkins University Press.

Sjoberg, S., & Imsen, G. (1988). Gender and science education: I. In P. Fensham (Ed.), *Development and dilemmas in science education* (pp. 218-248). London: The Falmer Press.

Smail, B. (1987). Organizing the curriculum to fit girls' interests. In A. Kelly (Ed.), *Science for girls?* (pp. 80-88). Milton Keynes, UK: Open University Press.

Smith, W., & Erb, L. (1986). Effect of women science career role models on early adolescents' attitudes toward science and women in science. *Journal of Research in Science Education, 23,* 667-676.

Spear, M.G. (1984). Sex bias in science teachers' ratings of work and pupil characteristics. *European Journal of Science Education, 6,* 369-377.

Stage, E.K., Kreinberg, N., Eccles (Parsons), J., Becker, J.R. (1985). Increasing the participation and achievement of girls and women in mathematics, science, and engineering. In S.S. Klein (Ed.), *Handbook for achieving sex equity through education* (pp. 237-268). Baltimore, MD: Johns Hopkins University Press.

The Task Force on Women, Minorities, and the Handicapped in Science and Technology. (1989, December). *Changing America: The new face of science and engineering* (Final report). Washington, DC: Author.

Tobin, K., & Gallagher, J. (1987). The role of target students in the science classroom. *Journal of Research in Science Teaching, 24,* 61-75.

Tobin, K., & Garnett, P. (1987). Gender related differences in science activities. *Science Education, 71.* 91-103.

Vetter, B.M. (1987). Women's progress. *MOSIAC, 18*(1), 2-9.

Whyte, J. (1986). *Girls into science and technology.* London: Routledge & Kegan Paul.

Widnall, S.E. (1988). *Voices from the pipeline.* Paper presented at the 154th National Meeting of the American Association for the Advancement of Science, Boston, February 11-15.

Science Education in the United States: Issues, Crises and Priorities. Edited by S. K. Majumdar, L. M. Rosenfeld, P. A. Rubba, E. W. Miller and R. F. Schmalz. ©1991, The Pennsylvania Academy of Science.

Chapter Forty-Two

FACTORS INFLUENCING THE UNDERREPRESENTATION OF WOMEN IN SCIENCE

CAROLYN F. MATHUR

Department of Biology
York College of Pennsylvania
York, PA 17403

More men than women become scientists in the United States today. This, coupled with decreased interest in science among America's youth, is causing government officials and others to express concern over the long term implications of a potential shortage of scientists. We need more scientists to meet the demands of our increasingly complex technological society. The disproportionately low numbers of women compared to men in many areas of science contributes to the magnitude of this problem. One way of filling this anticipated shortage of scientists is to specifically recruit more women into the sciences.

Another reason for attracting more women into science is to provide both women and men with equal opportunities and encouragement to study science and to enjoy productive scientific careers. Women have not always experienced such opportunities in the past. The purpose of this chapter is to describe the underrepresentation of women in science and to examine some of the factors associated with this problem.

DEFINING THE PROBLEM

Although women comprise 44% of all employees and 50% of the population, they make up only 15% of the scientists and engineers in this country.[1] In private industry only 9% of all the scientists are women. In education they make up approximately 23% of the scientists.[2] However, we are making some progress in narrowing the scientific gender gap. Numerous studies, articles, books, programs and courses concerning women and science have appeared recently, indicating greater awareness of the problem. In the last twenty five years the number of diplomas awarded to women in certain scientific fields increased. Bachelors degrees in biology earned by women increased from 28% in 1966 to 48% in 1986. In enginnering over the same period, the gain was from 0.4% to 13%.[3] We need to maintain this momentum in order to attract more women into science.

With this in mind, science educators should be aware of the special problems associated with women and science. Why do so few women become scientists? What could happen during the education of a woman to steer her away from science? What can be done to solve this problem? Widespread recognition of the problem is an important first step. Teachers are influential in a student's decision to become a scientist. Although education is not the only factor related to this issue, it is an important component of a complex system that results in few women in science.

HISTORICAL PERSPECTIVE

A brief historical background is helpful to give perspective concerning where women have been scientifically in the recent past. One hundred years ago women's position in science was very tenuous. Vera Ruben, a renowned astronomer and one of only 57 women elected to the over 2500-member National Academy of Sciences, describes early female astronomers as mainly underpaid, skilled technicians.[4] Lab directors realized that women could carefully and cheaply accomplish the tedious detail work that would involve higher labor costs if men were hired. Typically, women's accomplishments were unrecognized, and their work not given adequate financial or institutional support. Most women could study science only at women's colleges. As scientists, women faced limited opportunities for permanent jobs while simultaneously having a husband and family. They faced widespread beliefs that women were mentally and socially inferior to men and that science was an inappropriate career for women.[5] Women with scientific ambitions were considered abnormal. Strong European and American beliefs asserted that women belong "in the home" and that they should not challenge men as leaders of the family. Society encouraged women to participate in perceived "masculine" endeavors only during times of pressing economic need, such as occurred in World War II. This social support for working women was withdrawn later when jobs became more scarce and preference was given to men. Many of these attitudes still persist although they are not as widely held today as they were previously. Such attitudes influence women not to pursue a scientific career and contribute to the lower numbers of women in science.

FEMALE SCIENTIST

Another negative influence on a young girl or woman considering a scientific career is the misrepresentation of the female scientists's experience. A young woman's decision to become a scientist is strongly influenced by the barrage of information she receives through history books, television and other media. Popular history books and science textbooks make little mention of the scientific contributions of women.[6] Women's historical or current roles in science are not adequately covered. The problem of accurate biographical portrayals of women's experiences, not just in science but in general, is examined in a recent book by Carolyn Hielbrum.[7] She describes "the problem of living a life not scripted by others", and although not directly related to science, the dearth of realistic biographies of women is apparent in the sciences as in other areas. Women's lives are often presented according to the current societal view of what a woman should be instead of describing her real, human experiences. Men's views of women as objects, parts of nature, helpers or sources of inspiration to men tend to dominate descriptions of outstanding women's lives.[8] These narrowly-defined feminine role models may limit a young woman's perception of what opportunities are truly available to her.

DEFINING SCIENCE

The problems caused by these distorted and limiting presentations of women's capabilities are compounded by another common misunderstanding of exactly what is a scientist and what does he/she do? The non-scientist often has a narrow view of the typical scientist and how he/she conducts "science". They view a scientist as coldly analytical, objective, abstract, masculine, detached, unsociable, extremely intelligent and formidable! Such unbalanced and unrealistic representations are enough to discourage both young men and women alike from pursuing science as a career. The non-scientist's view puts little emphasis on the subjective, creative and intuitive nature of science, or the feelings of wonder, joy and elegance that one can experience as a scientist. Open cooperation, communication and teamwork are essential in promoting scientific advances. The diversity of scientific fields and levels of work available offers opportunities for many different styles of thinking and performing. Young people can choose from a wide range of careers in science and can contribute to science in a variety of ways within each field.

Difficulty in defining science occurs not only among non-scientists, but also among educators and professionals who study science. The "science" that we as scientists experience is often quite different from how we see it described in children's science texts or even in sociological journals that report studies of women scientists. The latter has led to some disagreement between gender theorists studying the women and science problem and some women scientists themselves who, in addition to their scientific pursuits, are expressing the experience of being a woman scientist from their perspective.[9] Two different schools of thought concerning women and science are emerging. One includes a more conventional, conservative view that men and women are not innately different and that they possess the same capabilities and range of cognitive styles over a very broad spectrum of possibilities. The low number of women in science, then, is caused primarily by social, cultural and historical forces. The second theory stresses the differences between men's and women's capabilities or cognitive styles. Women and men have different approaches toward defining and solving problems.[10] Along with this is the belief by some that science needs to change, that it is too narrow or "masculine" in its focus, and that only by changing and accommodating the "feminine" perspective can more women participate in science. Science will be better off if it broadens its basic tenets to allow for the femine approach.[11] Numerous articles express views intermediate between these two, but central to the controversy is the lack of a clear understanding of the female life experience and a clear definition of science.

A recent study[12] examined the concepts of female and science with a different emphasis. Instead of viewing women and science in narrowly defined contexts, the authors describe the different cognitive styles present in researchers, regardless of gender, and the different styles observed in the various scientific disciplines. There is a continuum of styles present in the researchers themselves and in the type of science they pursue. These differences, coupled with societal-dominated sex-role definitions, can result in mismatches when women with particular cognitive skills enter an inappropriate field of science.

Science styles range from problem-finding, unconventional, open fields such as sociology or psychology, to problem-solving, conventional, more narrowly defined fields such as chemistry. Since science is a more unconventional career for a female, the unconventional cognitive style may be more common among women scientists. The higher proportion of women in these fields supports this observation. Women with conservative styles, best suited for problem-solving fields such as chemistry, would be unlikely to enter the science field at all due to their conservative personal style which would bend to society's pressures and channel them into more socially acceptable careers. These ideas incorporate a wide range of behaviors in men, women and scientists and describe how these universal behaviors can be influenced by societal pressures to result in the under-representation of women in science.

INNATE GENDER DIFFERENCES

Apart from the problem of finding a clear definition of femininity and science, many other factors are frequently cited as contributing to the marginality of women in science. A common recurrent debate, which I alluded to previously, is the issue of "nature vs. nurture" in explaining male-female performance differences. This concept has fascinated people for quite some time. In the late nineteenth century, scientists attempted unsuccessfully to explain women's "inferiority" by comparing total brain volumes or frontal lobe development in men and women.[5] Explanations offering innate or natural causes for women's inferiority occurred commonly throughout the twentieth century. It is against such a background that many resist the current theories that try to relate differences in male and female brain anatomy to possible behavioral differences. Research by Witelson describes anatomical differences in the corpus callosum - the area of the brain that connects right and left hemispheres.[13] Since previous studies indicate there may be a relationship between hemispheres and different types of thinking, these findings are intriguing. However, a variety of factors can influence brain anatomy and physiology. Hormonal and biochemical effects, as well as individual experiences themselves, are important in shaping each individual's brain. Exactly how such differences relate to subsequent behaviors are highly speculative. Since males and females have different social experiences from an early age, these differences in brain anatomy are not helpful in explaining some of the disparities that exist. Many in the field are against such studies when they are conducted in the context of explaining differences in male and female behaviors.[14] These findings tend to become sensationalized and misrepresented by the press and thus detract from other, more meaningful considerations. In spite of these concerns, such anatomical differences are of academic interest, and studies in this area should continue as long as the limits of their relatedness to the issue of women's marginality is recognized. There are obviously many differences between men and women that we are just beginning to understand. Acquiring more knowledge in this area can only be beneficial.

TESTING DIFFERENCES

Of particular importance to educators are the testing differentials that frequently occur between the sexes. Some studies of test results in math and science indicate equality between the sexes until junior high, at which point boys start to perform better than girls in a variety of standardized achievement and aptitude tests. [3,15] The magnitude of the differences increases when comparing tests of higher difficulty or graduate school entrance tests. For more than 20 years in the SAT tests, males have scored about 50 points higher than females.[3] In spite of scoring lower on these entrance exams, women tend to get better grades than men in the upper division schools.[16] Thus these exams are not good relative indicators of performance when comparing men and women in college. Lower test scores not only reduce women's self-esteem, but they also decrease women's opportunities for scholarships or acceptance at more prestigious schools. Some suggest that the SATs are biased against women, although the Educational Testing Service that administers the exam states that they take special precautions to avoid gender-biased questions.[16]

A study by Hanna[17] of eighth grade children from 20 different countries compared their mathematical abilities in 5 different areas. Mathematics ability is correlated with success in science. She found slight gender differences in 2 of the 5 areas tested, but the variability among countries was greater than any sex differences. These results suggest that gender differences in test performance are not due to inherent differences between the sexes but rather are due to environmental, social or educational differences, such as exist between countries.

Another study compared incoming freshmen men and women of matched ability with respect to their retention in science after one year.[18] All the students declared science as their major as they entered college, but by the end of the freshmen year, 69% of the men and 50% of the women remained as science majors. The factors that were important in maintaining the science major differed in the two groups. For women, having educated parents, a strong desire for control, high math SATs and a need to interact with others were significant for retention. For men, high grades in science courses and a strong commitment to the major before entering school were important. Thus male and female science students approached their commitment to science in very different ways. We need to recognize and better understand these differences in order to attract more students into the sciences. The underrepresentation of women in science will not change if such differential rates of attrition continue.

SOCIAL FACTORS

The impact of the environment is probably most important in influencing whether or not a young woman will choose science as her career. There are a host of factors that detract girls away from science. It is in these areas that educators can have the greatest impact since we are usually influential in a girl's life at a time when these critical decisions are made.

Negative social pressures exert themselves in subtle ways. An example of the subtlety is illustrated by the concept of "microinequities", which are brief, negative social encounters between individuals that appear insignificant at the time they occur.[19] However, frequent exposure to these brief encounters, such as many women experience over the years, tends to have a powerful, cumulative, negative effect on attracting, training and retaining women in science. These differential influences start from infancy onward, when we stress competition, achievement and problem solving in boys, and caretaking, commitment and supportive behavior in girls.[20]

Exposure to positive role models or supportive adults early in childhood is important in attracting women to science and mathematics.[21] Yet some suggest that teachers prefer boys over girls, as indicated by the observation "that girls get called on less often in class and are interrupted more often, and get more diffuse responses and less eye contact from the teachers".[3] In one study young women's conversations were not as dominant or significant as men's, with men interrupting women 96% of the time.[22] Not getting encouragement to speak or express opinions indicates lack of support and results in lost self-esteem - two recurrent themes frequently described in explaining women's reduced participation in science. Possibly as a result of this, women have a greater need than men to work in collaboration with others, or as a part of a team.

Teenage peers exert negative social pressure on girls who are interested in or have exceptional abilities in math and science. In addition, young women believe that they must be excellent to be science majors, and usually only women with very high math SAT scores choose science careers. Women seem to have unrealistically low estimations of their science and math capabilities. This is not true for young men who enter the sciences possessing varying abilities in math.[17] The high female attrition rate in the sciences is also related to the impression by some that women must work harder than men as science majors.

FEMININITY AND SCIENCE

Another phenomenon occurring in women who are scientists is their dissociation from femininity in the context of their scientific careers. A woman may say "I don't think of myself as a woman, I think of myself as a scientist". Women appear forced to deny that they are women in order to survive in science. An equivalent statement from a male scientist would seem inappropriate, yet in the context of today's social climate concerning women's roles, the statement from a woman is understandable.

Teenage women considering science as a career are acutely aware of this social dilemma. They are justifiably concerned about their femininity and how they can manage a career in science. Marriage and family are important to most women. Some scientists and educators view women as less committed than men to their career because of constraints imposed by family responsibilities and their spouse. The most significant barrier to women's advancement is not their families per se, however, but rather their limited mobility in relocating to better positions, etc.[23] This situation is less common in younger couples as more and more women are relocating families to advance their careers.

Men and women seldom openly admit to views that women are subordinate to men and that they belong primarily in their sex role rather than in a professional role. Such views are common, however, causing many women to reject science as a career, or to decrease their personal aspirations for themselves as their career progresses. We need to "help women feel that being a scientist is appropriate, normal and not unfeminine".[17] It is not necessary to deny femininity to be a scientist and thus be forced to make a choice between the two.

PROFESSIONAL OBSTACLES

Women who remain in science in spite of all these adversities may face more difficulties as they enter their college and graduate training and subsequent careers. Changing career paths for women in male dominated fields have been documented.[24] The attrition rate for women in the sciences is very high. Women drop out or lower their expectations not only during the education years but also at the career or professional level. Women are "channeled" into women's work. They may start their careers in a male-dominated area such as research, and then change course over time toward a more female-dominant area such as teaching.[12] Women give up their goals in order not to inconvenience others. Possible contributing factors to this channeling include social sex stereotyping, family pressures, lack of mobility, discrimination, lack of professional recognition, poor self image, lack of cooperation from others and financial constraints. Therefore, attracting women into science is not enough. Even if the educational system manages to provide society with a higher proportion of female scientiests, social changes in other areas are needed before women can achieve true parity with men in the sciences.

SOLUTIONS

Awareness of this gap between men and women scientists inspired an increased number of programs and associations to promote female participation in science and math and to encourage women to advance in their scientific careers. Examples of programs for girls in mathematics are EQUALS at the University of California, The Futures Unlimited Project at Rutgers University and Women and Mathematics (WAM) program of the Mathematics Association of America.[20] For women already in the field there is the Project on the Status and Education of Women of the Association of American Colleges, the forgivable loan program for doctorates for women and minorities at California State University in Los Angeles, the visiting professorship grants from the National Science Foundation and the American Women in Science organization, to name a few.

ROLE OF EDUCATORS

Educators in science can personally promote mathematics and science as careers for interested girls and women. Early intervention is critical, so teachers at the elementary and junior high level are important in attracting women into these fields. Young women and girls may require extra encouragement because of the many problems already described.

Educators should present science to students in a more realistic and doable setting in order to attract more women and men into the field. Students are discouraged by the unrealistically high standards they perceive as necessary for entry into the sciences. They may envision science as existing in a sterile, coldly objective vacuum instead of as the exciting but frequently fallible human endeavor that it truly is. As educators we should stress the importance of communication, networking and human interaction so vital to the growth and dissemination of information in science.[18] Women would respond favorably to this type of presentation and their increased participation in science would benefit the sciences since women tend to do well in these areas.

Europeans also are moving towards formally promoting the involvement of women in science. Some foreign scientists believe their young women have had greater social permission to pursue science in the past.[25] Others disagree, stating that women's dominant role in domestic areas creates a restrictive environment for scientific women in other countries similar to what we experience here.[26]

SUMMARY

Women are underrepresented in the sciences when compared to their numbers in other fields. This disparity needs to be recognized by society as a whole, and in particular by parents, teachers and professionals at all levels. Factors contributing to this problem include narrowly-defined roles of women in society, and unrealistic definitions of science or descriptions of what scientists actually do. These combine to create an exceptional strain on women trying to visualize themselves realistically as a female scientist.

Other highly controversial factors that were descibed through the years to explain women's lower status include actual physiological or innate differences between the sexes. Explored frequently in the past, this idea is still under scrutiny now although it is scientifically difficult to relate any gender-based physical differences in the brain to human intellectual performance. Less controversial factors include women's social environment from infancy through adulthood, including lack of role models during childhood or lack of strong support for a career in science from peers or adults.

Microinequities, or frequent exposure to negative comments or pressures from a variety of sources, tends to have a cumulative effect on a women's self-esteem and sense of ability to accomplish anything in the sciences. Lower test scores for women, lower social acceptability for women in the sciences, lower salaries compared to men, lack of recognition of women's scientific achievements both historically and presently and unrealistically low expectations by women of what they can achieve in science all contribute to fewer women in science.

Despite these problems women are entering the sciences in greater numbers in recent years. Educators can play an important role in accelerating the entrance of young women into science since they are influential in young women's lives at a time when career decisions are made. Widespread awareness of particular problems associated with women in the sciences will help decrease the disparity that exists.

REFERENCES

1. Sims, C. 1990 Jan 7. The Overlooked. *The New York Times.* Sect.4A: EDUC 23.
2. Vetter, B. 1986. The Last Two Decades. *Science 86.* 7(6):62.
3. Berger, J. 1989 Aug 6. All in the Game. *The New York Times.* Sect.4A: EDUC 23.
4. Rubin, V. 1986. Women's Work. *Science 86.* 7(6):58-65.
5. Cole, J.R. 1979. *Fair Science: Women in the Scientific Community.* Columbia University Press.
6. Kyle, S.A.P. 1980. Curriculum resources: women in science, medicine and technology. *Dissertation Abstracts.* 41(10-A):4269.
7. Hielbrum, C.G. 1988. *Writing a Woman's Life.* Ballantine Books, New York.
8. Hine, H. 1981. Women and science: fitting men to think about nature. *Int. J. Women's Studies.* 4:369-72.
9. Ruskai, M. 1990 Mar 5. Why women are discouraged from becoming scientists. *The Scientist.* p. 17-19.
10. Gilligan, C. 1982. *In a Different Voice.* Harvard University Press, Cambridge, MA.
11. Women and Science. 1989. *Hood on the Issues.* (1): Hood College, Frederick, MD.
12. Bar-Haim, G., and J.M. Wilkes. 1989. A cognitive interpretation of the marginality and underrepresentation of women in science. *J. Higher Ed.* 60(4):371-387.
13. Kolata, G. 1989 Aug 6. Mind Blowing? *The New York Times.* Sect.4A: EDUC 25.
14. Epstein, C.F. 1989. *Deceptive Distinctions: Sex, Gender and the Social Order.* Yale University Press, New Haven, CT.
15. Benbow, C.P. and J.C. Stanley. 1983. Sex differences in mathematical reasoning ability: more facts. *Science.* 222:1029-1031.
16. Mandula, B. 1990. Is the SAT unfair to women? *Assoc. Women Sci. Newsletter.* 19(1):8-10.
17. Hanna, G. 1989. Mathematics achievement of girls and boys in grade eight:

results from twenty countries. *Educ. Stud. Math.* 20:225-232.

18. Ware, N.C., N.A. Steckler and J. Lesermen. 1985. Undergraduate women: who chooses a science major? *J. Higher Ed.* 56(1):73-84.

19. Project on the status and education of women. 1982. The classroom climate: a chilly one for women? Association of American Colleges, Washington, D.C.

20. Church, M. 1989 May 14. Bringing women into science. *The News and Observer*, Raleight, NC, p. 7D.

21. Fabricant, M., S. Svitak and P.D. Kinschaft. 1990 Feb. Why Women succeed in mathematics. *Mathematics Teacher.* 83:150-154.

22. Pfeiffer, J. 1985. Girl talk-boy talk. *Science 85.* 6(1):58-63.

23. Marwell, G., R. Rosenfeld and S. Spilerman. 1979. Geographic constraint's on women's careers in academia. *Science.* 205:1225-1231.

24. Widnall, S. 1988. AAAS Presidential Lecture: Voices from the Pipeline. *Science.* 241:1740-1745.

25. Yost, E. 1959. *Women of Modern Science.* Dodd, Mead and Co., New York, p. vii.

26. Haley-Oliphant, A.E. 1985. International perspectives on the status and role of women in science. pp. 169-192. In: J.B. Kahle (Ed>) *Women in Science.* The Falmer Press, Philadelphia.

Science Education in the United States: Issues, Crises and Priorities. Edited by S. K. Majumdar, L. M. Rosenfeld, P. A. Rubba, E. W. Miller and R. F. Schmalz. ©1991, The Pennsylvania Academy of Science.

Chapter Forty-Three

MINORITIES AND SCIENCE EDUCATION

MELVIN A. JOHNSON
Dean, College of Arts and Sciences
Central State University
Wilberforce, OH 45384

INTRODUCTION

As we move closer to the twenty-first century, a strong sense of accomplishment is shared by all who have initiated, witnessed or somehow benefitted from the achievements our nation has reached in areas of social, economic, and technological advances. As a nation we have become a super-power and have maintained a "cutting-edge" status in our progress towards scientific and technological excellence. However, in spite of our technological and scientific advances, we would be remiss if we failed to address the critical and frequently asked question: Why has the U.S. ignored, compromised, or otherwise neglected its minority citizenship in an unwise effort to maintain the status quo in the fields of science and technology?

The answer, it appears, is as complex as the question itself. None-the-less, the question demands the utmost consideration and most cautious reply. Interestingly enough, this question, perhaps in variation, has been asked many times. Also, interesting as well are some of the responses, or in many cases, assessments, which have been offered in addressing the concern over minority participation and inclusion in the scientific and technology fields. The aim of this chapter is to 1) address the aforementioned question from an educational perspective, tempered with a historical backdrop 2) review some of the existing programs designed specifically to offset the minority shortfall in the scientific and technological fields and 3) to propose recommendations that may help offset this shortfall.

HISTORY

On March 17, 1981, the author, who at that time was president of the National Institute of Science, presented the keynote address at the thirty-eighth annual joint meeting of the National Institute of Science and Beta Kappa Chi Honorary Scientific Society. The title of the speech was "The Black Scientist-Past, Present, and Future" and the two aforementioned organizations to whom it was delivered are composed of minority scientists and students (predominantly black) from historically black colleges and universities throughout the country.

In that address, the author pointed out that "Negroes were invited to America in 1619, but it was not until 1876 that Edward Bouchet, a black American, received the first Ph.D. in the sciences (physics)". The achievements of other black scientists of the past were mentioned; namely, Benjamin Banneker, George Washington Carver, Ernest Just, Percy Julian, and Hubert Crouch. Also the exploits of several black physicians of the past were outlined—William A. Hinton, Charles Drew and Daniel Hale Williams.

Although the first American black to earn a Ph.D. degree in the natural sciences received it in 1876, by 1900, there was only one other American black who had earned a doctorate in science. By 1930, there were only 13; by 1943, only 119 in all natural science fields! However, tremendous strides have been made in terms of an increase in the percentage of Blacks in all natural sciences as well as an increase in the variety and levels of contribution.

It is a fact that historically black colleges and universities (HBCUs) still provide the greatest opportunities for black students pursuing a degree in the natural sciences. This is certainly ironic—even though these institutuions had 32 percent of the black student enrollment in the U.S. in 1970, they received less than 1 percent of the funds for the improvement of college science programs.[2] In spite of these negative circumstances, better educational opportunities have been developing for the black students in the natural sciences in the traditionally black colleges.

Finally, as an over-all review, *The Negro Education Review*[3] indicates that prior to 1950, the greatest percentage of research output by Black scientists was made by holders of the M.D. degree. Many of these researches also held the Ph.D., Sc. D., Dr. P.H., or Pharm, D. degree.

Jay[1] estimated that approximately 850 Blacks had earned the doctorate in the natural sciences through 1972 (Anthropology and Psychology not included). An examination of another survey done by the Ford Foundation in 1968 and 1969, indicated that only 0.8 percent of 37,456 Ph.D. degrees awarded by 63 American graduate schools of arts and sciences were received by blacks.

It has appeared that over the years black students have been poorly prepared in the natural sciences; however, efforts are being made to correct this deficit.

*A Brief Survey of Science and Mathematics Preparation
at the Elementary and Secondary School Levels*

What are the facts and statistical data concerning the education of young black students in the sciences and mathematics? Since the late 1960's the National Assessment of Educational Progress has conducted surveys of student proficiencies in several content areas, namely science and mathematics, at the 9-, 13-, 17 year old age levels. Figure 1 shows a comparison of the results of scores in science between black and white students at the three age levels[5].

In the science skills assessment, the overall means for *nine-year-old* blacks in 1986 was about 36 points lower than whites, 196.2 versus 231.9. Since 1978, the mean for blacks has risen from 174.9 closing what was then a 55 point difference.

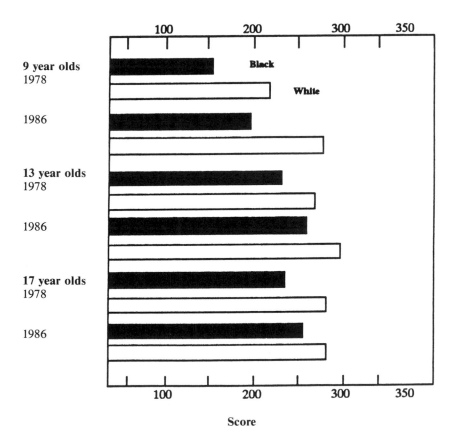

FIGURE 1. Science Assessment Scores for black and white students, by age level: 1978 and 1986.
Note: Score is based upon a 0-500-point scale.

In *thirteen-year-old* blacks, differences in scores have also narrowed. In 1978 blacks showed means 43 points lower than those of whites; in 1986 the difference was 33 points. For the most recent year, their respective scores were 226.1 and 259.2.

The *seventeen-year-old* groups show the largest variance in means for blacks and whites. In 1986, blacks scored 252.8, and this was more than 45 points lower than whites. In 1978, the difference was more than 57 points.

Figure 2 shows a comparison of the results of scores in mathematics between black and white students at the three age levels. In mathematics skills assessment, blacks scored below whites in all three age levels. However, blacks have improved in the last several years and the difference in scores from that of the white students is markedly less.

The *nine-year old group,* in 1986 showed a mean difference of 25 points between blacks and whites, 201.6 versus 226.9. This difference has diminished since

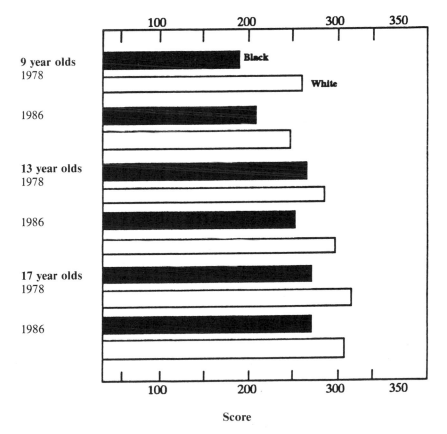

FIGURE 2. Mathematics Assessment Scores for black and white students, by age level: 1978 and 1986. Note: Score is based upon a 0-500-point scale.

the School of Science in the amount of $4,000. Also, students may qualify for scholarships, which are based on merit not need.

Evaluations are conducted by questionnaires given before and after the program. There is also follow-up communication with students and their high schools to monitor college plans. Profiles of student applicants are compared with successful Purdue minority graduates. Their progress at Purdue will be tracked along with a control group of minority science students who did not participate in the summer program.

The Conference on Pre-College Education of Minorities in Science and Engineering at the New Jersey Institute of Technology

National reports underscore the need to initiate action plans to deal with the lack of minority and female representation in the science and engineering fields. In addressing the aforementioned concerns, a "Special Incentive Program in Science and Computer Science for Selected High School Students" was initiated at Ramapo College of New Jersey. The program was designed to counter such problems as those mentioned above. It functioned as a mini-magnet school for the sciences[8].

The program, initiated in 1984, sponsored a four-day course format. The offerings covered a myriad of science including: chemistry, biology, biochemistry, physics, psychology, math and computer sciences.

Financial backing for the programs was made possible through grants by the New Jersey Department of Higher Education. Approximately $98,000 was awarded for the program from 1984 through 1987. The program's success was evidenced by the increasing number of participants from 60 students in 1984, of whom 27% were minorities and 44% were women, to 104 students in 1986, of whom 33% were minorities and 51% were women, to 107 students in 1987, of whom 38.3% were minorities and 55.1% were women.

The program's objectives were made clear: to expand and enrich students' cognitive skill and mastery of science contents, and to elevate and motivate their interest in science and /or mathematics/computer science. Effort was also exerted to involve students in a collegial setting, exposing them to faculty and college students who served as positive role models.

A recruitment plan was outlined and initiated to maximize the potential for readily identifying minority students for the program. The primary target group was students from inner-city areas and schools with rapid turnovers in populations of ethnic and minority groups.

The long range benefits which are expected to be derived from this kind of program include an increased number of minorities and women who will enter the fields of science and engineering and hence our national resource of technically educated people will expand.

UNDERGRADUATE EDUCATION

Instruction in college science and mathematics classes must be: (1) current, (2) relevant, (3) well-prepared, and (4) state-of-art with reference to laboratory experiments.

All colleges and universities must continue to maintain and improve their assessment programs. These programs can provide proficiency examinations, for example, the Junior English Proficiency Examination (required for all students at Central State University). Students upon reaching their junior year must take this examination and pass it. This is now one of the requirements for graduation.

It is very important that the reading, writing and speaking skills of black students be improved and assessed. The assessment program offers other services such as mock tests for the Graduate Record Examination, the Medical College Admission Test, and the Dental Aptitude Test.

Since 85% of black science graduates are from HBCUs, then more federal funding is needed for the science programs. Currently two major federally funded programs are enhancing science education at HBCUs. They are 1) Minority Access to Research Careers (MARC), and 2) The Minority Biomedical Research Support Program (MBRS). The MARC program is a special training support program of the National Institute of General Medical Sciences of the National Institutes of Health. One of the major goals of this program is to strengthen both science curriculum and biomedical research. The MARC program supports the research training of third and fourth year honors undergraduates science and mathematics which in turn will assist them to prepare for successful entry into graduate programs. Two other objectives of this program are to (1) support doctoral and post-doctoral training and (2) provide for visitation of outstanding science scholars at minority institutions to enhance research. Currently 60 minority colleges are supported by the MARC progrm, 35 are HBCUs. 85% of the minority graduates who were MARC students as undergraduates go into post-baccalaureate programs and 300 students have been supported by the predoctoral fellowship program.

The Minority Biomedical Research Support Program (MBRS Program), another division of the National Institute of General Medical Science, National Institutes of Health provides release time for faculty from teaching duties to engage in research and financial support for students who collaborate on these projects under the guidance of their faculty mentors. This program provides undergraduate students with in-depth experience in biomedical research.

Presently, the MBRS program supports 49 HBCUs plus six other schools with predominantly black student enrollment (total of 55)[9]. About 16,000 students have been supported by the program since 1972. In 1987, a survey was made of 1,078 minority health professionals and terminal degree recipients in the natural sciences. All of the recipients (as undergraduates) participated in the MBRS program; of these 773 were M.D.'s, 254 were Ph.D's and 51 were recipients of the D.D.S. degree.

The MARC and MBRS progams have greatly enhanced the science education of minority undergraduate students. The two programs come together each year and present a joint MARC-MBRS symposium. At this symposium, students from both programs deliver research presentations which they are conducting at their respective institutions. In addition, the symposium serves as an important site for universities to recruit black students for graduate and professional schools.

A recent "college program" involving the University of Chicago and twelve campuses of the Big Ten universities has been initiated[10]. This program, which is an Institutional Cooperation Consortium selects minority students from a number of colleges and universities including several black colleges and universities. This program is not confined to science students, but seeks to encourage students to pursue graduate education by exposing them to different research experiences.

Two other organizations, namely, the National Institute of Science and Beta Kappa Chi Honorary Scientific Society have chapters at a number of HBCUs.

The National Instititute of Science (NIS) is composed primarily of black science and mathematics members throughout the country. It was founded in 1943 and serves to provide an avenue for the exchange of science information and the presentation of scholarly research papers by science students and faculty members of the HBCUs. The membership includes scientists in the disciplines of Biology, Chemistry, Earth Science, Physics, Mathematics, Experimental Psychology and Science Education.

Beta Kappa Chi Scientific Honor Science had its beginning at Lincoln University, Pennsylvania with a group of black scientists and mathematicians. The purpose of this organization was to encourage undergraduate students to achieve in the sciences. Its membership has grown to fifty-five chapters and over 11,000 initiates. The national meeting is held jointly with the National Institute of Science. In addition, the organization is a member of the Association of College Honor Society.

GRADUATE EDUCATION

Data from the National Science Foundation[11] indicates enrollment of blacks in graduate science and engineering programs rose about 6% between 1982 and 1988. In 1988, about 12,300 blacks or 4% of all students were enrolled in graduate studies in science and engineering fields. By way of comparison, enrollment of whites rose about 7% over the 6 year period.

The field distribution of blacks and whites differs substantially. Blacks are more likely than whites to be enrolled in social science programs. In 1988, about

87% of blacks were in graduate programs in science fields: about one out of every two of these enrollees were in social science.

The number of doctorates awarded to blacks in science increased between 1978 and 1988 from 437 to 487 (2.4% of the total in 1988).[11] This increase, however, masks two very different trends. While the number of blacks on temporary or permanent visas earning Ph.D.s increased, the number of U.S. citizens represented a reduction in 1988. This number was down from 64% a decade earlier. On the other hand, the proportions of temporary black residents rose from 29% to 37%. For black U.S. citizens, declines in Ph.D. study were most evident in the physical, mathematical, life, and social sciences. Degrees in these four fields dropped from 169 to 113 over the 10-year period. Engineering doctorates however, increased from 9 to 10. In 1988, blacks comprised about 1.8 percent of new doctorates awarded to U.S. citizens. This percentage is alarmingly low and must be increased.

National Science Foundation (NSF) Minority Graduate Fellowships

The National Science Foundation Minority Graduate Fellowship Program began in 1978 as an experimental mechanism designed to increase the number of minority scientists and engineers. Institutional selection was used as the nominating mechanism, and, in 1979, the program was redesigned as a national competition to carry out the broadened concept of support of graduate study for minorities.

In 1980, the number of applicants to the Minority Fellowship Program was 404, 612 in 1985, and increased to 739 by 1988.[11] By field, about two-fifths of the applicants were in either engineering, mathematics, or the physical sciences; slightly less than one-third each were in the behavioral and social sciences or life and medical sciences. Of the 739 applicants in 1988, about 10% (75) were offered new awards. An additional 29% (214) received honorable mentions. In 1980, the fraction of applicants receiving new awards was 14% (55) of the 404 applicants. One-third (130) of the applicants also received honorable mentions.

These fellowships have certainly been a step in the right direction for the support of black graduate students. The author has served three times as a panelist for reviewing the National Science Foundation Minority Graduate Fellowship applications.

CONCLUSION

Talented minority students will be needed in the coming century to contribute to scientific and technological discoveries. These students need to be supported by programs at the local, regional and national levels.

We must remove any negative perceptions about science and scientists and send a "you can succeed' message to our minority students. Let us get the minority community involved so that children, teachers, and parents all can be motivated to take on the challenge of scientific careers in the future.

RECOMMENDATIONS

The following are several important recommendations which will help to enhance the production of more qualified and outstanding black scientists:
1. Establish enrichment programs, grades K-12.
2. Improve instruction and undergraduate science and mathematics courses, especially at the introductory level.
3. Develop communication packages to inform black leaders and parents about educational opportunities in science.
4. Maintain constant state-of-the-art curricula and articulations between two year and four year colleges and/or universities.
5. High school counselors should encourage more black students to enroll in college preparatory courses and/or curricula.
6. Science teachers at the secondary school level should make science courses interesting and relevant.
7. More majority colleges and universities should use national meetings involving minority institutions, such as, the Annual MBRS- MARC Syposium, NIS- Beta Kappa Chi Joint Annual Meeting for recruiting academically sound black students for graduate programs in the sciences.
8. Media should increase success stories concerning minority scientists. This provides students at all levels-elementary, high school, and college with contemporary role models.

REFERENCES

1. Jay, James M. 1971. Negroes in Science: National Science Doctorates, 1876-1969. Ballamp Publishers, Detroit, Michigan.
2. National Science Foundation. Resources Supporting Scientific Activities at Predominantly Black Colleges and Universities. NSF 82-11; NSF 84-332.
3. Richardson, F.C. 1976. A Quarter Century of the Black Experience in the Natural Sciences, 1950-1974. *Negro Education Review.* 27:135-154.
4. Schorr, L.B. 1988. Within Our Reach: Breaking the Cycle of Disadvantage. Anchor/Doubleday Publishers, New York.
5. National Science Foundation. 1990. Women and Minorities in Science and Engineering, pp 35-36.

6. Johnson, David. 1990. Strengthening and Enlarging the Pool of Minority High School Graudates Prepared for Science and Engineering Career Options, pp. 75-93: Strengthening American Science and Technology. Career Communications Group Inc., Publishers, Baltimore, MD. pp. 121.
7. Johnson, Irene H. 1989. The Minority Pre-Collegiate Science Programs at Purdue University. Pre-College Education of Minorities in Science and Engineering Conference Proceedings. pp. 1-6. New Jersey Institute of Technology Press.
8. Borowitz, Grace B. 1989. Special Incentive Program in Science and Computer Science for Selective High School Students. Pre-College Education of Minorities in Science and Engineering Conference Proceedings. pp. 12-16. New Jersey Institute of Technology Press.
9. Gonzales, Ciriaco Q., Ph.D., Director, Minority Biomedical Research Support program, National Institute of General Medical Sciences, (Personal communication on June 22, 1990).
10. Wycliff, Don 1990. "Science Careers Are Attracting Few Blacks", The New York Times National. pp. A1, A14.
11. National Science Foundation. 1990. Women & Minorities in Science and Engineering. pp. 40-41.

Science Education in the United States: Issues, Crises and Priorities. Edited by S. K. Majumdar, L. M. Rosenfeld, P. A. Rubba, E. W. Miller and R. F. Schmalz. ©1991, The Pennsylvania Academy of Science.

Chapter Forty-Four

SCIENCE EDUCATION, THE NATIONAL ECONOMY, AND INTERNATIONAL COMPETITIVENESS

JOHN H. MOORE

Director
International Institute
George Mason University
4001 North Fairfax Drive
Suite 450
Arlington, Virginia 22203

The most fundamental asset of any nation is its human resources. This has always been true, of course, but the relative importance of a well-educated, technically capable labor force is growing and will continue to grow as economic success increasingly depends on technology.

Dependence on technology has already grown significantly. The competitive edge has shifted to emphasize the knowledge content of goods. No longer does the possession of abundant supplies of natural resources provide assurance of international competitiveness. Indeed, in the industries that define international competition (such as computers, electronics, motor vehicles, aerospace, and pharmaceuticals), knowledge has become the most important component. Not only do the existing products in these areas embody high degrees of knowledge and technology, there is intense competition to develop new products of ever-increasing sophistication. The creation of these new products can take place only through the application of newly created knowledge.

To compete in this world obviously requires people who can create that knowledge—highly talented, highly educated people to do the fundamental investigations that produce new knowledge. Nobody can specify the number of such people needed for basic and applied research. But we do know that the number of U.S. citizens entering the sciences and engineering and completing graduate work to the doctoral level has declined significantly in the last twenty years. They have been replaced by foreign nationals; in some fields of science and engineering, foreign nationals now receive forty to sixty percent of all doctorates awarded by U.S. universities. This is a new phenomenon: twenty years ago, foreign nationals were a district minority among our graduate students.

Why has this influx of foreign nationals occurred? Why has the number of U.S. nationals declined so significantly? The phenomenon has not been fully analyzed, so only hypotheses can be advanced. It may reflect, in part, rising incomes abroad, which make the outstanding educational opportunities available in the U.S. more easily attainable by foreign students. It almost certainly indicates that the attractiveness of careers in these fields is low relative to the alternatives available to young Americans. It may reflect something deeper—a lack of interest in pursuing knowledge, a deadening of curiosity about the world that leads students to other fields. It may suggest a lack of preparation in the schools for the sciences that carries with it a corresponding failure to instill in students the intellectual spark that is the heart of scientific careers.

Whatever the reasons, the facts are clear. Should they be a matter of concern for public policy? If young people make informed individual career choices, and these choices reflect real alternatives revealed by efficient markets, there is no basis for concluding that the outcome is undesirable from a social point of view. Accordingly, the conclusion about public policy depends on whether choices are informed and on the efficiencies of the markets involved. Both of these are complex issues, requiring much more analysis than can be devoted to them here. Two points—the public goods aspect of research, and the fact that most basic research in the U.S. is conducted in educational institutions which themselves have public goods aspects—lead to a preliminary conclusion that the market would undervalue academic research careers. It is less evident that students make uninformed career choices, or at least that there is any bias in information available to students that would steer them toward or away from scientific careers. In this regard, however, poor preparation, in the senses both of sheer knowledge of the basic mathematics and science needed for college level work and of the intellectual curiosity that leads people to research, could be said to bias choice. That is, more students would enter scientific career paths if they had better preparation; from that point of view, they are making uninformed choices.

If fewer young people are entering these careers than would be the case in the absence of public goods problems and if preparation is inadequate, the conclusion is that the stock of human capital available for scientific and engineering research is smaller than it would otherwise be. In turn, other relevant things the same, the flow of new knowledge that can be incorporated into new products is smaller than it would otherwise be. Of course, domestic firms competing on international markets, where the knowledge content of goods is increasingly the margin of competition, have an interest in maintaining the flow of new knowledge. But once again, there are significant public goods problems: especially in fundamental research (but also in much applied work), the productivity of research depends on open communication of results, which makes exclusive capture of their value impossible.

Relative to corporations from other nations, this situation would place American firms at a competitive disadvantage if the foreign corporations had access to better flows of new knowledge. Here lies another dilemma, for the pool of basic scientific knowledge is essentially an international pool, precisely because of its public goods nature and the general acceptance in the scientific community of the broad dissemination of results. An increase in research effort in the United States produces results that are available to firms throughout the world; it is alleged, for example, that Japanese firms have profited greatly by using scientific results produced in American laboratories to develop new products. If more American research is to benefit American firms differentially, either the distribution of research results must be restricted (which almost certainly would produce retaliation and would be counterproductive) or American firms must somehow take better advantage of the results produced in American laboratories. Of course, proximity gives domestic firms a built-in advantage in this respect. From that perspective, increases in research in U.S. universities should produce potential benefits for domestic corporations.

From this perspective, a flow of young people to science and engineering careers that is reduced by the factors noted above is suboptimal and should be a concern of public policy. Of course, there is considerable government intervention in these markets in the form of fellowships, support for research, and the like. The public policy question is whether the existing effort is sufficient in the sense of producing an adequate flow into these fields. No quantitative answer to that question is possible, but available estimates of the social rate of return to research suggest that marginal investment in basic research produces high net benefits. This result is consistent with a suboptimal inflow of research personnel. This is likely correct even if foreign nationals are considered an integral constituent of the domestic supply. The last point is the subject of much debate as well, since developments entirely beyond domestic control could result in rapid and significant reductions in the numbers of foreign nationals coming to the United States for their education and then staying on in the labor force. Of course, any such reduction would produce market adjustments, but the adjustment would be lengthy and costly.

An inadequate flow of young people to research careers should be a matter of public policy concern in and of itself for the reasons just developed. But if, as is widely believed, part of the reason for the inadequate flow is poor preparation in the nation's schools, additional concerns are raised. International competition, as already argued, is increasingly fought on the basis of knowledge and technology. Producing that knowledge is the domain of the science and engineering research effort. But translating it into products must be done by practicing engineers and a technologically capable workforce. If preparation of young people for entry into professional research career paths is inadequate, it is also inadequate for young people who will be the practicing engineers, technicians, and factory floor employees of the future. There can be no doubt that the nature of production is shifting from traditional manufacturing to much more demanding methods. With that shift, there is a shift in the skills needed in the general workforce. From a wide range of anecdotal evidence, it seems clear that the nation's schools are failing to produce the knowledge and technical sophistication needed by young people for work in modern, productive enterprises. Even more, perhaps, than the apparent under-investment in research career preparation, this failure is a matter of serious concern for public policy.

Wherever the emphasis is placed, the general conclusion is the same: competitiveness requires a significantly deeper and broader technical capability in the nation's labor force. Meeting this need requires increased efforts at every level of education in the United States.

Science Education in the United States: Issues, Crises and Priorities. Edited by S. K. Majumdar, L. M. Rosenfeld, P. A. Rubba, E. W. Miller and R. F. Schmalz. ©1991, The Pennsylvania Academy of Science.

SCIENCE EDUCATION IN THE UNITED STATES: EDITORS' REFLECTIONS

PETER A. RUBBA[1], E. WILLARD MILLER[2], ROBERT F. SCHMALZ[3], LEONARD M. ROSENFELD[4], and SHYAMAL K. MAJUMDAR[5]

[1]Associate Professor of Science Education
[2]Professor of Geography, Emeritus
[3]Professor of Geology
The Pennsylvania State University
University Park, PA 16802

[4]Assistant Professor of Physiology
Jefferson Medical College
Philadelphia, PA 19107
and
[5]Professor of Biology
Lafayette College
Easton, PA 18042

"Crisis" is the most appropriate word to use in describing the present status of science education in the U.S. This is a crucial time for science education, kindergarten through the post-doctoral level. We are at a turning point. The actions we take, or decide not to take, will determine whether positive or negative consequences follow.

We are all familiar with the dire consequences that are predicted to follow; this nation's ability to sustain its economy as well as compete in a world economy are called into question by the adequacy of our science education. The general level of scientific literacy among our citizens; the quality of school science instruction; the number of individuals, especially women and minority group members, entering science careers; inadequate funding for science education at all levels; the rapid growth of knowledge in scientific fields; the balance between teaching and research at the college level . . . are often pointed to as the critical issues whose resolution will turn things around.

Over the past decade there have been initiated a number of reforms designed for the resolution of these issues. The reforms have mostly included stricter regulations, such as science teacher standards that require greater science content background, teacher pre-certification and certification testing, and an increase in minimum science course requirements for graduation from high school. The reforms also have included increased teacher salaries, stiffer academic requirements within science courses, the introduction of Science, Technology and Society (STS) into school science, increased use of standardized tests, the funding of some special programs designed to keep women and minority group members in the pipeline, an expansion in the number of "general studies" college science courses, and calls for greater emphasis on teaching at the college level.

The authors of the chapters in this volume have discussed the application of these and other reforms, and the science education issues they are intended to resolve. Rather than summarize those discussions in this, the final chapter, we prefer to extend them by exploring an alternative perspective.

The implementation of reforms and the call for further reforms have been made under the presumption that science education in the United States could be fixed by replacing some of the machinery and "retuning" the system. The failure of these reform efforts to affect significant change, however, suggests to a growing number of people from inside and outside of education (including business and industry) that we have interpreted the problems in science education, indeed the problems with our entire educational system, much too simply. The failure of our initiatives to "re-fit" education is seen in some quarters as evidence that no amount of "retooling" and "retuning" will improve the productivity of an otherwise outmoded educational system; that nothing short of "reconstructing" education will work. The recent successful restructuring of the nature of work in leading American industries, such as General Motors, Hewlett-Packard, IBM, John Deere, Motorola, and Xerox (Cook, 1990), is frequently held high as an example, with parallels drawn between the changes made in industry and those that are needed in schooling.

Over most of this century schooling has been conceived as a manufacturing process in which the raw material (youngsters) is operated upon by the educational process (machinery), some for a longer period than others, and turned into finished products. Youngsters learn in lock-step or not at all (frequently not at all) in an assembly line of sequenced classes, from workers (teachers) who run the instructional machinery. A curriculum of mostly factual knowledge is poured into the products to the degree they can absorb it, using mostly expository teaching methods. The bosses (school administrators) tell the workers how to make the products under rigid work rules that give them little or no stake in the process. Quality testing is the last step in this educational assembly line, but it is typically completed only on the 40% or so of the products that reach the end of the high school assembly line and desire a transfer to the college assembly line. If spotted, bad products are culled-out during the manufacturing process. When the clutter of discarded products accumulates to the point where it cannot be ignored, they are sent off for expensive repair.

American society has held and continues to hold tightly to the assembly line model of education, as evidenced by our efforts over the last decade to save the system. Yet, it is not clear the model worked well even in the so called "hay day" of American education, the mid 1960s. While we proudly remember that SAT scores peaked, that a substantial number went on to college, we also need to be reminded that a significant percentage of students failed to graduate. But high drop-out rates were less worrisome then; the 1960s were a different time economically for the United States. We were THE world economic power, and an unskilled high school drop out could still find a decent-paying job and buy a house. Today, it is a much different world. The United States is the world's largest debtor nation. Its industrial might has been ravished by other developed and developing nations. There is little economic security in a college degree, let alone a high school diploma.

Albert Shanker (1990), president of the American Federation of Teachers, argues that,

> Our persistent educational crisis shows that we've reached the limits of our traditional model of education. Given our present and foreseeable demographics, economic, social, and educational circumstances, we can expect neither greater efficiency nor more equity from our education system. (p. 345)

Shanker continues the argument by suggesting that the failure of the traditional assembly line model of education is due mainly to the absence of three conditions from today's turbulent American society, on which the model is dependent:

> . . . a cohesive family and social structure; a willingness to accept educating the vast majority of children to only a low level (and then pushing them out or letting them drop out) and a small minority to a high level; and a large supply of well-qualified teachers . . .

> Times have changed. *A Nation Prepared: Teaching for the 21st Century* [prepared by the Task Force on Teaching as a Profession, Carnegie Forum on Education and the Economy, 1986] puts the situation starkly: "If our standard of living is to be maintained, if the growth of a permanent underclass is to be overted, if democracy is to function effectively into the next century, our schools must graduate the vast majority of their students with achievement levels long thought possible for only the privileged few." (p. 347)

As has always been required of professionals, today's high tech and service jobs require individuals with the abilities to work cooperatively with colleagues, to process large quantities of information from a variety of sources, to make responsible decisions, and above all, to continue to learn. The trend in business, industry and the service sectors is to move to participative or "quality" work teams, in which group members perform a broad range of tasks that formerly were the responsibility of a number of employees at a number of levels working under the close supervision of a boss(es). Computer technology is the basic tool used to manage and process information. The need for middle management has been eliminated. Teams set goals, determine procedures, manage time and tasks, cooperatively solve problems, take responsiblity for their performance, determine the composition of the team, etc. (Cook, 1990, Wirth, in press).

Wirth (in press) cites, William Duffy, Director of Research in Work Innovation at General Motors as follows on the future of the assembly line:

GM used to boast that the production line had been designed so that every task could be learned in fifteen minutes or less; efficiency could be secured by engineering work tasks so small that any idiot could do them. If morale and workmanship were poor, the answer was to step up supervision and control. GM is now convinced, he said, that a model based on "increased control by supervision of a reluctant work force which produces shabby products is not viable for survival."

The new Saturn Motor Division of GM, which was eight years and $3.5 billion in the making, is a jointly managed enterprise of General Motors and the United Auto Workers. "Nearly 3,000 union members and GM managers fully share decision making on every issue from how teams of workers perform their jobs to the crafting of budgets and profitability models" (Cook, 1990, p. 51).

The need for higher order thinking skills is not limited to the work place, all citizens need to be able to make informed, responsible decisions on science and technology-related societal issues as contributing members of society and as these issues impinge upon their personal lives. Yet, within the assembly line model, these are capabilities developed only by a few.

Shanker (1990) attributes Jack Bowsher, former director of education at IBM, as having said,

. . . that if IBM were producing results comparable to those of our schools—that is, if 25% of their computers were falling off the assembly line before they reached the end and if 90% of the completed ones didn't work 80% of the time—the last thing in the world the company would do would be to run that same old assembly line an additional hour each day or an extra month each year. Instead, IBM would rethink the entire production process. (p. 347)

Similarly, former Secretary of Education Lauro Cavazos recently declared the reform efforts of the 1980's to have failed, "We tried to improve education by imposing regulations from the top down, while leaving the basic structure of the school untouched . . . Obviously that hasn't worked" (as cited by Wirth, in press).

The research in cognitive psychology completed over the past 20 years also shows the assembly line model to be untenable. Eminent researchers like David Ausubel, Joseph Novak, Rosalind Driver and John Clement have shown learning to be an active interpretive process by which learners interact with the external world and construct meaning from the interactions. Prior knowledge learners bring with them dramatically affects the understanding that is constructed (Watson and Konicek, 1990).

Viewed from the constructivist epistemology (please see the chapters in Part II this volume by Staver and Gates for a more complete discussion), teachers are not the workers in an industrial analogy. The learners are the workers. The products are the understandings constructed by the learners. Teachers are the managers who assess and analyze learners' understandings, organize learning experiences that will allow learners to confront inconsistencies in their understandings, and provide support and guidance as more appropriate understandings are constructed. Computer technology is used to help manage the process, gather and analyze information on learners and their understandings, as well as to deliver instruction. From this perspective, education is a participative model.

As with American industry, the change from an assembly line model to a participative model in education will not come easily. Most of us went to traditional schools; our conceptions of schools and colleges are limited to the status quo. Many people continue to believe the traditional model can be reformed, if not by making a few modifications to tighten-up the system, then by adopting a successful model from Europe, or even Japan. However, this suggestion fails to recognize that education is a social institution. The application of European models would require significant changes American society will not accept; mainly, the centralized control of education by the federal government and development of a national curriculum.

But fundamental change may be necessary. Rhodes (1988) recommends that we look for guidance to Copernicus and Galileo, who also asked society to look at reality from a wholly different paradigm or point of view:

> What it takes to make such a quantum leap is everyone's acceptance of a new or different fundamental belief, which can then serve as a reference point for a different perspective. (p. 28)

Kuhn's (1970) study of the "paradigm shifts" that result in scientific revolutions suggests that if a paradigm shift occurs in education, it will occur rather quickly:

> Just because it is a transition between incommensurables, the transition between competing paradigms cannot be made a step at a time, forced by logic and neutral experience. Like the gestalt, it must occur all at once (though not necessarily in an instant) or not at all (p. 150).Unfortunately, as Kuhn also notes, persuasion is the most effective course of action for social and political change (p. 92).

It is frightening to think that as the crisis in education deepens and proposals are made, some to implement further reforms and others for reconstruction;

as competing camps clash, one defending the old model and others seeking to institute new ones; the future of science education may not be decided on educational merits and social need. Let us hope that the educational system we will need to keep the United States scientifically and economically competitive—that will provide the cadre of scientists and well-trained technical support people, and the supporting base of scientific and technologically literate citizens—is not beyond our vision or our will.

REFERENCES

Task Force on Teaching as a Profession. (1986). *A nation prepared: teaching for the 21st century.* New York: Carnegie Forum on Education and the Economy.

Cook, W.J. (1990). Ringing in Saturn. *U.S. News and World Report,* October 22, 1990, 51-54.

Kuhn, T.S. (1970). *The structure of scientific revolutions.* (2nd ed.). Chicago: University of Chicago Press.

Rhodes, L.A. (1988). We have met the system—and it is us! *Phi Delta Kappan,* September 1988, 28-30.

Shanker, A. (1990). A proposal for using incentives to restructure our public schools. *Phi Delta Kappan,* January 1990, 345-357.

Watson, B. and Konicek, R. (1990). Teaching for conceptual change: confronting children's experience. *Phi Delta Kappan,* May 1990, 680-685.

Wirth, A.G. (in press). Issues in the reorganization of work: implications for education. *Theory Into Practice.*

Subject Index